Urban Hydrogeology Studies

Urban Hydrogeology Studies

Editor

C. Radu Gogu

MDPI • Basel • Beijing • Wuhan • Barcelona • Belgrade • Manchester • Tokyo • Cluj • Tianjin

Editor
C. Radu Gogu
Technical University of Civil Engineering
Bucharest, Romania
Vice-Chair of the Groundwater Management Subgroup, International Water Association
London, UK

Editorial Office
MDPI
St. Alban-Anlage 66
4052 Basel, Switzerland

This is a reprint of articles from the Special Issue published online in the open access journal *Water* (ISSN 2073-4441) (available at: https://www.mdpi.com/journal/water/special_issues/Urban_hydrogeology).

For citation purposes, cite each article independently as indicated on the article page online and as indicated below:

LastName, A.A.; LastName, B.B.; LastName, C.C. Article Title. *Journal Name* **Year**, *Volume Number*, Page Range.

ISBN 978-3-0365-4737-4 (Hbk)
ISBN 978-3-0365-4738-1 (PDF)

© 2022 by the authors. Articles in this book are Open Access and distributed under the Creative Commons Attribution (CC BY) license, which allows users to download, copy and build upon published articles, as long as the author and publisher are properly credited, which ensures maximum dissemination and a wider impact of our publications.

The book as a whole is distributed by MDPI under the terms and conditions of the Creative Commons license CC BY-NC-ND.

Contents

About the Editor vii

Constantin Radu Gogu
Urban Hydrogeology Studies
Reprinted from: *Water* **2022**, *14*, 1819, doi:10.3390/w14111819 1

Stephen Foster
Global Policy Overview of Groundwater in Urban Development—A Tale of 10 Cities!
Reprinted from: *Water* **2020**, *12*, 456, doi:10.3390/w12020456 5

Susie Mielby and Hans Jørgen Henriksen
Hydrogeological Studies Integrating the Climate, Freshwater Cycle, and Catchment Geography for the Benefit of Urban Resilience and Sustainability
Reprinted from: *Water* **2020**, *12*, 3324, doi:10.3390/w12123324 13

Youcef Boudjana, Serge Brouyère, Pierre Jamin, Philippe Orban, Davide Gasparella and Alain Dassargues
Understanding Groundwater Mineralization Changes of a Belgian Chalky Aquifer in the Presence of 1,1,1-Trichloroethane Degradation Reactions
Reprinted from: *Water* **2019**, *11*, 2009, doi:10.3390/w11102009 37

Fabrice Rodriguez, Amélie-Laure Le Delliou, Hervé Andrieu and Jorge Gironás
Groundwater Contribution to Sewer Network Baseflow in an Urban Catchment-Case Study of Pin Sec Catchment, Nantes, France
Reprinted from: *Water* **2020**, *12*, 689, doi:10.3390/w12030689 59

Mitja Janža, Joerg Prestor, Simona Pestotnik and Brigita Jamnik
Nitrogen Mass Balance and Pressure Impact Model Applied to an Urban Aquifer
Reprinted from: *Water* **2020**, *12*, 1171, doi:10.3390/w12041171 79

Carly M. Maas, William P. Anderson Jr. and Kristan Cockerill
Managing Stormwater by Accident: A Conceptual Study
Reprinted from: *Water* **2021**, *13*, 1492, doi:10.3390/w13111492 95

Sadaf Teimoori, Brendan F. O'Leary and Carol J. Miller
Modeling Shallow Urban Groundwater at Regional and Local Scales: A Case Study in Detroit, MI
Reprinted from: *Water* **2021**, *13*, 1515, doi:10.3390/w13111515 113

Weicheng Lo, Sanidhya Nika Purnomo, Bondan Galih Dewanto, Dwi Sarah and Sumiyanto
Integration of Numerical Models and InSAR Techniques to Assess Land Subsidence Due to Excessive Groundwater Abstraction in the Coastal and Lowland Regions of Semarang City
Reprinted from: *Water* **2022**, *14*, 201, doi:10.3390/w14020201 137

Cristian-Ștefan Barbu, Andrei-Dan Sabău, Daniel-Marcel Manoli and Manole-Stelian Șerbulea
Water/Cement/Bentonite Ratio Selection Method for Artificial Groundwater Barriers Made of Cutoff Walls
Reprinted from: *Water* **2022**, *14*, 376, doi:10.3390/w14030376 165

Floris Cornelis Boogaard
Spatial and Time Variable Long Term Infiltration Rates of Green Infrastructure under Extreme Climate Conditions, Drought and Highly Intensive Rainfall
Reprinted from: *Water* **2022**, *14*, 840, doi:10.3390/w14060840 177

Alina Radutu, Oana Luca and Constantin Radu Gogu
Groundwater and Urban Planning Perspective
Reprinted from: *Water* **2022**, *14*, 1627, doi:10.3390/w14101627 . . . . **191**

About the Editor

C. Radu Gogu

Radu Constantin Gogu, Full Professor, is specialized in water resources engineering, with international experience achieved in different countries: Switzerland, Belgium, Spain, Greece, and Romania. He has a Ph.D. in Applied Sciences from the University of Liege (Belgium), and he worked as a Senior Researcher at the Swiss Federal Institute of Technology (ETH, Zurich) and the Polytechnic University of Catalonia (Barcelona, Spain). He received a Research Grant of Excellency, "Ramon y Cajal", from the Spanish Ministry of Research.

At present, he is the Deputy Chair of the Groundwater Management Specialist Group—the International Water Association (IWA), and his expertise area includes urban water, groundwater resources, environment protection, natural hazards, and risk assessment. His current focus is on urban groundwater studies.

Radu Gogu initiated and developed the Bucharest city urban groundwater model, which is currently used as a city-scale background for the implementation of sustainable drainage systems.

Among his achievements can be mentioned the design of the groundwater database and the prototype hydrogeological maps of the Walloon region, the design and development of the first geospatial database concept for active volcanoes (ETHZ), the design and development of the Barcelona hydrogeological spatial database, the development of software instruments for water quality, and 3D geological analysis.

Editorial

Urban Hydrogeology Studies

Constantin Radu Gogu [1,2]

1. Groundwater Engineering Research Centre, Technical University of Civil Engineering, Bucharest, bd. Lacul Tei 124, 020396 Bucharest, Romania; radu.gogu@utcb.ro or radu.constantin.gogu@gmail.com
2. Groundwater Management Group, International Water Association (IWA), London E14 2BA, UK

1. Introduction

Urbanization is a pervasive phenomenon of our time, and sustainable urban development is one of the greatest challenges faced by the contemporary world. The subsurface plays a range of roles in such developments through the complex processes of urbanization, including building developments, constructing roads, providing water-supply, drainage, sanitation and even solid-waste disposal.

For most cities, the groundwater system represents a 'linking component' between various elements of the urban infrastructure. Since urban processes have an influence on groundwater and groundwater conditions have an impact on the urban infrastructure, groundwater systems thus exhibit a close relationship with the processes of urbanization, and this continuously changes with the urban development cycle. Consequently, cities around the world face issues related to urban hydrogeology, requiring attention at least as much as those provided by other planning-related problems in urban areas.

Urban groundwater problems are usually predictable. However, they are not predicted early enough, as action usually responds to emergencies rather than to planning. Consequences resulting from a lack of accurate and detailed knowledge of the underground environment and the interaction between the urban groundwater and urban infrastructure are faced by cities across the entire world in economic, environmental, social, legal and political terms.

The lack of data and planning, as well as poor communication between the scientific community and city managers, exacerbate the difficulties of solving urban hydrogeology problems. To provide the necessary understanding, experts have to use robust datasets of urban fabric, infrastructure networks, groundwater and geothermal energy systems at the city scale. Furthermore, this knowledge and understanding must also be accessible to urban planning processes.

In recent decades, progressive advances in the scientific understanding of urban hydrogeological processes and the groundwater regimes of a substantial number of cities have been documented. This extensive array of subsurface challenges which cities have to contend with lies at the core of the sustainability of the urban water cycle. This is threatened by the increasing scale and downward extent of urban subsurface construction, including utilities (cables, sewage, drainage), transportation (tunnels, passages) and storage (cellars, parking lots, thermal energy). The cumulative impact of this subsurface congestion on the surrounding geology, and especially the groundwater system, has to be constantly studied and addressed.

In this volume, key connections amongst the urban hydrogeology activities are identified consistent with scientific results and good practices in relationship to subsurface data and knowledge of sub-surface systems. The volume supports a useful dialogue between the providers and consumers of urban groundwater data and knowledge offering new perspectives on the existing research themes.

2. Summary of This Special Issue

The opening article [1] defines a context for policy development and proposes steps for integrating groundwater into the urban water and land management context. On the basis of a strategic assessment of the major trends perceived in 10 world spread developing cities, Professor Stephen Foster analyzes the benefits both to water users and to the broader community of groundwater use. Associated risks considering resource sustainability, built infrastructure, public-health hazards and the economic distortion of water-sector investments are evidenced. This is developed further in the article of Mielby and Henriksen [2] that outlines the crucial need of using cross-disciplinary methodologies to develop adaptive design and resilience in urban planning and management. On the basis of several hydrogeological studies performed in Odense, Denmark, the authors identify needed improvements in standardized data and modelling to promote strategic planning and decision making in order to diminish the environmental consequences of city infrastructure developments.

The main directions of the current research worldwide in urban hydrogeology are well balanced in this Special Issue. Three papers [3–5] describe different research experiences in quantitative urban groundwater studies with case studies in Detroit, USA, and Nantes, France, the third one being a laboratory study. The qualitative urban groundwater nucleus of this volume is represented by three other papers of three distinct cases studies that describe distinct groundwater contamination problems. These are located in Ljubljana, Slovenia [6]; North Carolina, USA [7]; and Belgium [8].

A study [3] in the Great Lakes Basin in Detroit, USA, provides an improved understanding of urban groundwater focusing on two approaches: the regional and local areas. A regional groundwater model that encompasses four major watersheds has been defined to outline the large-scale groundwater characteristics. On this basis, the local-scale model has been developed to analyze the local urban water budget with subsequent groundwater simulation. Both regional- and local-scale models can be further used to evaluate and mitigate urban environmental risks.

Using experimental data recorded on a small urban catchment, a model has been built to simulate the groundwater seepage into the sewer network in Nantes, France [4]. The experimental analysis revealed a strong correlation between groundwater levels and sewer base flow. The authors mentioned that the total groundwater volume drained by the sewer system represents 42% of the total rainfall annually.

The third [5] article focuses on the effects of artificial groundwater barriers made of cutoff walls. The authors underline the necessity of building effective, strong, flexible and low-permeability cutoff walls. Their study proposes a methodology to choose optimum construction materials. Based on laboratory results, their approach focuses on assessing the viscosity and representation of the Water–Cement–Bentonite components and presents an improved method for materials composition.

Ljubljana, the capital of Slovenia, considered the greenest city of the world, is the chosen study area [6] to model groundwater nitrogen loads coming from agriculture and leaky sewer system. The estimated total nitrogen load has been used to simulate the distribution of nitrate concentration into the aquifer using a groundwater contaminant transport model. The model, quantifying the impact of pressures from different contamination sources, can be further used in groundwater quality management activities.

In cold urban environments, unplanned natural raised wetlands represent one solution to reduce the impact of road salt contamination in surface water and groundwater. A study [7] was conducted in northwestern North Carolina, USA, on the capacity to control the timings and reduce peak concentrations of road salt into a stream of an accidental raised wetland. The study, based on the modelling of multiple meltwater and summer storm event scenarios, indicates that accidental wetlands improve stream water quality, and they may also reduce peak temperatures during temperature surges in urban streams.

Abandoned industrial sites represent a strong environmental threat in many areas worldwide. A study case [8] has been developed in Belgium to investigate the hydrochemical processes controlling groundwater mineralization through the characterization of the backfill and groundwater chemical composition. The analysis focuses on groundwater pollution, due to a mixture of chlorinated solvents with mainly 1,1,1-trichloroethane (1,1,1-TCA) at high concentrations.

Nowadays, remote sensing techniques are increasingly used to identify subsidence due to groundwater overexploitation. The volume also has a paper dedicated to the monitoring and modelling of subsidence as consequence of groundwater overexploitation in Semarang City, the capital of Indonesia [9]. A study integrating numerical modeling and synthetic aperture radar interferometry (InSAR) is presented. The models, simulating the hydromechanical coupling of groundwater flow and land subsidence, describe groundwater management measures to reduce the rate and affected area of subsidence.

Green infrastructure is one of the most important responses to urbanization, and a large-scale study based on a set of infiltration experiments [10] carried out in three swales completes this volume. In Holland, in the municipality of Dalfsen, a research project analyzed infiltration rate variation under extreme climate events. The study contributes to a better understanding of infiltration processes in this type of drainage system.

Urban planning should consider both above-ground and underground infrastructure development, including accurate groundwater management. Radutu et al. [11] highlight the need for accurate studies to properly discriminate the phenomena and processes generating subsidence. Satellite remote sensing used with a methodology characterizing the study area, demonstrates a major concordance between the groundwater level and the vertical displacements. With the purpose of understanding the connection between the cities' development and urban groundwater, the article reviews the urban plans in Romania, analyzes the strategic and planning framework of Bucharest city and discerns the role of groundwater as one of the main subsidence-triggering factors.

The future of our cities will increasingly need to rely on the sustainable use and reliable management of urban groundwater. This volume addresses this for groundwater and geotechnical specialists, civil engineers, infrastructure developers, land-use planners and geodetic experts. In addition, city managers and experts involved with various sectors of municipal utilities and environmental departments could use this Special Issue to improve their understanding of urban hydrogeology as a basis for accurate subsurface management.

Funding: This research received no external funding.

Conflicts of Interest: The author declares no conflict of interest.

References

1. Foster, S. Global Policy Overview of Groundwater in Urban Development—A Tale of 10 Cities! *Water* **2020**, *12*, 456. [CrossRef]
2. Mielby, S.; Henriksen, H.J. Hydrogeological Studies Integrating the Climate, Freshwater Cycle, and Catchment Geography for the Benefit of Urban Resilience and Sustainability. *Water* **2020**, *12*, 3324. [CrossRef]
3. Teimoori, S.; O'Leary, B.F.; Miller, C.J. Modeling Shallow Urban Groundwater at Regional and Local scales: Case Study in Detroit, MI. *Water* **2021**, *13*, 1515. [CrossRef]
4. Rodriguez, F.; Le Delliou, A.L.; Andrieu, H.; Gironas, J. Groundwater Contribution to Sewer Network Baseflow in an Urban Catchment- Case Study of Pin Sec Catchment, Nantes, France. *Water* **2020**, *12*, 689. [CrossRef]
5. Barbu, C.S.; Sabau, A.D.; Manoli, D.M.; Serbulea, M.S. Water/Cement/Bentonite Ratio Selection Method for Artificial Groundwater Barriers Made of Cutoff Walls. *Water* **2022**, *14*, 376. [CrossRef]
6. Janza, M.; Prestor, J.; Pestotnik, S.; Jamnik, B. Nitrogen Mass Balance and Pressure-Impact Model Applied in an Urban Aquifer. *Water* **2020**, *12*, 1171. [CrossRef]
7. Maas, C.M.; Anderson, W.P.; Cockerill, K. Managing Stormwater by Accident: A Conceptual Study. *Water* **2021**, *13*, 1492. [CrossRef]
8. Boudjana, Y.; Brouyere, S.; Jamin, P.; Orban, P.; Gasparella, D.; Dassargues, A. Understanding Groundwater Mineralization Changes of a Belgian chalky aquifer in the Presence of 1,1,1- Trichloroethane Degradation Reactions. *Water* **2019**, *11*, 2009. [CrossRef]
9. Lo, W.; Purnomo, S.N.; Dewanto, B.G.; Sarah, D.; Sumiyanto. Integration of Numerical Models and InSAR Techniques to Assess Land Subsidence due to Excessive Groundwater Abstraction in the Coastal and Lowland Regions of Semarang City. *Water* **2022**, *14*, 201. [CrossRef]

10. Boogaard, F.C. Spatial and Time Variable Long Term Infiltration Rates of Green Infrastructure under Extreme Climate Conditions, Drought and High Intensive Rainfall. *Water* **2022**, *14*, 840. [CrossRef]
11. Radutu, A.; Luca, O.; Gogu, C.R. Groundwater and Urban Planning Perspective. *Water* **2022**, *14*, 1627. [CrossRef]

Perspective

Global Policy Overview of Groundwater in Urban Development—A Tale of 10 Cities!

Stephen Foster [1,2,3]

1. Groundwater Management Specialist Group, International Water Association, 2521 Den Haag, The Netherlands; DrStephenFoster@aol.com
2. Urban Groundwater Network, International Association of Hydrogeologists, P.O. Box 4130, Goring, Reading RG8 6BJ, UK
3. Department of Earth Sciences, University College London, London WC1E 6BT, UK

Received: 18 December 2019; Accepted: 4 February 2020; Published: 8 February 2020

Abstract: Urbanisation is the predominant global phenomenon of our time. This overview provides an assessment of the trends in both public and private use of groundwater for urban water-supply in 10 developing cities and their policy implications, which is based on the global experience during 2001–2012 of the World Bank—Groundwater Management Advisory Team (a multi-disciplinary team of groundwater specialists working long-term for the World Bank, with special funding principally from the Netherlands Government supplemented by the United Kingdom and Denmark), together with subsequent follow-up enquiries. The strategic assessment analyses both the benefits to water users and the broader community of groundwater use, and the associated risks in terms of (a) compromising resource sustainability, (b) impacting the built infrastructure, (c) public-health hazards arising from widespread groundwater pollution and (d) the economic distortion of water-sector investments.

Keywords: urban development; developing cities; groundwater policy; integrated management

---

1. Objective of This Overview

In the course of providing data analysis and investment advice to potential World Bank-funded urban water-supply and water-resource projects worldwide, the World Bank—Groundwater Management Advisory Team (GW-MATe) collected extensive evidence of fast-increasing dependence upon groundwater (for public and/or private water-supply) with both benefits and hazards to water users and the broader community, in the 10 developing cities that are the focus of this paper (Table 1). The detailed results of the GW-MATe technical surveys and policy diagnostics are available elsewhere [1,2] and thus the primary objective of the present paper is to provide a concise global overview of the results of that work, without entering into detail about each of the 10 cities nor about the methodologies employed.

Table 1. Summary of groundwater concerns and management approaches in 10 cities assessed.

CITY Country	POPULATION * Utility Water-Supply	MAIN ISSUES AND CONCERNS	MANAGEMENT APPROACH
BANGKOK Thailand [3]	8 million; 1200 Ml/d (5%–10% groundwater)	Municipal groundwater use constrained but massive private waterwell construction into the 1990s, causing land subsidence and other impacts	Water utility groundwater use drastically reduced; groundwater resource regulator empowered to control private waterwell use

Table 1. Cont.

CITY Country	POPULATION * Utility Water-Supply	MAIN ISSUES AND CONCERNS	MANAGEMENT APPROACH
LUCKNOW India [1]	3 million; 500 Ml/d (50% groundwater)	Declining water-table/well yields; in-situ sanitation causing high nitrate levels in a shallow aquifer; uncontrolled private waterwell drilling	Accidental/incidental conjunctive use by the water-service utility since 1973 but groundwater use requires adequate planning and new protected sources
AURANGABAD India [1]	1 million; 150 Ml/d (0% groundwater)	Poor utility water-service led to excessive private waterwell drilling in an aquifer of limited storage and polluted by universal in-situ sanitation	Supplementary water source needed; water-service utility needs to embrace private waterwell use (30+ Ml/d) and prioritise the sewerage system
LIMA Peru [2]	8 million; 1100 Ml/d (30% groundwater)	Declining groundwater levels/reserves and increasing saline intrusion from excessive water utility/industrial abstraction in the 1980s–1990s	The government commissioned a water utility who successfully controlled public/private waterwell use and implemented conjunctive management
BUENOS AIRES Argentina [1]	4 million; 500 Ml/d (20% groundwater) (reduced since 1990)	Utility/industrial waterwell use reduced in the late 1980s (due to pollution fears); strong water-table rebound caused groundwater flooding	Multi-stakeholder task-force on urban groundwater needed; some water-utility pumping must be maintained for water-supply/drainage purposes
NATAL Brazil [1]	1 million; 280 Ml/d (100% groundwater)	In-situ sanitation causing groundwater nitrate pollution widely to 50+ mg/L; seawater intrusion concerns	Water utility prioritising sewage for aquifer protection and harmonising private waterwell use (45 Ml/d)
RIBEIRÃO PRETO Brazil [4]	1 million; 500 Ml/d (100% groundwater)	Groundwater levels have fallen by 50+ m since 1970, with increased pumping costs and some pollution risks	Institutional agreement on urban waterwell drilling ban; improved monitoring and protection measures for public water-supply sources
ADDIS ABABA Ethiopia [5]	3 million; 400 Ml/d (50% groundwater)	Severe water-supply shortage being partly met by new deep wellfield development	Improved monitoring and management of limited groundwater resources needed to avoid conflicts
DAR-es-SALAAM Tanzania [5]	5 million; 350 Ml/d (10% groundwater)	Surface-water sources decreasing in reliability; major private waterwell use (80+ Ml/d) but aquifer widely polluted by in-situ sanitation	Developing major new municipal wellfields in rural areas with favourable groundwater conditions to the south
NAIROBI Kenya [5]	4 million; 500 Ml/d (10% groundwater)	Surface-water sources decreasing in reliability; major private waterwell drilling (90–280 Ml/d) but groundwater levels have fallen by 50+ m	Metering/charging of private waterwells, with a moratorium on drilling and licensing new waterwells

* approximate figure for water-service area (political boundaries render census data inappropriate).

2. Context for Policy Development

2.1. Drivers of Urban Groundwater

Dependence on groundwater for public and/or private water-supply is a fast-increasing phenomenon in developing cities [6–9], which is occurring in response to population growth, accelerating urbanisation, higher ambient temperatures, increasing per capita usage and reduced security of river intakes (due to increased pollution). This global trend has been facilitated by the reducing (and generally modest) cost of waterwell construction [10]. Today many cities—from across the EU and USA to Brazil, China, India, Mexico, Nigeria, Pakistan and Vietnam—exhibit a high level of dependence on groundwater for urban water-supply.

2.2. Major Issues Relating to Increased Groundwater Use

Those urban centres underlain and/or surrounded by high-yielding aquifers (e.g., Bangkok, Lima, Lucknow, Natal and Ribeirão Preto/Table 1) usually have better public water-service levels. However, there are rarely sufficient groundwater resources within urban areas themselves to satisfy the entire water demand of larger cities, and resource sustainability often becomes an issue with serious localised aquifer depletion, potentially causing quasi-irreversible side-effects (such as induced seepage of contaminated water, land subsidence and coastal saline intrusion) [1,11].

This situation is often exacerbated by groundwater quality deterioration due to inadequately controlled urban sanitation, given the close connection of wastewater handling, disposal and re-use with underlying shallow groundwater (e.g., Lucknow and Natal/Table 1). It is thus critical that groundwater quality is systematically monitored, and the risks posed by pollution carefully assessed and acted upon [10].

The use of groundwater is not restricted to cities with ready access to high-yielding aquifers, but also occurs widely where the public water-supply is imported from distant surface-water sources of low reliability and high cost (e.g., Aurangabad, Dar-es-Salaam and Nairobi/Table 1). In this situation, private in-situ waterwell construction often mushrooms, and self-supply from groundwater widely represents a significant proportion of water 'actually received' by users (widely 20%–30% of total water-supply and well above 50% in extreme cases).

3. Urban Interactions with Groundwater

Urbanisation modifies the groundwater flow and quality regime in underlying shallow aquifers by:

- substantially increasing recharge rates—since the reduction of infiltrating rainfall through land-surface impermeabilisation almost always is more than compensated for by water-mains leakage, infiltrating pluvial drainage and the incidental 'return' of wastewater from in-situ sanitation and sewer leakage, especially where the municipal water-supply is 'imported from external sources'.
- greatly increased subsurface contaminant load, mainly from in-situ sanitation (in developing cities), and to lesser degree from sewer leakage and casual spillage/disposal of industrial and community chemicals.

These modifications are in continuous evolution. The processes combine to present numerous threats and some benefits for urban groundwater resources [8]—with the threats being more severe where shallow aquifers are used for private in-situ self-supply. In one sense, groundwater systems under cities represent 'the ultimate sink' for urban pollutants (with nitrates from wastewater and some synthetic hydrocarbons being very persistent), but in practice the extent to which the contaminant load impacts groundwater will vary widely with the vulnerability of the aquifer system.

In turn, man-made modifications to groundwater regimes can seriously impact urban infrastructure, with the impacts becoming more serious under unstable groundwater conditions:

- falling water-table due to excessive groundwater extraction, leading to land subsidence, building damage and increased surface-water flood risk (e.g., Bangkok/Table 1);
- water-table rebound, if groundwater use is abandoned for some reason such as serious pollution (e.g., Buenos Aires/Table 1), causing basement damage, groundwater flooding to transportation routes, malfunction of septic tanks and excessive inflows to deep collector sewers.

All of these problems can be very costly and surprisingly persistent—and while most are 'predictable' in theory, they are rarely 'predicted' in practice because of the widespread absence of an institution charged with the oversight and management of urban groundwater.

4. Integrating Groundwater into Urban Water and Land Management

Groundwater is far more significant in the water-supply of fast-developing cities than is widely appreciated, and is also the 'invisible link' between various facets of the urban infrastructure. Regretfully, many organisations concerned with urban water-supply and environmental management have a poor understanding of groundwater as a component of the urban water-cycle and one that always needs to be integrated into decisions on urban infrastructure planning and investment, whatever its status.

4.1. Conserving Groundwater Reserves for Water-Supply Security

In future, it will be important for the major groundwater storage of most aquifers to be used to improve urban water-supply security, as part of a strategy to cope with the pressures generated by urban growth and climate change.

First and foremost, there is widely a need to introduce effective demand management measures to constrain unnecessary use and to reduce 'unaccounted for water', and these need to be accompanied by managed aquifer recharge (from roof-drainage soakaways, permeable pavements and injection of excess surface water via lagoons or large wells) to maximise urban aquifer replenishment. All of this will require a 'resource culture' to be cultivated within water utilities.

Where existing groundwater development is excessive and beginning to cause land subsidence that may threaten the stability of urban infrastructure, there is a need to employ risk assessment techniques to identify potential problems [12,13].

It is important that schemes of conjunctive use of groundwater and surface-water sources (Figure 1) be pursued imaginatively, since most so-called 'conjunctive use' currently practised in developing cities amounts to a 'piecemeal coping strategy' rather than the more 'optimised approaches' that have been successfully implemented in Lima and Bangkok (Table 1).

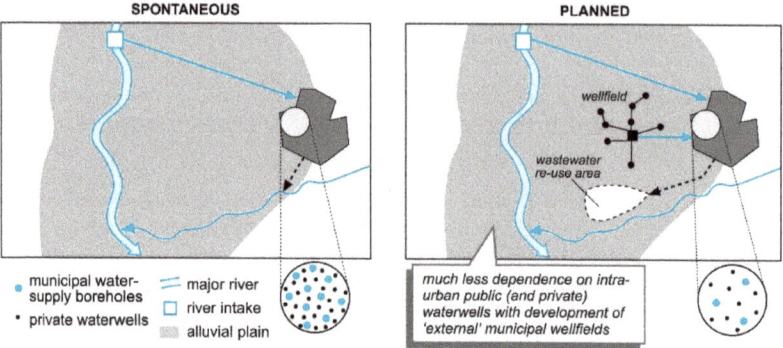

Figure 1. Planned and spontaneous conjunctive use for urban water-supply.

The development of external wellfields for utility water-supply (e.g., Addis Ababa, Dar-es-Salaam and Ribeirão Preto/Table 1) is another much needed policy response for the strategic use of groundwater. Investment in such wellfields and water delivery-mains requires insurance through declaring their 'capture areas' as drinking-water protection zones [14,15]. This often encounters administrative impediments related to fragmented powers between the municipalities comprising 'metropolitan areas', and improved governance arrangements and economic incentives need to be explored to overcome this problem.

Given the continuous evolution of groundwater use in 'urban aquifers', and some hydrogeologic uncertainty in predicting their precise behaviour, it is desirable to adopt an 'adaptive management approach' to urban groundwater resources. This should be based on continuous monitoring of groundwater levels and quality, and guided by a (periodically updated) numerical aquifer model.

4.2. Incorporating Groundwater into Sanitation and Drainage Planning

The groundwater–sanitation nexus is extremely relevant in developing nations, where extensive in-situ sanitation presents a significant groundwater quality hazard (Figure 2). It may be decades before the full extent of pollution becomes fully apparent because contamination of large aquifers is a gradual and insidious process. Thus, it is critically important to recognise the incipient signs of groundwater pollution through shallow aquifer monitoring and to put in place groundwater protection measures.

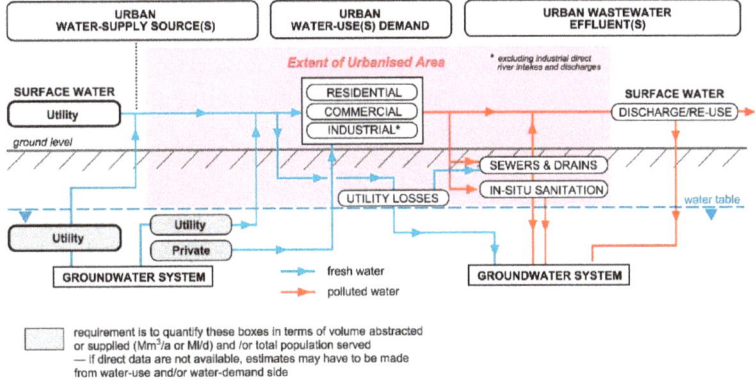

Figure 2. Groundwater relations within the urban water-cycle.

In most settings (except for shallow and vulnerable aquifers), there will be sufficient attenuation capacity to eliminate faecal pathogens from percolating wastewater, but the hazard increases markedly with inadequate waterwell construction and/or poor septage management, which are typical of anarchical fast-growing cities. However, troublesome levels of nitrogen compounds and dissolved organic carbon will arise to varying degrees, depending on the population density served by in-situ sanitation. For municipal water-supply, the problem is often dealt with by dilution through mixing, but this requires a secure source of high-quality water and also has absolute limitations because some wastewaters contain a wide array of pharmaceutical residues and industrial chemicals.

An integrated approach is required to reduce the cost and improve the security of the urban water infrastructure. There are numerous practical measures that can be taken to improve the sustainability of groundwater use, including:

- prioritisation of recently-urbanised districts for sewerage cover to protect their good-quality groundwater from gradual degradation;
- establishment of groundwater protection zones around utility waterwells favourably located to take advantage of parkland;
- imposition of better controls for the handling and disposal of industrial effluents and solid wastes to reduce the risk of aquifer pollution.

Groundwater pollution can be reduced by deploying dry (eco-sanitation) units, in which urine is separated from faeces, with both being recycled. The latter installations are recommended for new urban areas overlying shallow aquifers, but they are not the universal solution to groundwater contamination because large-scale retro-installation in existing dwellings is very costly.

An important corollary is making better use of increasing wastewater resources generated by wider sewer coverage—this could be through reuse for amenity and agricultural irrigation that is spatially planned and appropriately controlled so as to minimise public health hazards, including incidental pollution of groundwater in potable use [16].

4.3. Private In-Situ Groundwater Use—Reducing Risks and Improving Benefits

Private capital investment to access groundwater in situ for urban self-supply often mushrooms during periods of inadequate utility water-supply [1,17]. This is essentially a 'coping strategy' by multi-residential properties, and commercial and industrial enterprises. Water users tend to take their supply from multiple sources according to their availability and relative cost, with much more expensive tankered water as the last resort.

Private self-supply from groundwater is then likely to continue by many users as a 'cost-reduction strategy' when the availability of the utility supply improves. Although the 'economy of scale' can be poor, the cost of water-supply from this type of source often compares favourably with the tariffs implied by full cost-recovery for new utility surface water-supply schemes. Whether private residential groundwater use presents a serious threat to users will depend on the type of anthropogenic pollution (or natural contamination) present and the type of water-use concerned.

An emerging policy question is under what circumstances the risks or inconveniences of private residential self-supply from urban groundwater justify attempts to ban the practice. Many private waterwells are, at best, unregulated and at worst illegal. In the longer run, this is counterproductive for the private user and the public administration but can be regularised by management interventions such as:

- using advances in geographical positioning and data capture to locate waterwells;
- registering commercial/industrial users and residential use for apartment blocks/housing estates, and charging for groundwater use by waterwell pump capacity or metering discharges to sewers;
- issuing water-quality use advice and health warnings to private waterwell operators, and if pollution is severe, declaring sources unsuitable for potable use;
- gaining civil-society commitment through effective participatory mechanisms and incentives for 'self-monitoring'.

4.4. Groundwater in Decentralised Urban Water-Service Provision

Given the escalating rates of urbanisation globally, new paradigms for water-service provision need to be found. 'Closed-loop' operations for suburbs of 10,000–50,000 population on a 'decentralised basis' would appear to be an attractive alternative, either managed by the water utility or a community organisation. Such systems can be operated to minimise infrastructure costs, energy use and water losses, since they reduce the distance between household use and water treatment [18]. They can also promote energy and nutrient recovery by converting the current liability of wastewater treatment into assets, through urine/faeces separation and re-use of the former as a fertiliser and the latter for energy generation.

The natural drought resilience and quality protection of deep waterwells means that they are well suited to be the water-supply source for decentralised closed-loop water-service systems. Since these systems treat wastewater as a resource, their installation should substantially reduce the urban subsurface contaminant load, and thus one major groundwater pollution threat. Nevertheless, it will also be necessary to put more local attention and effort into on-the-ground inspection and control of other forms of urban land and groundwater contamination (such as gasoline stations, small-scale motor shops, dry-cleaning laundries, etc.) to prevent the pollution and loss of key waterwell sources.

5. Filling the 'Institutional Vacuum' in Urban Groundwater

There is a clear need for groundwater considerations to be integrated into decision-making on infrastructure planning and investment, but institutional capacity to take the lead is often limited because:

- water-resource agencies rarely have the capacity to cope with urban development dynamics;
- urban water utilities are often 'resource illiterate' despite increasing groundwater dependence;

- urban land and environment departments have a poor understanding of groundwater.

As a consequence, the strategic importance of urban groundwater is not yet reflected by sufficient investment in management and protection of the resource base. Urban groundwater tends to affect everybody but is often the responsibility of 'no body'. Municipal, provincial or state, and national governments need to find the political will and practical means to:

- constrain groundwater demand and limit abstraction by socio-economic and regulatory measures, so as to avoid excessive aquifer depletion and degradation;
- encourage the spread of groundwater abstraction for municipal water-supply over larger land areas, through the declaration of special conservation and/or protection areas;
- plan urban sanitation better and regulate the storage of industrial chemicals and handling of industrial effluents in the interest of much improved groundwater quality protection.

Governments, from national to local level, need to seek realistic policies and effective institutions to proactively manage urban groundwater resources. They will require political leadership and the engagement of major stakeholder groups. In pursuing these policies, it will be essential to make a critical appraisal of the actual and desirable roles of national/state water-resource and environment agencies, municipal water-service utilities, and municipal government offices responsible for land-use and industrial licenses—as well as the avenues of consultation and communication between them. This is required to address split institutional interests and responsibilities, and identify a mechanism for long-term periodic review and systematic action on urban groundwater.

Moreover, the dynamics of urban development and its relationship with groundwater are such as to merit the formation of a 'cross-sector urban groundwater consortium' (or standing committee) of all major stakeholders and regulatory departments/agencies. Such consortia should be tasked with communicating groundwater issues at the political and executive levels, and must be empowered and financed to define and implement a 'priority action plan'. They must also be provided with a sound technical diagnosis from an appropriate group of institutional and university specialists.

Funding: The writing of this review received no external funding.

Acknowledgments: The author would like to acknowledge stimulating interaction with the following people on urban groundwater issues: Ricardo Hirata (Brazil), Ken Howard (Canada), John Chilton (UK) and Radu Gogo (Romania), and the anonymous reviewers of this paper who made valuable suggestions for its improvement.

Conflicts of Interest: The author declares no conflict of interest.

References

1. Foster, S.; Hirata, R.; Misra, S.; Garduño, H. *Urban Groundwater Use Policy–Balancing the Benefits and Risks in Developing Nations*; GW-MATe Strategic Overview Series 3; World Bank: Washington, DC, USA, 2011. Available online: www.un-igrac.org (accessed on 1 September 2019).
2. Foster, S.; van Steenbergen, F.; Zuleta, J.; Garduño, H. *Conjunctive Use of Groundwater and Surface Water—From Spontaneous Coping Strategy to Adaptive Resource Management*; GW-MATE Strategic Overview Series 2; World Bank: Washington, DC, USA, 2010. Available online: www.un-igrac.org (accessed on 1 September 2019).
3. Buapeng, S.; Foster, S. *Controlling Groundwater Abstraction and Related Environmental Degradation in Metropolitan Bangkok, Thailand*; GW-MATe Case Profile Collection 20; World Bank: Washington, DC, USA, 2008. Available online: www.un-igrac.org (accessed on 1 September 2019).
4. Foster, S.; Hirata, R.; Vidal, A.; Schmidt, G.; Garduño, H. *The Guarani Aquifer Initiative—Towards Realistic Groundwater Management in a Transboundary Context*; GW-MATe Case Profile Collection 9; World Bank: Washington, DC, USA, 2009. Available online: www.un-igrac.org (accessed on 1 September 2019).
5. Tuinhof, A.; Foster, S.; van Steenbergen, F.; Talbi, A.; Wishart, M. *Appropriate Groundwater Management Policy for Sub-Saharan Africa in Face of Demographic Pressure and Climatic Variability*; GW-MATe Strategic Overview Series 5; World Bank: Washington, DC, USA, 2011.

6. Howard, K.W.F. *Urban Groundwater—Meeting the Challenge*; IAH Selected Paper Series 8; Taylor & Francis: Oxford, UK, 2007.
7. Taniguchi, M.; Dausman, A.; Howard, K.; Polemio, M.; Lakshmana, T. Trends and sustainability of groundwater in highly stressed aquifers. In Proceedings of the Symposium HS. 2 at the Joint Convention of the International Association of Hydrological Sciences (IAHS) and the International Association of Hydrogeologists (IAH), Hyderabad, India, 6–12 September 2009; Taniguchi, M., Dausman, A., Howard, K., Polemio, M., Lakshmana, T., Eds.; IAHS Press: Wallingford, UK, 2009.
8. Foster, S.; Hirata, R. Groundwater use for urban development: Enhancing benefits and reducing risks. *Water Front* **2011**, 21–29.
9. Foster, S.; Bousquet, A.; Furey, S. Urban groundwater use in Tropical Africa-a key factor in enhancing water security? *Water Policy* **2018**, *20*, 982–994. [CrossRef]
10. Resilient Cities and Groundwater. IAH Strategic Overview Series. Available online: https://iah.org/wp-content/uploads/2015/12/IAH-Resilient-Cities-Groundwater-Dec-2015.pdf (accessed on 1 September 2019).
11. Morris, B.L.; Seddique, A.A.; Ahmed, K.M. 2003 Response of the Dupi Tila Aquifer to intensive pumping in Dhaka-Bangladesh. *Hydrogeol. J.* **2003**, *11*, 496–503. [CrossRef]
12. Lyu, H.M.; Sun, W.J.; Shen, S.L.; Arulrajah, A. Flood risk assessment in metro systems of mega-cities using a GIS-based modeling approach. *Sci. Total Environ.* **2018**, *626*, 1012–1025. [CrossRef] [PubMed]
13. Lyu, H.M.; Shen, S.L.; Zhan, A.; Yang, J. Risk assessment of mega-city infrastructure related to land subsidence using improved trapezoidal FAHP. *Sci. Total Environ.* **2019**, 135310. [CrossRef] [PubMed]
14. Foster, S.; Hirata, R.; D'Elia, M.; Paris, M. *Groundwater Quality Protection—A Guide for Water-Service Companies, Municipal Authorities and Environment Agencies*; GW-MATe Publication; World Bank: Washington, DC, USA, 2007. Available online: www.un-igrac.org (accessed on 1 September 2019).
15. Sun, R.; Jin, M.; Giordano, M.; Villholth, K. Urban and rural groundwater use in Zhengzhou—China: Challenges in joint management. *Hydrogeol. J.* **2009**, *17*, 1495–1506. [CrossRef]
16. Foster, S.S.D.; Chilton, P.J. Downstream of downtown–urban wastewater as groundwater recharge. *Hydrogeol. J.* **2004**, *12*, 115–120. [CrossRef]
17. Gronwall, J.T.; Mulenga, M.; McGranahan, G. *Groundwater and Poor Urban Dwellers—A Review with Case Studies of Bangalore and Lusaka*; IIED Human Settlements Working Paper 25; International Institute for Environment and Development: London, UK, 2010.
18. Vairavamoorthy, K.; Tsegaye, S.; Eckart, J. Urban water management in cities of the future: Emerging areas in developing countries. *Water Front* **2011**, 42–48.

© 2020 by the author. Licensee MDPI, Basel, Switzerland. This article is an open access article distributed under the terms and conditions of the Creative Commons Attribution (CC BY) license (http://creativecommons.org/licenses/by/4.0/).

Article

Hydrogeological Studies Integrating the Climate, Freshwater Cycle, and Catchment Geography for the Benefit of Urban Resilience and Sustainability

Susie Mielby * and Hans Jørgen Henriksen

Geological Survey of Denmark and Greenland (GEUS), 8000 Aarhus C, Denmark; hjh@geus.dk
* Correspondence: smi@geus.dk; Tel.: +45-2055-5310

Received: 10 September 2020; Accepted: 19 November 2020; Published: 26 November 2020

Abstract: Today, there is an increasing need to understand how to link the management of the surface and subsurface to avoid disasters in many urban areas and/or reduce the likelihood of future risks. There is a need for thorough investigation of subsurface processes. This investigation should entail an analysis of water security, flood risks, and drought hazards in urban areas that may affect long-term sustainability and the ability to recover from disturbance, e.g., a capacity for resilience. In this context, as part of this analysis, potential biophysical and hydro-meteorological hazards need to be studied and subdivided according to geological, hydrogeological, man-made, and climatic origin, and by their characteristic temporal scales and site specific characteristics. The introduction of adaptive design and resilience in urban and suburban planning and management requires a shift towards more organic, adaptive, and flexible design and management strategies. This leads to the use of a complex cross-disciplinary methodology. We consider data collation, modelling, and monitoring designed to fit typical urban situations and complexity. Furthermore, implementation of strategic planning, decision-making to manage the consequences of future infrastructure and constructions are considered. The case studies presented are experiences from different hydrogeological studies performed in Odense, Denmark. Rising population and densification is affecting Odense, and there is risk of raised seawater level, groundwater, and surface-water flooding. The anthropogenic modification of subsurface structures and increased climate changes enhance the risk of hazards and the risk of coinciding impacts.

Keywords: urban; resilience; sustainability; hazards; subsurface; water cycle; land-use; infrastructure; planning; catchment; hydrogeology

---

1. Introduction

Under 'natural' conditions, the hydrological water cycle is in a state of balance. In numerous ways, however, basins are being artificially modified by land use changes and urban development. These modifications affect natural processes and can result in increased flood or drought risk, due to reduced storage and faster runoff, as well as numerous other impacts reducing the value of the ecosystem services that healthy aquifers and river basins can provide [1] (p. 31).

Cities and their residents can be particularly vulnerable to the negative impacts of man-made river basin processes and future changes resulting from climate change.

1.1. Urban Needs

Cities need to improve their understanding of how river basin management and nature-based measures can be hydrogeologically, ecologically, and socially constructed to create benefits in urban areas. There may also be competition in use of subsurface areas and water resources. These challenges are of increasing importance in light of the abovementioned issues of global population growth, increased urbanization, and potential climate change impacts [1] (p. 31), and planning and management must focus on the consequences in order better to protect current and future interests [2].

1.2. State of Focus on Implementation

The International Water Associations (IWA) initiative on river-basin connected cities has primarily been focused on surface activities and runoff. However, IWA notes that catchments are predominantly defined by surface topography and that groundwater systems are defined by geological basin geometry and climate. Consequently, the involvement of the groundwater, with its 3D complexity, is seen as a possibility for improving the planning of water resilience and sustainability [1] (p. 39). As the subsurface is an important constituent of the physical environment of cities, better urban subsurface knowledge and communication of this knowledge to decision-makers have been the focus for the COST Action TU1206 Sub-Urban [3,4]. This COST Action TU1206 Sub-Urban, comprised a study of 26 cities (with participating geoscientists in 31 countries), to identify common knowledge and communication gaps between subsurface experts, urban planners, and decision-makers [5]. They stated that "the only possible way to bridge this gap is to provide the right type of subsurface information in the right format and at the right time and to ensure that decision makers and urban planners are able to understand and use this information to make decisions". The need is to better understand where in the planning hierarchy and processes different types of information are needed [6]. Other recent projects (NORDRESS [7,8]) have focused on societal security in relation to personal, community, infrastructural, and institutional resilience—with resilience being defined as "an integrated, learning-based approach to management of human-ecological interactions, with explicit implications for planning interventions and resulting design forms" [9]. This means that urban designs and interventions must be adaptive and resilient to a change that cannot be predicted with certainty or controlled completely. Resilience therefore, is a broader concept than simply managing infrastructure failure and damage levels (for example due to flooding) guided by a single discipline or expertise. Instead, meaningful community engagement in the planning and adaptive design process is necessary. Key issues include the ability to learn and to make collective decisions, and the ability for different actors to influence risk management. No single country, region or city has achieved complete competency on all four points [8].

For interacting systems like urban landscapes and groundwater the long-term sustainability, and health of landscape and their underlying groundwater has three major dimensions—environmental, economic, and social. Social-ecological research argues that these dimensions should not be addressed by 'silo' approaches. Instead, emphasis must be on system level characteristics such as resilience or adaptive capacity of human–technology–environment systems [10], where management and adaptation are adaptive, resilient, and reflective, honoring the catchment and subsurface that sustain the city [11].

1.3. Strategy for the Work Achieving Resilience and Sustainability

The implementation of Sustainable Development Goals (SDGs) implicitly calls for a system of models and plans, which target research toward something more durable (long lasting and longsighted), more diverse (with different properties and scales), and more robust (with the ability to deliver replicable solutions and assessments). SGD 6, among others, deals with access to clean water, prevention of pollution, integrated water resource management, and protection and reestablishment of ecosystems. SDG 11 deals with inclusive and sustainable cities, protection of culture and nature, reducing impacts of disasters, and increased access to green areas. SDG 13 deals with resilience against climate related risks and hazards, sharing of knowledge. SDG 17 deals with partnering for shared knowledge, development, technology, and monitoring.

We argue for the importance of providing and enhancing specific knowledge about all conceived needs and potential hazards of the water cycle from an early stage in order to avoid later surprises. It has been the tradition for many years to elaborate new models in order to solve emerging problems, which implies that the bases for planning and management rely on different data and thus are often not comparable. Therefore, a more systematic mapping and modelling of the landscape and urban subsurface must be based on the characteristics in the geographical area; the climate and predicted changes in that area, the degree of subsurface stability (abstraction and infiltration), land use, drainage, etc. But, the mapping and modelling must also consider the broader aspects of resilience connected to human-technology-environment systems [10–14].

1.4. Aim of the Paper

In this paper, we first focus on a study of geological and hydrological processes and man-made impacts including the relevant parts of integrated planning (such as construction, economy, green areas, occupation, infrastructure, water environment, etc.) [2]. Then, we consider the use of technical solutions for enhanced and cohesive decision and planning support. We mainly build our understanding on years of studies and experience. This is gained from existing hydrogeological models and monitoring projects in the Danish municipality of Odense, along with the selected 'sustainability' indicator (e.g., depth to shallow groundwater table), elaboration of a monitoring strategy, and the consequences tolerated for individual indicators inspired by the Monitor–Data–Indicator–Assess–Knowledge chain (MDIAK) (see Figure 1, modified after The European Environmental Agency (EEA), [15]). Furthermore, we build on the understanding of the person–community–infrastructure–institutional societal resilience framework developed in NORDRESS [7,8] in order to minimize risks from hazards and integrate scientific knowledge.

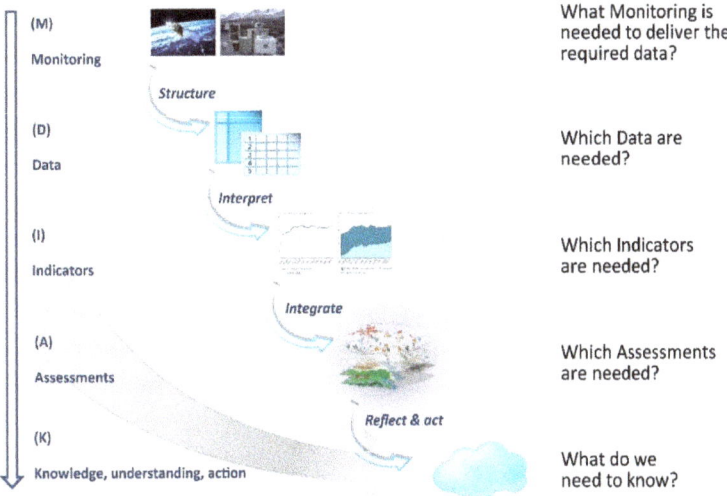

Figure 1. The Monitoring–Data–Indicator–Assess–Knowledge (MDIAK) reporting chain leading to knowledge, understanding, and action (on the right-hand side the same chain can be used to design assessments, communicate results and support the monitoring).

Odense City is used as example. Odense City has a risk of impacts from man-made activity and, due to its position near the bottom of Odense River catchment and the coast, it is at risk of groundwater flooding, surface-water flooding, and raised sea-level. This implies, associated with climate change modifications, that there is an increased possibility of areas with various coinciding hazards.

2. Analysis of Natural and Man-Made Processes in Regard to Urban Studies

To be able to perform knowledge-based decisions in urban planning, all important natural and anthropogenic processes must be taken into account. These processes have to be characterized and analyzed if their impacts are costly or represent a threat to health, cultural heritage, or ecosystems.

Key elements that have to be considered include climate, the water cycle, and natural and man-induced hazards. Table 1 compares the size and extension of natural and man-made hazards in the urban landscape. In addition, common relationships between the climate and man-made impacts have to be studied. The use of data, modelling, and monitoring have to be considered for improved planning and management.

Table 1. Estimated spatial extension of urban processes related to climatic, geological, and man-made hazards (capital bold letters indicate the main spatial regime). Modified after The PanGeoProject [16].

Hazard Driver	Processes	Local Unit (Construction)	City Quarter	City Landscape	Metropole	Hydrological Basin
Meteorologically Induced event	Climate Change, Cloudbursts, Droughts, Sea level Rise /Storm Surges			x	x	X
Geologically Induced Deep Ground Movement	Earthquakes, Tectonic Movements, Salt Tectonics, Volcanic Inflation/Deflation			x	X	x
Geologically Induced Ground Instability	Land Slide, Soil Creep, Ground Dissolution, Collapsible Ground, Running Sand/Liquefaction		X	x		
Geologically Induced Ground Movement	Shrink-Swell Clays, Compressible Ground	X	X			
Man-made Ground Instability	Shallow Compaction, Peat Oxidation, Groundwater Abstraction, Mining, Underground Construction, Artificial Ground, Oil and Gas Production, Surface Excavation	X	X	x		

2.1. Natural Hydrological Processes Impacting Urban Freshwater Cycle

The hydrological cycle is based on an equilibrium state between precipitation, evaporation, surface runoff, infiltration, and groundwater flow. Any modification in the hydrological catchment results in changes towards a new state of equilibrium. Precipitation, wind, and temperature are important meteorological drivers of the hydrological cycle that have to be considered. In Denmark, increasing precipitation has been observed within the last century and further climate changes are predicted in years to come that will increase water levels, facilitate cloudbursts with floods, and produce dryer summers with droughts. Moreover, a rise in sea level is expected.

Increased precipitation intensity can nearly instantaneous induce streamflow alterations but no corresponding impact on groundwater is detected. Whereas, long-lasting meteorological events, such as droughts or wet seasons, can be associated by similar changes in groundwater levels identified by the regional monitoring and need to be taken into account in urban planning.

2.2. Geological Processes in Urban Areas

The landscape form and ground permeability of near-surface layers largely control runoff and infiltration. A terrain characterized by steep slopes combined with high clay content in the upper soil layers usually results in a large amount of surface runoff. On the contrary, infiltration into the ground increases, if the land-slope is gentle and near-surface layers have a higher permeability. Thus, geological mapping provides important knowledge for the characterization of an area (See Tables 1 and 2).

Table 2. Natural hydrological processes in the freshwater cycle affecting urban planning (capital bold letters indicate impact magnitude).

Hydrological Processes	Groundwater Re-/Discharge	River Basin	Groundwater System
Meteorology (net precipitation)		X	x
Surface runoff	X	X	
Soil and Unsaturated zone percolation	X		X
Stream flow		X	x
Groundwater flow	x	X	X

Urban geology drastically influences city development, as well as its freshwater cycle [4] (pp. 39–41). Within the urban subsurface, the occurrence of even smaller thicknesses of organic deposits or sensitive clays can cause geotechnical problems for the foundations of buildings and cultural heritage. Such problems increase with changes in the groundwater level, as can be seen in e.g., Denmark and The Netherlands. Another example, relating to some of the Mediterranean countries, is tectonic and volcanic activity released forces resulting in uplift, changing flow directions in streams and aquifers.

2.3. Anthropogenic Deposits Resulting from Urban Evolution

Urban anthropogenic deposits are a consequence of urban growth, land-use changes, and technological development. These deposits consist of buildings, bricks, pipes, tunnels and fills and they reflect anthropogenic history. These deposits need to be judiciously studied as they impact traditional hydrogeological modelling. To understand and interpret urban geology, it is necessary to include a mapping procedure for the anthropogenic elements, and the developments over time [17] have to be taken into consideration. For example, the vast majority of Danish cities have developed in the past 200 years (see Table 3). Consequently, the city center will have a sequence of subsurface layering affected by all historical activities, while only recent ones will occur in the youngest parts of the city.

Man-made activities are able to affect groundwater levels, superficial groundwater runoff, and surface runoff. In Table 4, examples of man-made urban impacts affecting the hydrological cycle within the last 200 years are summarized and classified. In addition to the originally targeted purpose of constructions, unwanted and often unforeseen impacts can also occur (e.g., blocking and intersection of aquifer flow [18,19]). Lowering of the groundwater tables can result, simultaneously with declining groundwater quality due to infiltration from contaminated sources—for example, in Odense [20,21] and Copenhagen [20–23] pollution from surface water and saltwater intrusion, have occurred, and pyrite-oxidation with increased sulphate concentrations in groundwater related to the unsustainable drawdown of groundwater levels near wellfields [22,23]. Furthermore, lowering of the water table in several cities has resulted in a compaction of the subsurface, inducing a land subsidence as a result (e.g., in Shanghai [24]). Stopping or reducing groundwater abstraction can result in groundwater level and quality changes, as has been seen in Odense [20,25].

Table 3. Activities in Danish urban areas; historical evolution within the last 200 years (Compilation of the evolution of urban areas based on Cartographic maps from Odense).

Man-Made Action	Main Activity Years
Densification of the cities.	–Now
Anthropogenic deposition (especially during the first 100 years), along with urbanization.	1840–1940
Increasing pollution related to the industrialization process.	1840–1980
Moving of rural population into the cities	1840–Now
Piping of natural streams and draining wetlands.	1870–1980
Increasing fortified areas and infrastructural constructions.	1900–Now
Increasing groundwater abstraction of groundwater for households and industry, until groundwater pollution from industries and human activities reduced groundwater abstraction.	1940–1970 1970–Now
Subsurface construction with the technological development.	1940–Now
Introduction of Danish Planning Act, intended to subdivide cities into separate areas for trade, industry, and dwellings.	1970–Now
Initiating extensive cleaning of industrial sites and landfills.	1982–Now

Table 4. Examples of man-made urban impacts affecting the hydrological cycle within the last 200 years (Table based on authors experience from Odense and other places).

Man-Made Activity	Urban Impacts
Increased pavement extension, especially within the past 200 years.	Modifying infiltration and evaporation, as well as increasing surface runoff.
Establishing large construction works (buildings, pipelines, tunnels, roads, etc.)	Inducing modification of surface runoff and of groundwater flow levels.
Groundwater abstraction	Affecting both groundwater flow and level.
Draining of wetlands	Increasing surface runoff.
Land surface draining	Lowering of groundwater level.
Laying water pipes and sinks in trenches filled with sand and gravel.	Local change of upper groundwater flow direction and velocity (convergence with 'channels' and sinks).
Sustainable Drainage Systems (SuDS) established to reduce excess rainwater.	Local increase of ground water level.

The land surface near shallow groundwater stratum can also be affected by regulation of ditches and watercourses, which affect surface runoff. Abstraction in the upper end of the catchment can affect runoff in the summer seasons, which will be of special importance for the ecosystems.

2.4. Climatic Impacts on the Hydrogeological Catchment Related to Anthropogenic Activity

When considering the effect of man-made activities in a catchment, it is also necessary to include the impact of expected climate change. Climate changes can result in a change of sea level due to melting polar ice, and modifications of the precipitation and evapotranspiration regimes in terms of quantity and distribution. Identification of potential impacts on existing buildings, cultural heritage areas and ecosystems is the main focus of climatic-change studies It is important to provide a common understanding of the likelihood of impacts from other man-made processes. Mitigation and adaptation will typically require a coherent sequence of assessments in order to clarify:

- Where in the catchment climate change will have an (unacceptable) effect on surface runoff and groundwater levels?

- Where is it technically possible to reduce and delay these unacceptable impacts?
- Where in the catchment should potential adaptation and mitigation solutions have an acceptable influence on groundwater levels and surface runoff?

In Denmark, implementation of SuDS (Sustainable Drainage Systems) is commonly used to separate rainwater from grey water. The challenge of using SuDS in these areas is to infiltrate excess water into the subsurface. If infiltration is possible, it must be checked that raising groundwater levels in the infiltration area and surrounding land are acceptable. An example is Vinkælderrenden/Skibhuskvarteret in Odense; after a detailed mapping and hydrological modelling, it was realized that the groundwater level elevated unacceptably in the neighboring areas. Therefore, an alternative solution had to be found.

2.5. Influences of Human Settlements

For millennia, our ancestors have accessed water bodies for drinking water, fishing, and easy transport. Many cities were, therefore, founded near rivers and near the coastline, and thus at the bottom of hydrological catchments. Activities in the entire upstream catchment thus require consideration, with its boundary often situated at considerable distance from the city and its peri-urban areas—because man-made activities at the top end of the catchment can have downstream impacts. The management of rural areas situated upstream of cities is usually done with a different perspective with agriculture being the main focus. In general terms, the draining of agricultural land, watercourse straightening, and crop cutting, together with increasing paved areas, all tend to increase surface runoff and flood risk in downstream urban areas. Extreme flooding events are expected to increase with predicted climate change with more warm and wet winters, and more intense cloudburst events in the summer. A further complication arises, when catchment areas cross authority boundaries and the management of these upstream areas is the responsibility of another authority. Thus, the management structure and relative position of a city and its boundaries are of importance for determining resilience and sustainability.

2.6. Essential Added Knowledge for City Management and Planning

Traditionally, urban planning and management have been based on thematic maps containing information about constructions, infrastructure, occupation, and land-use. Since urban areas also have a buried infrastructure (cultural and archaeological sites, subsurface constructions etc.), such topics as aquatic ecosystems, aquifer vulnerability, thermal and mineral resources, ground stability, and waste disposal all require hydrogeological consideration for hazard identification and risk management. For hydrogeological planning and management purposes, there is a need for a regularly repeated process (for each planning period) that also includes relevant urban issues, rather than site-specific projects covering only one specific problem. There is a need for stepwise knowledge building—starting with: existing maps and data sets, new data capture and data collation, and basic hydrogeological model building, assessments, and use of knowledge in planning (Figure 2). This requires building relevant and conceptualized information—for example, temporal and geographical scale, history, demography, infrastructure, land-use, and hydrogeology—to be represented in three dimensions and visualized with integrated GIS at a higher level than previously.

The process contributes to a large amount of data and flexible use of selected information is needed in order to avoid starting over again whenever new situations arise. Thus, there is a need to determine which information should be made digitally available, and how this can be performed by which digital processes. The evaluation of existing, digitally-available hydrogeological data and models from previous studies, and the evaluation of supplementary data, are important parts of the analytical process and working methods for determining increased resilience and sustainability.

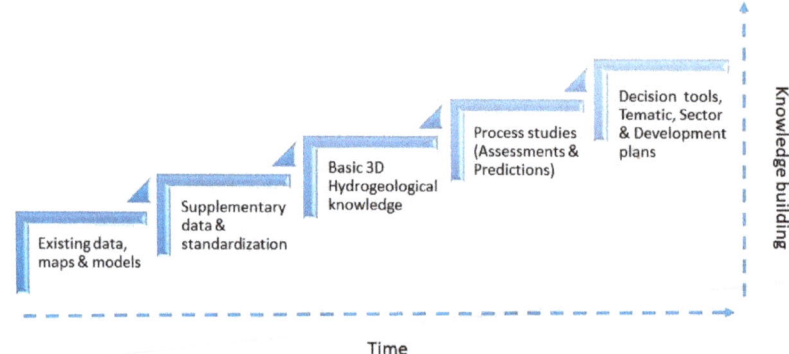

Figure 2. Urban hydrogeology system with step-wise knowledge building process: from screening existing data to development of decision-making tools and plans.

Urban management and planning, involving natural and anthropogenic processes requires a systemic focus on the following data:

- Basic hydrogeological information system (including climate, hydrogeology, surface-water catchment and sea level, and potential changes to them);
- Drinking water-supply and waste water plans;
- Urban infrastructure and environment (including issues to support infrastructure development, cultural, and environmental risk assessment);
- Planning hierarchy and its management.

What is important is that urban management and planning activities require a good understanding of current hydrogeological and hydrological conditions, a good comprehension of their evolution over a long time, and a reliable estimation on the consequences of potential changes expressed in terms of modelling scenarios. It involves natural and anthropogenic processes from two viewpoints, contributing to greater resilience and sustainability:

- A geoscientific urban perspective that includes new model building, monitoring, and assessment;
- An adaptive planning and governance perspective that handles the necessary information and its correct use.

3. Hydrogeology in Odense (Denmark)—Methods of Mapping and Assessment

Odense Municipality (Denmark) is a good basis for a practice example, since the water supply, the municipality, and the former county authority (the Danish counties were closed in 2007 as a result of a local (municipalities) and regional (counties) governmental reform) have had a long tradition of data collection and groundwater mapping. Therefore, data and experience are available for further knowledge building.

3.1. Subsurface Mapping and Assessments

It is important to use a robust hydrodynamic subsurface model (anthropogenic and geological) and relatable time series that can define indicators to assess the groundwater dynamics of selected aquifers. One of the keys to replicable scientific work in Odense is using common public databases with relevant, standardized, and correct datasets that are easy to access. This ensures that the same basis for decision-making and for information updating are used. The foundation for this work was the development of a 'new' city-scale 3D geological and hydrogeological model that provides a coherent and updated basis for the urban water cycle analysis and for management of the urban subsurface.

Depending on the specific situation, one must select hydraulic assessment models that are built on these datasets (see Figure 1). The connection with the work performed in knowledge-building (from data collation to hydrodynamic model) is important for the following hydrological assessments and monitoring. Dealing with hydrogeological modelling we differentiate between existing project situations and a new municipality model with a year-by-year updating. Figure 3 illustrates the four steps consisting of Subsurface data, Hydrodynamic subsurface modelling, Hydrological assessment and Monitoring.

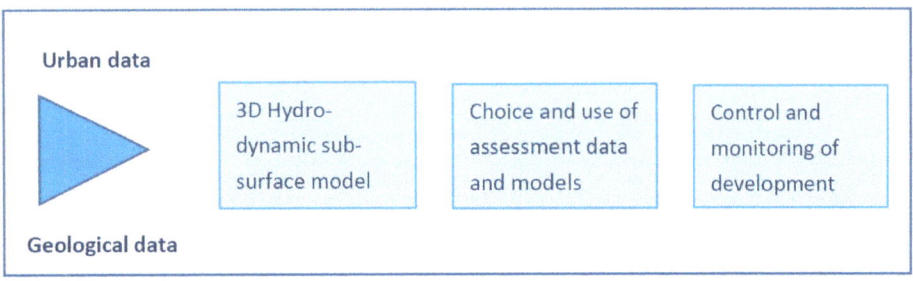

Figure 3. The four steps: Subsurface urban data, Subsurface modelling, Assessment, and Monitoring.

Requirements of a City Scale Concept Hydrodynamic Model Framework

First, a city-scale concept is required to allow the combination of anthropogenic and hydrogeological information. It is essential that the modelling, in selected areas, is based on detailed knowledge of natural and man-made processes and their potential impacts. When the modelling of subsurface knowledge is extended to the entire municipality systematic mapping rather than a project-based approach is mandatory. This is a cross-disciplinary process with several challenges:

- The established hydrogeological model is based on heterogeneous information from the existing boreholes, geophysics, maps, and models of varying quality and age;
- The anthropogenic man-made deposits beneath the cities differs in age, scale and origin from geological deposits. Collation of data and modelling on anthropogenic deposits are related to the urban history and deviates very much from hydrogeological processes;
- A workflow for interlinking and merging different kinds of models has to be designed;
- A definition of due care in regard of the maintenance of the modelling is necessary, and has to be defined from the outset;
- A design of information output has to be defined.

3.2. Development of 'New' City Scale Hydrogeological 3D-Model

In Odense, the national geological survey, the municipality, and the local water utility found in 2012 that most geological and hydrogeological mappings was in rural areas, but similar activity for the urban areas was lacking. Thus, a goal was established to produce one model at a municipality or city scale, which, through sustained maintenance and development, could be improved year-on-year with the introduction of new geological and hydrogeological data and information on anthropogenic changes. The focus was put on groundwater and surface water, the connection between deep and shallow aquifers, the secure use of existing models all relevant data, large-scale overview mapping, and detailed small-scale mapping, and based on the selected typical work situations (see Table 5 and Figure 4). A hydrogeological 3D municipality model project was established through a pilot study ('Development of a 3D Geological/Hydrogeological model as basis for the urban water cycle' [25–31]). The pilot-project was completed in 2015 with a report [25–31], and the developed model was selected as one of the examples of good practice of urban modelling by the European COST Action TU1206 Sub-Urban project ([6,32] (pp. 64–74)).

Table 5. Adaptive 'real' planning and governance projects chosen for design of the 'new' urban hydrogeological model in Odense.

Management and Planning Situation	Study	Test Location
Early project evaluation for major construction planning, green/blue solutions, and prospects for geothermal plants	Potential areas for infiltration and runoff	Odense new University Hospital
Adaptive planning project for re-urbanization after the abandonment of a four-tracked road through the heart of the city	Collection and use of old and new geotechnical data and visualization	Thomas B. Thriges Gade
Well-field study of the effect of stopping urban groundwater abstraction and potential use of ditches for surface run-off	Mapping of existing low-lying areas with poor possibilities of surface runoff	Sanderum Tørvehave
Residential area project to increase the water infiltration into soil due to high surface runoff	Collection and use of new detailed data and modelling (see [25])	Vinkælderrenden—a part of City quarter Skibhuskvarteret.

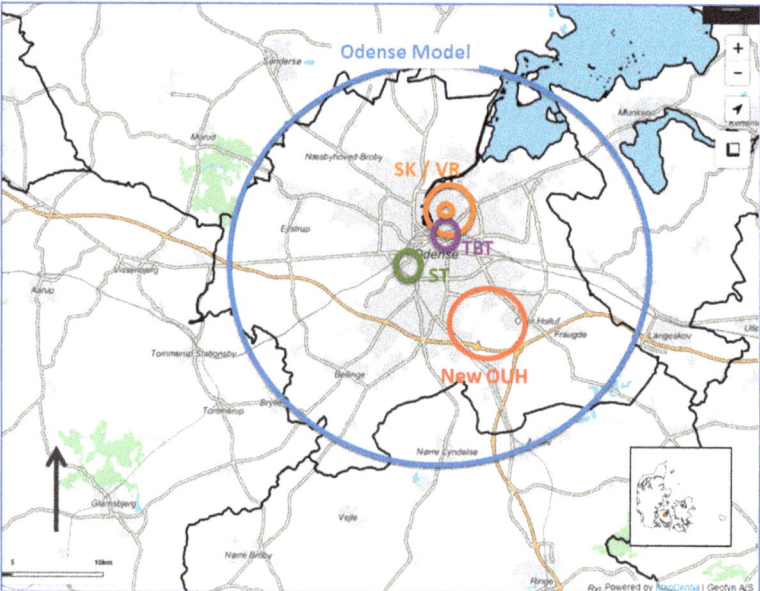

Figure 4. Location of projects used for development of a conceptual design of The Odense Model. (SK/VR) Skibhuskvarteret/Vinkælderrenden, (TBT) Thomas B. Thrigesgade, (ST) Sanderum Tørvehave, (New OUH) New Odense University Hospital (for descriptions see Table 5).

The following four case-studies illustrate several important technical aspects for the Odense urban subsurface and hydrogeological model:

The **first** case-study focuses on the design of a hydrodynamic model area for the entire municipality level area and some separate detailed areas. Analysis has been performed on the current and potential planning and management activities (see Table 5) that were compared with existing data and models. The analyses searched for optimal solutions. This model should be able to operate on a municipal level and in selected areas (constructions and up to city-quarter, see Table 2) with a detailed discretization.

Consequently, the Municipality model had to be built on existing digitized data and include the following elements: deep and shallow aquifers, deep and shallow geology, man-made deposits, and constructions. Figure 5 shows a view of the resultant conceptual model.

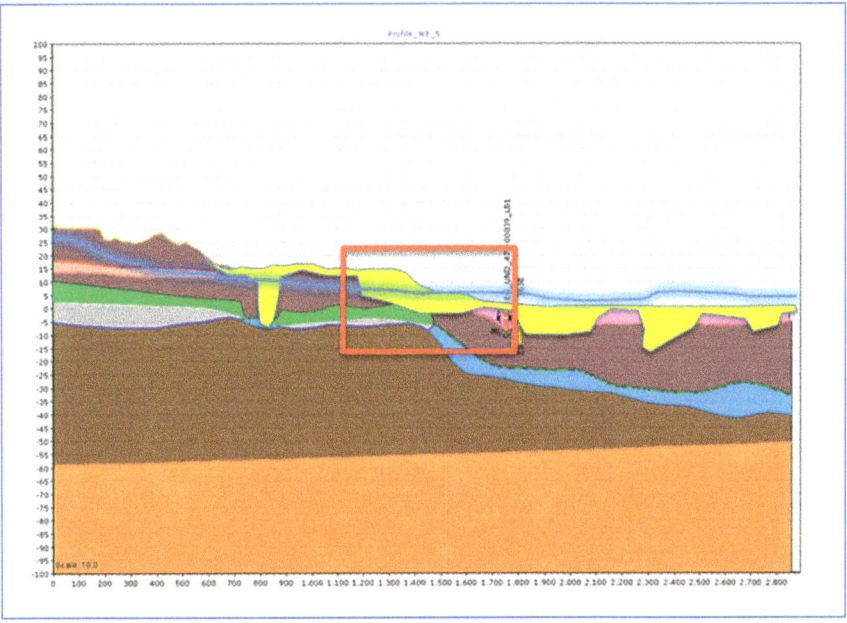

Figure 5. Conceptual frame for the hydrodynamic municipality model [25,29]: Yellow layer indicates anthropogenic origin; blue line is estimated hydraulic groundwater level, and red box indicates a detailed modelling area. Scale in meters.

The **second** case-study focuses on hydrogeological modelling beneath cities. The existing data from Odense city revealed areas with only limited or old data (see Figure 6). The lack of sufficient site-specific data [33] caused difficulty in making sound decision-making and assessing uncertainty, and led to the need for gathering supplementary data. The operation of assembling data, especially of geotechnical and cadastral types, proved to be time consuming, and data updating and modelling required professional administrative governance.

Modelling of the urban anthropogenic layer represents the **third** case-study. This layer consists of bricks, buildings, constructions, pipes, roads, excavations, urban fill, and many other materials of different age. The modelling process includes cadastral data coming from the municipality, data characterizing the sewer system and water supply network (from the local water utility, VCS Denmark), as well as borehole logs and geophysical surveys (from GEUS and VCS Denmark).

The information is frequently updated as result of urban dynamics. A rule based data handling procedure was chosen to process and update systematically the huge amount of data—GeoScene3D Software (www.GeoScene3D.com) being selected for data processing. Finally, a detailed voxel model for the anthropogenic layer of the Thomas B. Thrigesgade (Odense City) has been obtained, which is illustrated in Figure 7.

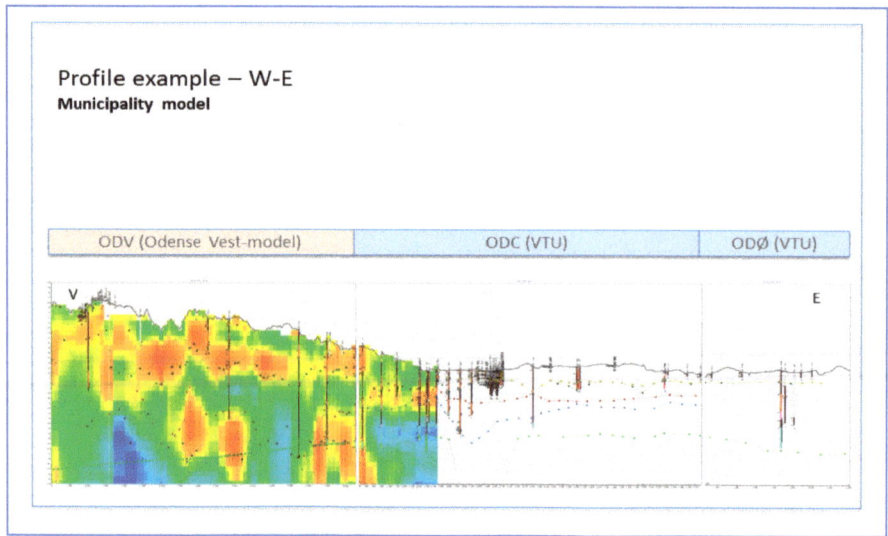

Figure 6. Variations in data density in the subsurface information in Municipality model on west–east profile [25,29]. The western side of the urban area is shown with a comprehensive groundwater mapping illustrated by the 'colored' results of the geophysical survey (Odense Vest model). The city center is located in the middle profile and has many shallow boreholes, and on the eastern side limited groundwater interests are illustrated by restricted data availability.

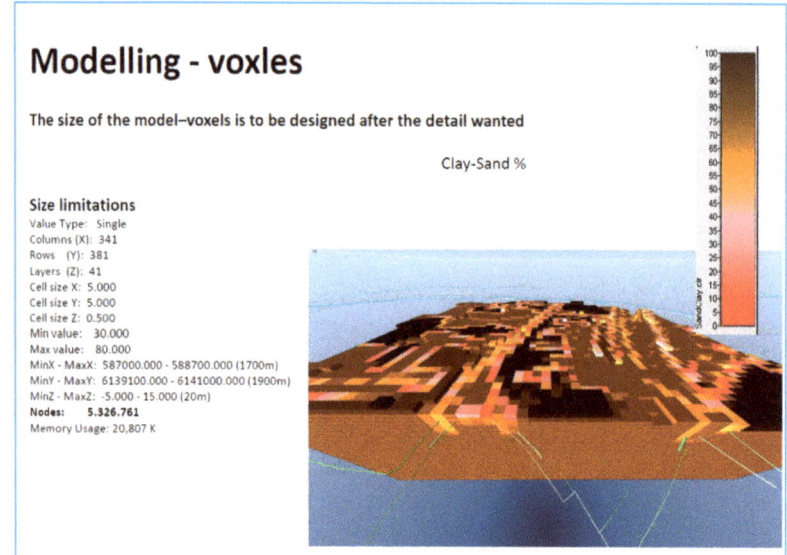

Figure 7. Visualization of detailed voxel modelling of an urban area [25,30] with buildings, and transecting boxes for roads and, pipes converted to clay-sand % and combined with different soil types.

Defining the procedure of working with different models and model scales stays as the **fourth** case-study. Urban modelling often requires the use of information from other sources in order to augment model information or regional trends. To combine geological models and anthropogenic models a common workflow is required. Figure 8 illustrates the anthropogenic data, the resulting anthropogenic layer and the combined anthropogenic-geological model. Figure 9 shows a workflow for merging a municipality model and a detailed anthropogenic model.

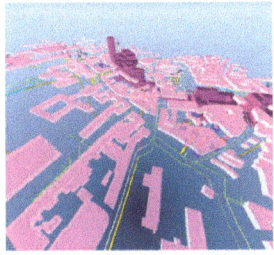  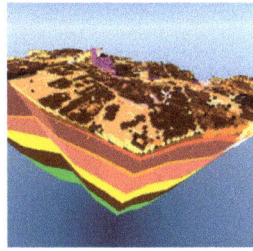

Figure 8. Visualization of Thomas B Thriges District (in dark purple). On the left is anthropogenic data; in the middle the anthropogenic layer; on the right the combined geological and anthropogenic model [25,30].

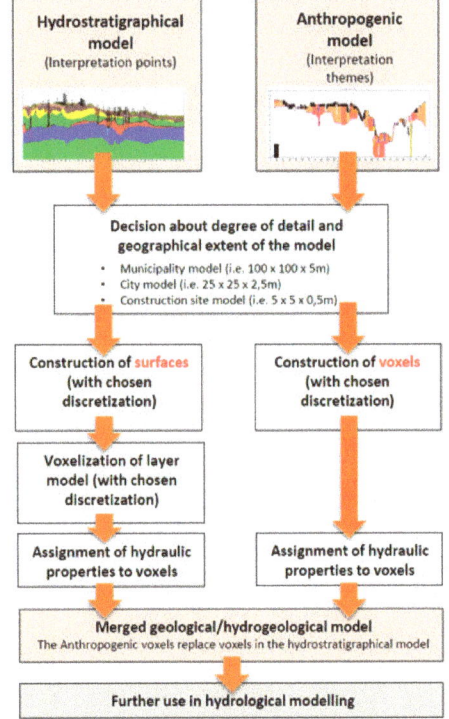

Figure 9. Workflow for the combined geological and anthropogenic model [25] of the urban area.

3.3. 'New' Approach for Monitoring Groundwater Levels in Odense

Monitoring of shallow groundwater levels offers important information in urban hydrogeological studies, especially if they are part of long time-series. The monitoring data can offer a valuable picture of the spatial and temporal evolution of groundwater levels and urban groundwater dynamics. The monitoring of complementary parameters, such as water temperature and quality, can offer valuable additional knowledge.

In Denmark, observation of groundwater levels mainly takes place in three types of monitoring wells. At the national level, monitoring is normally performed in boreholes outside urban areas and is used to study natural processes (e.g., changes in winter net precipitation on groundwater level) in areas that are not affected by abstraction and other man-made influences. The second level relates to licenses for groundwater abstraction, and the monitoring wells are situated near wellfields. Finally, the third level relates to the collation of additional data for hydrogeological surveys, for example for detailed modelling of groundwater levels due to the infiltration of excess rainwater (as seen in Vinkælderrenden in Odense) [34]. Groundwater monitoring for the preservation of ecosystems, and the cultural heritage is not a standard procedure.

In Odense City, comprehensive monitoring networks, which were established in the mid-1960s for wellfields, were later abandoned. Today, this information would have been useful for detailed urban studies, but a valuable picture of the evolution of groundwater levels since the 1950s is given by the national monitoring system (See Figure 10).

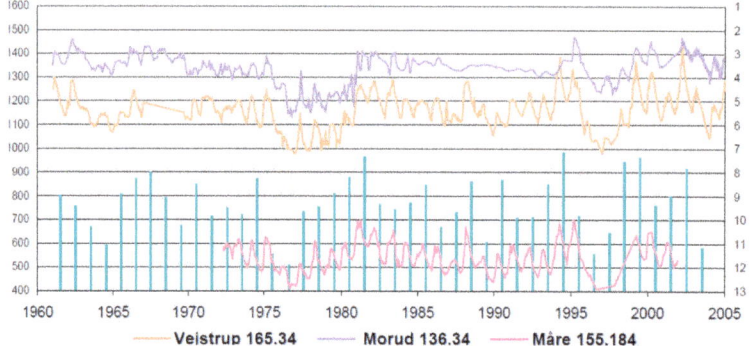

Figure 10. Illustration of development of groundwater level in surroundings of Odense. The figure shows annual precipitation (mm on left-hand axis, blue columns) compared with groundwater depth in the three national monitoring wells (in m below ground level on right-hand axis) (Vejstrup 165.34, Morud 136.34 and Måre 155.84) [35]. Data from Geological Survey of Denmark and Greenland and Danish Meteorological Institute, processed by the local water utility (VCS Denmark).

Earlier urban monitoring focused on the impacts of groundwater abstraction on neighboring wells and drinking water quality, but today there is a tendency to focus on the impacts of changes in groundwater abstraction patterns and their impacts of subsurface constructions and infrastructure.

The frequency of monitoring in national monitoring wells has been improved in time. In early days, this monitoring was performed by discrete observations (just a few times per year) but this was later improved to monthly readings, and today, data-loggers measure groundwater levels remotely up to several times a day (for studying the relation between shallow aquifers water-level and wetlands).

3.4. Urban Groundwater Studies on the Interaction of Hydrological Events

A master study of the Nordic Centre of Excellence on Resilience and Societal Security (NORDRESS) [36] illustrated that the forced infiltration of rain-water from paved areas (as investigated in Skibhuskvarteret), leads to a marked increase in groundwater levels close to the infiltration basins. The local geology appears to exert a major influence on this phenomenon, since high hydraulic conductivity is needed to 'spread' the infiltrated water. Increases of groundwater level of up to 1 m, are modelled in the Odense area due to increased winter precipitation, and an increased sea-level of a similar magnitude occurs near the Odense River and Fjord (due to global warming and sea level rise and storm surges), which challenges water management activities for climate change adaptation and risk reduction.

Another NORDRESS master study [37] evaluated previous assessments of Odense flooding risks and established that without adaptation work, flood damage at Odense will increase from 3.9 million of Danish kroners in 2010 to 9.3 million of Danish kroners by 2100. The study [37] revealed a projection of 239 million of Danish kroners for the expected increase in damage costs resulting from climate change by the year 2100 and highlighted the need for a cost–benefit analysis of potential adaptation options.

The need for adaptation is further challenged by the expected increase of maximum runoff of the Odense River. Several reports [37,38] suggest a 40% increase in the daily maximum 100-year return period runoff event in Odense River by year 2100. The NORDRESS study [37] highlighted that the inundation resulting from concurrent events from storm surges and river flooding could result in damage costs of some 833 million of Danish kroners for publically owned properties. This does not include critical infrastructure damage costs nor the potential interruption of commerce due to inaccessibility.

In Odense (Figure 11 [37]), the interacting impacts from different hazards are important for urban planning strategy. It is critical to manage disaster risk reduction, climate change adaptation, and water-supply security in an integrated and adaptive way.

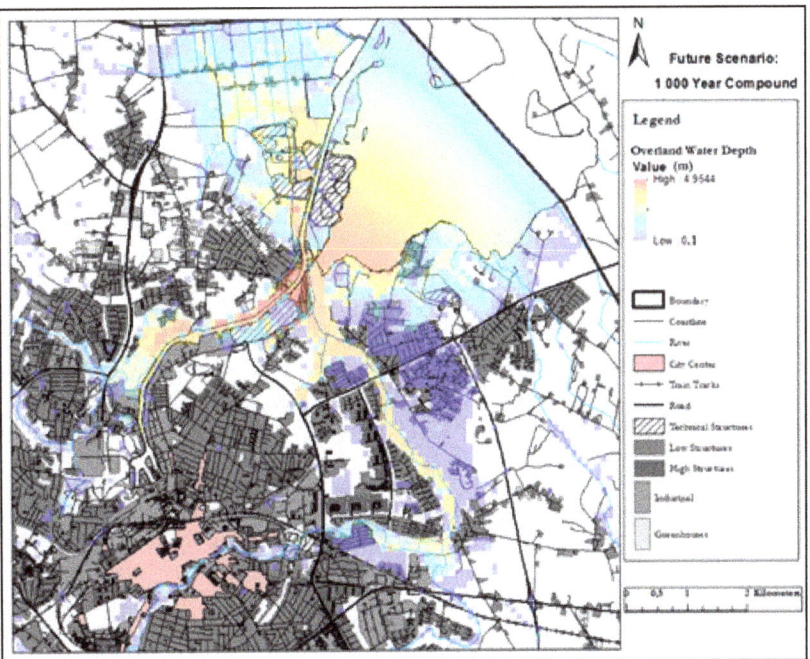

Figure 11. Scenario of potential climatic coincidence of increased runoff in Odense River and increased sea level in Odense Fjord. Resultant flooding (yellow to red) is observed along downstream part of Odense River [37].

3.5. 'New' Concept of Adaptive Planning and Governance

Classical modes of planning were grounded in rationalistic, authoritarian planning model where politically described goals were translated into plans implemented by the administration. Contrary, to this adaptive planning and governance is grounded in neo-institutionalism, social-science based approaches which consider governance, collective action, informal rules and the redefinition of spatial planning toward "place" governance. Instead of focus on scientific-rational development plans, in the new concept, focus is on communicative determination of needs, the joint development of plans in participatory processes and the activation of citizens to participate in spatial planning [39]. Adaptive planning and governance therefore focus on polycentric governance e.g., cross-sectoral and multi-level coordination with focus on learning capacity and adaptability of plans and actor networks. The main advantage of the new concept compared to the old classical mode of planning is that it enhances adaptability and capacity for self-organization and learning [40]. Learning here is understood in different types e.g., reflexive social learning, instrumental, political, and symbolic learning [39] which can be regrouped into institutional and individual learning [41].

As in many other cities, Odense has an increasing need to use its subsurface efficiently, in order to balance conflicting uses and avoid natural and man-made hazards, reduce the likelihood of future climate change risk, and to safeguard its ecosystem services. The city needs to be able to adapt, plan and manage according to transparent governance principles at the water catchment, city, and street dimensions. This means that adaptation plans and measures should be prepared, implemented, and followed-up regularly [11]. City management cannot wait for additional measuring, collating, and processing fundamental information before acting. Thus, leadership is important. From the outset, this work must be systematic and well described to a level fit-for-purpose.

3.5.1. Towards a More Resilient and Sustainable Urban Area

In Odense [21,25], historical development has illustrated on several occasions the consequences of man-made activities years previous, which were unforeseen or negative.

In the COST Action TU1206 Sub-Urban Project [42], it was important to make a correct inventory of problems (screening the problem to solve it), to have proper mapping (screening of required data and processes), and then to be able to make use of this mapping. Thus, hydrogeological and geological characterization had to be based on geoscientific skills and expertise with a focus oriented towards the required scale, depth, land-use, and timing, with upscaling of the expected impacts on surface-water and groundwater. Planning and management had to be based on mapping of the history of urban development, as well of its geology.

3.5.2. Strategic Elements for Management of Urban Subsurface

The strategic elements in the European Environmental Agency's Monitor-Data-Indicator-Assess-Knowledge (EEA MDIAK) decision chain are given in Figure 1. This contains useful tools for integration and management of the urban subsurface. Focus is put on parts of planning, modelling, monitoring, and mapping that deviate from traditional hydrogeological techniques for rural areas—and they include city planning (construction, city quarter, city landscape, metropole, and state), the planning of a drinking water-supply (well–field and local abstraction, and future development), and integrated resource planning.

Modelling enables us to study the relevant processes and forecast the consequences for the environmental planning. Relevant issues and data must be integrated to a level that fit the individual case of each area, as no model fits all cases. The catchment scale must be decided upon, since models need to be very detailed for urban areas (e.g., for modelling constructions and city quarters). The coherence between hydrogeological modelling, hydrological modelling, and monitoring is important because their methodologies complement each other.

It is not possible to manage properly if there are too few data compared to problems solved. Therefore, monitoring is the first important step towards achieving a more resilient and sustainable urban area. For example, it is necessary to study, for dry summers, how and when the processes evolve and register their impacts (Figure 1). Selected parameters have to be mapped in terms of their time of appearance, frequency and duration, and size. Choice of the limiting value for each indicator is important. Monitoring is thus a key issue for management [43–47].

3.5.3. Operationalizing Resilience

An investigation of social resilience in the NORDIC countries, including Denmark [8] shows that there is a general focus on the co-production of knowledge, with an emphasis on mapping resilience, producing territorial level indicators of adaptive capacity, vulnerability, and social learning as part of good and transparent governance. These factors can facilitate adaptive urban design and resilience to natural hazards by allowing flexible responses and adaptation to changing contexts, and acting as potential drivers for increased societal resilience.

Enhanced information sharing at the local, national, international and cross-border levels were identified to facilitate information sharing [8]. However, dealing with uncertainty and especially ambiguity as part of knowledge co-production is still in its infancy. This means that that there is not yet full acknowledgement of subjectivity and proportionality, and there are multiple ways of knowing, especially when it comes to handling emergent groundwater management challenges. In many smaller Danish municipalities with only one or few persons dealing with groundwater management (and because municipalities are the authority responsible for climate change adaptation and urban planning [47]), co-production capacity may be limited.

As part of environmental governance, and as described with these examples from Odense, adaptive planning and governance of the urban subsurface and water resources need to be coordinated with the other types of planning, e.g., ecosystems, infrastructure, housing, cultural heritage protection, etc. This is especially in order to address the environmental challenges and problems at hand, which are often wicked or complex and thus has a very high degree of uncertainty. Furthermore, similar to spatial planning projects or adaptive measures dealing with environmental issues, such require policy responses that are cross-sectoral, rethinking of norms, territories (or places), and policy levels.

3.6. Use of Data Outcomes

A nationwide comprehensive groundwater mapping (in years 1998–2015) based on airborne geophysical surveys, boreholes, mathematical/physical groundwater models, chemical analyses, etc., were performed as basis for local action plans for protecting the drinking water resources. The collated data and results contribute to the national hydrogeological databases and is used as valuable 3D hydrogeological input to the nationwide water resource model, DK-model [48–50].

The proposed workflows in this paper are based on comprehensive data collations and 3D modelling exercises that are not dealt with earlier. Besides, the work processes also generate a considerable amount of data to be synthesized in the strategic planning and decision-management. For this purpose, automatized data handling with the use of indicators, remote sensing, modelling, and monitoring ease the access to visualizing valuable information in the urban area and its subsurface.

As described earlier (e.g., in Section 2) urban areas require more detailed and exact mapping due to more intensive land use and impacts on the hydrogeological catchment. Typically, in urban planning and management purposes we are looking for places suitable for specific land-use/cover characteristics such as nature based solutions (e.g., extended green/blue areas), areas with negative impacts due to man-made or natural induced hazards or results of combined land-use just causing a negative evolution.

Typical indications for critical impacts are unwanted changes in land-use/cover or terrain levels, unwanted changes in groundwater level or groundwater depth below terrain, or unwanted changes of the frequency of flooding. Mapping of change detection can be based on results of assessments or monitoring of groundwater levels in boreholes, but it can also be detection of changes in specific features in cartographic maps, geophysical or satellite information, etc.

The synthesizing of outcomes can be performed as a part of the data collation, as satellite land-use/landcover mapping, as change detection on land/geology and as a mapping of the impacts on the groundwater or surface-water. GIS-suitability analysis (or impact analysis) exploits unacceptable and acceptable differences between normal and changed situation. A part of this analysis is weighted overlay of GIS-layers.

4. Results

This paper focuses on the urban hydrogeological studies that have to be implemented in the activity of cities planning, decision making and management in order systematically to support their resilience and sustainability. Besides focus is on improvement of the following topics: assessment of sustainability and resilience, the importance of subsurface conditions, hydrogeological/hydrological conditions and the need of comprehensive planning.

4.1. Natural and Man-Made Processes

The meaning of using sound knowledge of subsurface conditions is clear. The freshwater cycle with all its components and external influences plays an important role in decision-making. The location, amount, density, and spatial and temporal scales (Tables 1–4) highlight how much natural and man-made environments have changed and their impact.

In Odense, observations during the last few hundred years show a strong anthropogenic influence on the natural water system in urban areas, whose consequences are first understood many years after their implementation [25]. It is also observed that the effect of non-governed actions (groundwater abstractions, drainage, piping, infrastructures, etc.) likewise influence the hydrological system. These actions initially individually seem to be small or unimportant, but over a time-span of years, it is realized that they amount to a significant impact.

Improved quality and access to information improves urban hydrogeological information and results in improved decision-making and management to support the city resilience and sustainability.

The present conceptual urban model for Odense is based on selected typical projects in Odense (see Figure 4). Since the urban modelling in Odense started, several new projects have arisen covering a broad variety of natural and man-made processes (protection of wellfields, impacts of climate changes, city-quarters with SuDS, new big subsurface constructions etc.). These forewarn on groundwater flooding or sea level rise and confirm the need and rationale behind a more permanent and integrated city urban model solution. As example of development: As some of the wells in the city were abandoned a new well field west of Odense were established. This action induces a new situation with a mapping project aiming to protect the catchments of both the remaining wells in Odense and the new wells established in recreational surroundings outside the city. This new mapping is already under realization.

4.2. 'New' Urban Model Design

This paper concentrates on providing an evaluation of improvements in urban modelling from standardized data and modelling. Procedures includes modelling of the hydrogeological and anthropogenic layers and methodologies to work with different models and model scales. Additionally, an example of an assessment combining the climate related impacts of increased runoff from Odense River Catchment, a sea level rise in Odense Fjord and flooding in Odense River is provided.

A part of the process, development in future urban monitoring is important, but often affected by several artificial processes, and requires supplementary regional data and limiting values.

4.3. Integrated Urban Hydrogeological System

For the technical part of the solution, we have described a design to deal with hydrogeological and hydrological conditions (see Figure 12). The purpose is to contribute with data, model, assessment, and knowledge based decision support to the planning and management. Visualization of data outcomes, whether they appear just in collated form, as impact results of models or results of GIS analysis is an important part of the dissemination of the integrated system.

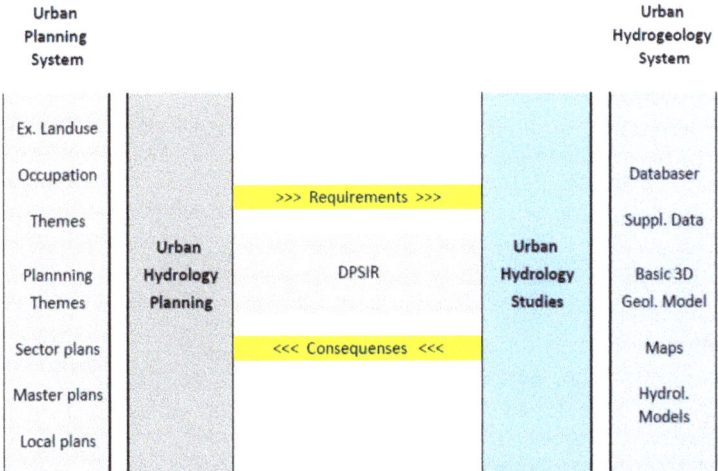

Figure 12. Interaction between Urban Planning System and hydrogeological studies in a City-Scale Hydrogeological Framework. Urban planning contributes with requirements, and hydrogeologists deliver consequences—in a continuous Driver-Pressure-State-Impact-Response (DPSIR) loop until a solution is found.

4.4. Planning for Increased City Resilience and Sustainability

Urban hydrogeological studies of the groundwater level evolution requires data to establish trends and assess limitations. Assessments must determine the climate changes forcing impacts and trends from monitoring data. The accuracy of urban area modelling is determined by data density, coverage, and connection to other models.

Planning and management must be based on the urban assessments and evaluate the evolution. To achieve a resilient and sustainable management of urban areas, the approach must be both strategic and operational. Operational resilience requires involvement, co-production, risk awareness and information, and must be built on well-defined processes, indicators and modelling.

5. Discussion and Conclusions

In the introduction of this paper, we argued that urban resilience and sustainability must be based upon our handling of the hydrogeological environment. This must be based on sound knowledge of natural conditions and man-made processes and their temporal and spatial regimes. Meteorological and geological hazards have different development periods. For example, response of surface runoff is much faster and more visible than that of deep groundwater (which makes it natural for city planners to be aware and react to the former but not the later).

We should encourage urban planning based on sufficient knowledge and analyses and knowledge of subsurface and surface-water cycles to achieve a high city sustainability and resilience. Adaptive planning and governance of the urban subsurface and water resources need to be coordinated with other types of planning (such as that of ecosystems, infrastructure, housing, and protection of cultural heritage).

An analysis of territorial governance in the Nordic Countries [8] reveals a difference in the management structures of Denmark and Norway, which are more centrally managed, while Finland is more guided. Municipalities in Denmark, Norway, and Sweden are ultimately responsible for risk and vulnerability assessments, while the municipalities in Finland and Iceland produce more specific assessments for limited risk areas.

It is fundamental that all the expected processes of natural and artificial changes are identified and described according to their importance. Impacts on selected 'sustainability' indicators require estimation, elaboration of a monitoring strategy, and a definition of the consequences tolerated for the individual indicators.

The upper hydrogeology in the Odense case study area consists mainly of glacial moraine layers consisting of clayey till and sandy aquifers that is intersected by systems of buried valleys. This complicated geology forms the basis for the abstracted groundwater for drinking water supply. Furthermore, climate change adaptation has to work on top of this complex subsurface when identifying suitable adaptation measures in city center area with existing infrastructure and when planning for new infrastructure in city center and peri-urban areas.

The elaboration of reliable information takes time, and thus it makes good sense to do this work continuously and regularly. Urban areas are a focus of concentrated activity, with numerous people involved, and thus are more intensively used. We need to learn more about the ground beneath them (including the interactions between geology, man-made ground, groundwater, surface water, and urban drainage systems) since the existing information for decision-making is not sufficient. Information on the subsurface needs to be organized and supplementary data needs collation and availability for others to use. Often, available data are site-specific, and it is important to differentiate between the required spatial and temporal scales. There is a lack of awareness and experience in mapping the effects of how subsurface usage develops and interacts in urban areas over time and space (e.g., sewage renovations, the use of drainage systems, start/stop groundwater abstractions, greening of cities, etc.).

This illustrates the complexity of urban hydrogeology, and the basis for sustainable and resilient planning and decision-making will require more systemic integration of the entire water cycle of our cities, founded on urban history and more continuous monitoring. Building knowledge based on coherent hydrogeological modelling and monitoring allows the possibility of cross-disciplinary and multi-project assessment of the relevant processes, but it also demands capturing information in a general physical framework and using indicators to place restrictions on development in specific areas.

In urban areas, there are three typical management and planning levels—Level 1 is small construction. Level 2 is large and expensive construction (e.g., New Odense University Hospital), and Level 3 is the city quarters (e.g., Skibhuskvarteret). In all cases, it is important that management and planning are based on all existing data and maps, since they cannot wait the long time needed for implementing new mapping. Furthermore, it is important that the data handling procedure is standardized and documents the evaluation (of the sufficiency of existing usable information and the need for supplementary data and inventories) to ensure that the final decision is adequately informed. Hydrogeological data handling and integration supporting decision-making enable the cities to reduce the impact of relevant hazards in an enhanced planning and management process. Catchment areas sometimes extend beyond the borders of the urban authority, and will require effective liaison with neighboring municipalities. This situation is likely to occur if the catchment is large compared to the governance area and is of importance for the resilience and sustainability.

A review of recent initiatives [8] regarding early warning and monitoring systems suggests that there is great potential for adding more participatory approaches. A benefit of participatory early warning and monitoring systems is the increased awareness about the risks related to natural hazards and improved preparedness and responses. The key to increased resilience and successful adaptation in response to problems with high groundwater levels is the implementation of efficient monitoring programs and communication strategies that deal with the entire water cycle in urban areas. A proper understanding of local conditions by, and effective engagement of, professional stakeholders and citizens will be important.

Author Contributions: Conceptualization, S.M.; Methodology, S.M. and H.J.H.; Investigation, S.M. and H.J.H.; —Original Draft Preparation, S.M.; Writing—Review and Editing, H.J.H. and S.M.; Project Administration, S.M. All authors have read and agree to the published version of the manuscript.

Funding: This research received no external funding.

Acknowledgments: This paper is partly built on the results of the project 'Development of a 3D geological/hydrogeological model as basis for the urban water cycle' [26] and the NORDRESS project ('Nordic Centre of Excellence for Resilience and Societal Security' funded by the Nordic Societal Security Programme, http://nordress.hi.is/, [51]). Further, this paper is partly based on the work from the COST Action TU1206 Sub-Urban—'A European network to improve the understanding and use of the ground beneath our cities' [4]. Many participants have contributed, with their own expertise and experience, to these three projects, and the authors wish to acknowledge them all for their contributions. We especially thank Professor Radu Gogu and Stephen Foster for valuable comments, and the contributors of illustrations to the cases used in this paper.

Conflicts of Interest: The authors declare no conflict of interest.

References

1. IWA International Water Association. Basin-Connected cities: Principles and Case Studies. In Proceedings of the IWA World Water Congress & Exhibition, Basin-Connected Cities Forum, Tokyo, Japan, 20 September 2018; p. 49. Available online: www.worldwatercongress.com (accessed on 19 November 2020).
2. Van Leeuwen, C.J.; Frijs, J.; van Wezel, A.; van de Ven, F.H.M. City Blueprints: 24 Indicators to Assess the Sustainability of the Urban Water Cycle. *Water Resour. Manag.* **2012**, *26*, 2177–2197. [CrossRef]
3. Venvik, G. The Ground beneath Our Cities. In Proceedings of the 54th ISOCARP Congress, Bodø, Norway, 1–5 October 2018; p. 12.
4. COST Action TU1206 Sub-Urban—A European Network to Improve the Understanding and Use of the Ground Beneath Our Cities. Available online: http://sub-urban.squarespace.com/ (accessed on 19 November 2020).
5. Van der Meulen, M.; Campbell, S.D.G.; Lawrence, D.; Lois Gonzáles, R.C.; van Campenhout, I.P.A.M. *Out of Sight Out of Mind? Considering the Subsurface in Urban Planning–State of Art.* COST Action TU1206 Sub-Urban Report; T1206-WG1-001. 2016, p. 45. Available online: https://static1.squarespace.com/static/542bc753e4b0a87901dd6258/t/570f706201dbae9b1f7af3bf/1460629696046/TU1206-WG1-001+Summary+report+Out+of+sight+out+of+mind.pdf (accessed on 19 November 2020).
6. Mielby, S.; Eriksson, I.; Campbell, D.; De Beer, H.; Bonsor, H.; Le Guern, C.; van der Krogt, R.; Lawrence, D.; Ryzynski, G.; Schokker, J. *Opening up the Subsurface for the Cities of Tomorrow. Considering the Access to Subsurface Knowledge-Evaluation of Practices and Techniques.* COST Action TU1206 Sub-Urban Report; TU1206-WG2-001. 2016, p. 119. Available online: http://sub-urban.squarespace.com/s/TU1206-WG2-001-Opening-up-the-subsurface-for-the-cities-of-tomorrow_Summary-Report.pdf (accessed on 19 November 2020).
7. Henriksen, H.J.; Roberts, M.J.; van der Keur, P.; Harjanne, A.; Egilson, D.; Alfonso, L. Participatory early warning and monitoring systems: A Nordic framework for web-based flood risk management. *Int. J. Disaster Risk Reduct.* **2018**, *31*, 1295–1306. [CrossRef]
8. Van Well, L.; van der Keur, P.; Harjanne, A.; Pagneux, E.; Perrels, E.; Henriksen, H.J. Resilience to natural hazards: An analysis of territorial governance in the Nordic countries. *Int. J. Disaster Risk Reduct.* **2018**, *31*, 1283–1294. [CrossRef]
9. Holling, C.S. 'Buzz'. The resilience of terrestrial ecosystems: Local surprise and global change. In *Sustainable Development of the Biosphere*; Clarck, W.C., Munn, R.E., Eds.; Cambridge University Press: Cambridge, UK, 1986; pp. 292–320.
10. Berkes, F.; Colding, J.; Folke, C. (Eds.) Navigating Social-Ecological Systems. In *Building Resilience for Complexity and Change*; Cambridge University Press: Cambridge, UK, 2002.
11. Lister, N.M. Insurgent ecologies: (Re) Claiming ground in landscape and urbanism. In *Mohsen Mostafavi and Gareth Doherty*; Ecological Urbanism; Lars Müller Publisher: Baden, Switzerland, 2016; pp. 550–561.
12. EEA European Environment Agency. *Climate Change Adaptation and Disaster Risk Reduction in Europe: Enhancing Coherence of the Knowledge Base, Policies and Practices*; EEA Report No. 15/2017; EEA European Environment Agency: Copenhagen, Denmark, 2017.
13. CODATA Task Group. Next Generation Disaster Data Infrastructure. In *Linked Open Data for Global Disaster Risk Research (LODGD)*; White Paper; CODATA: Paris, France, 2019; p. 26.
14. Makropoulos, C.; Savic, D.A. Urban Hydroinformatics: Past, Present and Future. *Water* **2019**, *11*, 1959. [CrossRef]
15. EEA European Environment Agency. *Digest of EEA Indicators*; EEA Report No. 8/2014; EEA: Copenhagen, Denmark, 2014; p. 44.

16. PanGeo Project 7FP. 2013. Available online: http://www.pangeoproject.eu (accessed on 26 November 2020).
17. Mielby, S.; Schokker, J.; de Beer, J.; Sandersen, P.B.E.; Pallesen, T.M. Byen og dens Undergrund-Beliggenhed, Udvikling, Klimaforandringer og Bygeologi. In *Geoviden-Geologi og Geografi 4*; Geocenter Danmark: Copenhagen, Denmark, 2017; p. 20. Available online: https://www.geocenter.dk/wp-content/uploads/2018/07/geoviden-4-2017.pdf (accessed on 2 December 2020). (In Danish)
18. Pujades, E.; López, A.; Carrera, J.; Vázquez-Suñé, E.; Jurado, A. Barrier effect of underground structures on aquifers. *Eng. Geology* **2012**, *145*, 41–49. [CrossRef]
19. Wu, Y.X.; Shen, S.L.; Yuan, D.J. Characteristics of dewatering induced drawdown curve under blocking effect of retaining wall in aquifer. *J. Hydrol.* **2016**, *539*, 554–566. [CrossRef]
20. Hinsby, K.; Troldborg, L.; Purtschert, R.; Corcho Alvarado, J.A. *Integrated Dynamic Modelling of Tracer Transport and Long Term Groundwater/Surface Water Interaction Using Four 30 Year 3H Time Series and Multiple Tracers for Groundwater Dating*; International Atomic Energy Agency: Vienna, Austria, 2006; pp. 73–98.
21. Laursen, G.; Mielby, S. Odense. TU1206 COST Sub-Urban WG1 Report, TU1206-WG1-011. 2016, p. 37. Available online: https://static1.squarespace.com/static/542bc753e4b0a87901dd6258/t/57330a2e7da24f10a9dfdd46/1462962771253/TU1206-WG1-011+Odense+City+Case+Study.pdf (accessed on 19 November 2020).
22. Gejl, R.N.; Rygaard, M.; Henriksen, H.J.; Rasmussen, J.; Bjerg, P.L. Understanding the impacts of groundwater abstraction through long-term trends in water quality. *Water Res.* **2019**, *156*, 241–251. [CrossRef] [PubMed]
23. Gejl, R.N. Assessing Sustainable Groundwater Abstraction: An Evaluation of Impacts on Groundwater Quantity and Quality. Ph.D. Thesis, Danish Technical University, Kongens Lyngby, Denmark, September 2019.
24. Zhang, Y.; Wu, J.; Xue, Y.; Wang, Z.; Yan, Y.; Yan, X.; Wang, H. Land subsidence and uplift due to long-term groundwater extraction and artificial recharge in Shanghai, China. *Hydrogeol. J.* **2015**, *23*, 1851–1866. [CrossRef]
25. Mielby, S.; Laursen, G.; Linderberg, J.; Sandersen, P.B.E.; Jeppesen, J. *Udvikling af en 3D Geologisk/Hydrogeologisk Model som Basis for det Urbane Vandkredsløb. Delrapport 1—3D Modellen som Basis for det Urbane Vandkredsløb*, Special ed.; GEUS: Copenhagen, Denmark, 2015; p. 64. Available online: https://www.geus.dk/media/7768/urban_vandkredsloeb_del1.pdf (accessed on 19 November 2020). (In Danish)
26. Mielby, S.; Jespersen, C.E.; Ammitsøe, C.; Laursen, G.; Jeppesen, J.; Linderberg, J.; Søndergaard, K.; Kristensen, M.; Hansen, M.; Jensen, N.-P.; et al. *Udvikling af en 3D Geologisk/Hydrogeologisk Model som Basis for det Urbane Vandkredsløb—Syntese Rapport*, Special ed.; GEUS: Copenhagen, Denmark, 2017; p. 52. Available online: https://www.geus.dk/media/7766/urban_vandkredsloeb_syntese.pdf (accessed on 19 November 2020). (In Danish)
27. Kristensen, M.; Sandersen, P.B.E.; Mielby, S. *Udvikling af en 3D Geologisk/Hydrogeologisk Model som Basis for det Urbane Vandkredsløb. Delrapport 2—Indsamling og Vurdering af Data*, Special ed.; GEUS: Copenhagen, Denmark, 2015; p. 82. Available online: https://www.geus.dk/media/7769/urban_vandkredsloeb_del2.pdf (accessed on 19 November 2020). (In Danish)
28. Laursen, G.; Mielby, S.; Kristensen, K. *Udvikling af en 3D Geologisk/Hydrogeologisk Model som Basis for det Urbane Vandkredsløb. Delrapport 3—Geotekniske data til Planlægning og Administration*, Special ed.; GEUS: Copenhagen, Denmark, 2015; p. 32. Available online: https://www.geus.dk/media/7770/urban_vandkredsloeb_del3.pdf (accessed on 19 November 2020). (In Danish)
29. Sandersen, P.B.E.; Kristensen, M.; Mielby, S. *Udvikling af en 3D Geologisk/Hydrogeologisk Model som Basis for det Urbane Vandkredsløb. Delrapport 4—3D Geologisk/Hydrostratigrafisk Modellering*, Special ed.; GEUS: Copenhagen, Denmark, 2015; p. 106. Available online: https://www.geus.dk/media/7771/urban_vandkredsloeb_del4.pdf (accessed on 19 November 2020). (In Danish)
30. Pallesen, T.M.; Jensen, N.P. *Udvikling af en 3D Geologisk/Hydrogeologisk Model som Basis for det Urbane Vandkredsløb. Delrapport 5 – Interaktiv Modellering af Antropogene Lag*, Special ed.; GEUS: Copenhagen, Denmark, 2015; p. 58. Available online: https://www.geus.dk/media/7772/urban_vandkredsloeb_del5.pdf (accessed on 19 November 2020). (In Danish)
31. Hansen, M.; Wiese, M.B.; Gausby, M.; Mielby, S. *Udvikling af en 3D Geologisk/Hydrogeologisk Model som Basis for det Urbane Vandkredsløb. Delrapport 6—Teknisk Håndtering og Lagring af Data og Modeller*, Special ed.; GEUS: Copenhagen, Denmark, 2015; p. 22. Available online: https://www.geus.dk/media/7773/urban_vandkredsloeb_del6.pdf (accessed on 19 November 2020). (In Danish)

32. Schokker, J.; Sandersen, P.; De Beer, H.; Eriksson, I.; Kallio, H.; Kearsey, T.; Pfleiderer, S.; Seither, A. 3D Urban Subsurface Modelling and Visualization—A Review of Good Practices and Techniques to Ensure Optimal Use of Geological Information in Urban Planning. COST Action TU1206 Sub-Urban Report; WG2.3-004. 2017, p. 92. Available online: https://static1.squarespace.com/static/542bc753e4b0a87901dd6258/t/58c021e7d482e99321b2a885/1488986699131/TU1206-WG2.3-004+3D+urban+Subsurface+Modelling+and+Visualisation.pdf (accessed on 19 November 2020).
33. Mielby, S.; Ammitsøe, C. Need for a hydrogeological management framework as a basis for the urban water resources? In Proceedings of the 9th annual meeting of Danish Water Forum, Copenhagen, Denmark, 29 January 2015.
34. Jeppesen, J. *Udvikling af en Urban-Hydrologisk Model til Simulering af nye Innovative LAR-Løsninger til Lokal HåNdtering af BåDe Regnvand og Grundvand (LARG)*; VTU-Project; Alectia: Aarhus, Denmark, 2014; Volume 29. (In Danish)
35. Mielby, S.; Christensen, J.C.; Greve, C.; Lauritsen, L.; Laursen, G.; Müller-Wohlfeil, D.-I. *Kortlægning af Grundvandsressourcerne–Status for Vandressourcekortlægningen 2005*; Miljø- og Arealafdelingen: Funen, Denmark, 2005; p. 151. (In Danish)
36. Hole, Ø. Hydrological Modeling of the Urban Environment in Odense and the Impact of Forced Infiltration and Climate Change. Master's Thesis, Copenhagen University, Copenhagen, Denmark, 2 September 2016; p. 77.
37. Negus, A.R.A. Building Resilience to Extremity and Climatic Changes Investigating the Phenomena of Compound Events. Case Study: Odense, Denmark. Master's Thesis, Copenhagen University, Copenhagen, Denmark, 14 October 2016; p. 177.
38. Henriksen, H.J.; Pang, B.; Olsen, M.; Sonnenborg, T.O.; Refsgaard, J.C.; Madsen, H. Klimaeffekter på ekstremværdi afstrømninger. In *Fase 2 Usikkerhedsvurdering*; Danmarks og Grønlands Geologiske Undersøgelse: Copenhagen, Denmark, 2014; p. 38. (In Danish)
39. Schmitt, P.; Wiechmann, T. Unpacking spatial planning as the governance of place. *disP Plan. Rev.* **2018**, *54*, 21–33. [CrossRef]
40. Folke, C.; Hahn, T.; Olsson, P.; Norberg, J. Adaptive governance of social-ecological knowledge. *Annu. Rev. Environ. Resour.* **2005**, *30*, 441–473. [CrossRef]
41. Schmitt, P.; Van Well, L. Revisiting territorial governance: Twenty empirically informed components. In *Territorial Governance across Europe: Pathways, Practices and Prospect*; Routledge: London, UK, 2016; pp. 221–237.
42. Mielby, S.; Eriksson, I.; Campbell, S.D.G.; Lawrence, D. Opening up the subsurface for the cities of tomorrow. The subsurface in the planning process. *Procedia Eng.* **2017**, *209*, 12–25. [CrossRef]
43. Madsen, H.; Arnbjerg-Nielsen, K.; Mikkelsen, P.S. Update of regional intensity-duration-frequency curves in Denmark: Tendency towards increased storm intensities. *Atmos. Res.* **2009**, *92*, 343–349. [CrossRef]
44. Van Loon, A.F. Hydrological droughts explained. *WIRES Water* **2015**, *2*, 359–392. [CrossRef]
45. Van Loon, A.F.; Stahl, K.; Di Baldasarre, G.; Clark, J.; Rangecroft, S.; Wanders, N.; Gleeson, T.; van Dijk, A.I.J.M.; Tallaksen, L.M.; Hannaford, J.; et al. Drought in a human-modified world: Reframing drought definitions, understanding, and analysis. *Hydrol. Earth Syst. Sci.* **2016**, *20*, 3631–3650. [CrossRef]
46. Bertule, M.; Bjørnsen, P.K.; Constanzo, S.D.; Escurra, J.; Freeman, S.; Gallagher, L.; Kelsey, R.H.; Vollmer, D. *Using Indicators for Improved Water Resources Management-Guide for Basin Managers and Practitioners*; UN Environment-DHI Centre: Horsholm, Denmark, 2017; p. 73.
47. Jørgensen, L.F.; Villholt, K.G.; Refsgaard, J.C. Groundwater management and protection in Denmark: A review of pre-conditions, advantages and challenges. *Int. J. Water Res. Dev.* **2016**, *33*, 23. [CrossRef]
48. Stisen, S.; Højberg, A.L.; Troldborg, L.; Refsgaard, J.C.; Christensen, B.S.B.; Olsen, M.; Henriksen, H.J. On the importance of appropriate precipitation gauge catch correction for hydrological modelling at mid to high latitudes. *Hydrol. Earth Syst. Sci.* **2012**, *16*, 4157–4176. [CrossRef]
49. Højberg, A.L.; Troldborg, L.; Stisen, S.; Christensen, B.B.S.; Henriksen, H.J. Stakeholder driven update and improvement of a national water resources model. *Environ. Model. Softw.* **2013**, *40*, 202–213. [CrossRef]
50. Henriksen, H.J.; Troldborg, L.; Højberg, A.L.; Refsgaard, J.C. Assessment of exploitable groundwater resources of Denmark by use of ensemble resource indicators and a numerical groundwater–surface water model. *J. Hydrol.* **2008**, *348*, 224–240. [CrossRef]

51. Nordic Societal Security Programme–The NORDRESS project (Nordic Centre of Excellence for Resilience and Societal Security. Available online: http://nordress.hi.is/ (accessed on 19 November 2020).

Publisher's Note: MDPI stays neutral with regard to jurisdictional claims in published maps and institutional affiliations.

 © 2020 by the authors. Licensee MDPI, Basel, Switzerland. This article is an open access article distributed under the terms and conditions of the Creative Commons Attribution (CC BY) license (http://creativecommons.org/licenses/by/4.0/).

Article

Understanding Groundwater Mineralization Changes of a Belgian Chalky Aquifer in the Presence of 1,1,1-Trichloroethane Degradation Reactions

Youcef Boudjana [1],*, Serge Brouyère [1], Pierre Jamin [1], Philippe Orban [1], Davide Gasparella [2] and Alain Dassargues [1]

1   Hydrogeology and Environmental Geology, Urban and Environmental Engineering, University of Liège, Sart Tilman B52, 4000 Liège, Belgium; serge.brouyere@uliege.be (S.B.); pierre.jamin@uliege.be (P.J.); p.orban@uliege.be (P.O.); alain.dassargues@uliege.be (A.D.)
2   AECOM, Maria-Theresiastraat 34 A, 3000 Leuven, Belgium; davide.gasparella@aecom.com
*   Correspondence: y.boudjana@doct.uliege.be or boudjanayoucef@gmail.com

Received: 22 August 2019; Accepted: 23 September 2019; Published: 27 September 2019

Abstract: An abandoned industrial site in Belgium, located in the catchment of a chalk aquifer mainly used for drinking water, has been investigated for groundwater pollution due to a mixture of chlorinated solvents with mainly 1,1,1-trichloroethane (1,1,1-TCA) at high concentrations. The observed elevated groundwater mineralization was partly explained by chemical reactions associated with hydrolysis and dehydrohalogenation (HY/DH) of 1,1,1-TCA in the chalky aquifer. Leaching of soluble compounds from a backfilled layer located in the site could also have influenced the groundwater composition. In this context, the objective of this study was to investigate the hydrochemical processes controlling groundwater mineralization through a characterization of the backfill and groundwater chemical composition. This is essential in the context of required site remediation to define appropriate remediation measures to soil and groundwater. Groundwater samples were collected for chemical analyses of chlorinated aliphatic hydrocarbons, major ions, and several minor ones. X-Ray Diffraction Analysis (XRD), Scanning Electron Microscopy (SEM) and a leaching test according to CEN/TS 14405 norm were carried out on the backfill soil. $\delta^{34}S$ and $\delta^{18}O$ of sulphate in groundwater and in the backfill eluates were also compared. Both effects influencing the groundwater hydrochemistry around the site were clarified. First, calcite dissolution under the 1,1,1-TCA degradation reactions results in a water mineralization increase. It was assessed by geochemical batch simulations based on observed data. Second, sulphate and calcium released from the backfill have reached the groundwater. The leaching test provided an estimation of the minimal released quantities.

Keywords: hydrochemistry; chalk aquifer; 1,1,1-trichloroethane; degradation; sulphate; backfill; leaching test

1. Introduction

Chlorinated aliphatic hydrocarbons (CAHs) are among the most common pollutants in industrial sites because of their intensive use as cleaning and degreasing products [1]. CAHs in groundwater is a major concern because of their harmful effect on human health [2]. They may undergo different natural degradation pathways in groundwater. Physicochemical and geochemical data help to identify actual degradation reactions that occur naturally in groundwater [3–5]. Degradation reactions of CAHs in groundwater influence the hydrochemistry and may modify the physicochemical conditions [6,7]. This explains the interest given to enhanced monitoring and investigations about physicochemical parameters and groundwater hydrochemistry in cases of CAHs pollutions [8]. Identification of redox conditions, electron donors and acceptors and the source of carbon within the groundwater contamination plume improve the assessment of destructive biodegradation of CAHs [9].

1,1,1-trichloroethane (1,1,1-TCA) is often detected with other CAHs at contaminated sites because of their common use in industry [10]. In groundwater, both biotic and abiotic natural degradation reactions of 1,1,1-TCA are possible [11,12]. Although abiotic degradation of 1,1,1-TCA by hydrolysis and dehydrohalogenation (HY/DH) occurs independently of redox conditions in groundwater [13], it influences the pH that would impact the groundwater mineralization. In this case, a hydrogeochemical investigation including major ions analysis allow a direct link between degradation reactions and mineralization changes, as long as no additional pollution source influencing groundwater mineralization can be identified.

On the other hand, artificial man-made ground (i.e., backfill soil) represent a potential threat to the quality of shallow groundwater in urban context (e.g., [14–17]) as unwelcome chemical compounds can be leached and contaminate groundwater. Then, a remediation of the groundwater quality, in order to be compliant with drinking water standards, can be a very difficult task [18].

In Belgium, the past intensive industrial activities have caused environmental problems, especially for soil and groundwater. Most cases of local pollution, including the use of uncontrolled backfilling, are a legacy of past practices when the question of the environmental and health consequences of human activities was scarcely considered [19].

At an abandoned industrial site in Wallonia (South of Belgium), a chalky aquifer intensively used for drinking water supply [20] has been locally contaminated by a mixture of CAHs. Local changes in groundwater mineralization have been observed compared to the background composition of groundwater in the aquifer. During the investigations of Palau et al. [21], an increase of Ca^{2+}, HCO_3^-, Cl^- and SO_4^{2-} was observed within the plume of dissolved CAHs. Sulphate concentrations even exceed the EU drinking water standard (250 mg/L) [22] in some places, reaching concentration levels as high as 5 times the background concentration. During the investigations of Palau et al. [21], mineralization changes has been first explained by 1,1,1-TCA degradation reactions by HY/DH. However, the presence of a backfill layer at the site has allowed to presume the leaching of different chemical compounds with water infiltration that may also affect groundwater composition.

In this context, the general objective of the investigation is to better characterize and to quantify the hydrochemical processes controlling groundwater mineralization, through a combined approach of groundwater chemistry investigation with backfill soil characterization. This is essential in order to avoid overestimation of degradation reactions through their effect on groundwater mineralization and to define appropriate remediation measures to soil and groundwater in the context of required site remediation.

More specifically, the aim of the study is (1) to improve the understanding of hydrochemistry in the chalky aquifer in a context of 1,1,1-TCA degradation by hydrolysis and dehydrohalogenation, and (2) to identify and quantify processes influencing the hydrochemical changes in groundwater in the study area. The used method is based on groundwater analyses, laboratory backfill soil characterization and the study of sulphate isotopic signature in groundwater, along with the one in backfill eluates.

2. Study Area

The study area is located around an industrial site where the subsurface is polluted by a mixture of CAHs. They have been detected with high concentrations, not only in the unsaturated part of the soil but also in the underlying chalky aquifer [21]. The aquifer is unconfined, made of Cretaceous chalk with an average thickness of about 30 m in the study area. A summary of the local geology is given in Table 1.

Table 1. Local geology in the study area.

Geological Material	Top of Formation (m Below Surface)	Bottom of Formation (m Below Surface)	Description	Comments
Backfill layer	0	1.5	Loamy and sandy soil, with recycled construction materials, shale and coal waste	Heterogeneous backfill only at the industrial site
Loess	1.5	4.8/10	Loess, sandy and clayey loess	Variable thickness, higher thickness in piezometers located out of the contamination site
Flint conglomerate	4.8/10	10/18	Flint conglomerate in loamy and/or sandy and/or clayey matrix	/
Chalk	10/18	-	White chalk with observed fractured chalk in boreholes	Locally, the bottom of chalk was not reached. Average thickness 30 m from regional data mapping

In this aquifer, the hydraulic conductivity values of the chalk formation varies from 10^{-8} m/s for the chalky matrix [23] to values as high as 10^{-4} m/s for fractured chalk [24]. Hydraulic conductivity values ranging between 3×10^{-6} and 3×10^{-4} m/s were obtained from pumping tests near the study site. The effective (transport) porosity varies between 1 and 2% [25–27]. The importance of the immobile water on solute transport through matrix diffusion was studied in this chalk [28], with immobile water porosity values ranging between 8 and 42% and first-order transfer coefficients ranging between 9.8×10^{-8} and 10^{-6} s^{-1} were determined [24].

The overlying loess layer induces a delay for the transfer of pollutants that infiltrate from the land surface with water towards the saturated zone. The mean transfer velocity through the unsaturated zone was estimated at 1 m/year [23]. Locally, at the study site, depth to groundwater varies spatially between 17.16 and 28.60 m (in March 2013) with interannual fluctuations that can reach 5 m. Groundwater flow direction is from the South-East towards the North-West direction.

Since the first detection of the contamination in 1987, many investigations have been carried out with installation of a network of 30 wells to delineate and monitor the groundwater pollutants plume, and, more recently, the establishment of a combined venting-air sparging remediation (see here under). Various aliphatic organochlorines have been detected in groundwater, including mainly 1,1,1-TCA, 1,1-dichloroethene (1,1-DCE) and trichloroethene (TCE). The maximum concentrations recorded in the campaign of Palau et al. [21] in March 2013 were 7400 µg/L for 1,1,1-TCA, 4200 µg/L for 1,1-DCE and 2000 µg/L for TCE.

Between May 2013 and August 2016, a combined venting-air sparging program was undertaken to remediate the source zone of CAHs. Air sparging consists of injecting air into the saturated zone of an aquifer. Along the pathways of air bubbles towards the unsaturated zone, contaminant stripping occurs by volatilization of the chlorinated aliphatic hydrocarbons. A total mass of around 800 kg was extracted from the loess layer and the unsaturated upper chalk during this operation. For dissolved CAHs in groundwater, the ongoing monitoring indicates that the plume is currently shrinking. In addition to CAHs contamination, the monitoring program has highlighted significant increase in groundwater mineralization change compared to regional background groundwater composition in the chalk aquifer.

Figure 1 shows Stiff diagrams from groundwater samples taken in the plume in March 2013 (orange) compared to samples taken around before the contamination (blue). The highest concentrations of calcium, bicarbonate, chloride and sulphate are observed near the pollution source and they progressively decrease to the natural background as we move downgradient from the industrial site.

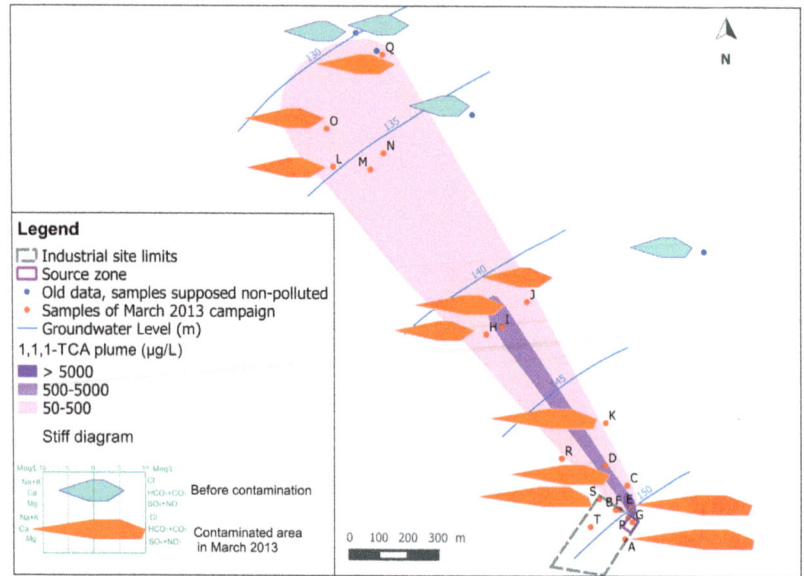

Figure 1. Stiff diagrams for groundwater samples: comparison between the contaminated area in March 2013 (orange diagrams) and the groundwater quality before contamination (blue diagrams) in the aquifer. The 1,1,1-TCA concentrations are taken Palau et al. [21].

3. Groundwater Quality Investigations

In March 2017, around 7 months after the end of the remediation, a new sampling campaign was carried out in the scope of this research, on the existing monitoring network to assess the groundwater quality and investigate other potential sources resulting on mineralization changes. In addition, analyses results were used to perform hydrogeochemical simulations in PHREEQC [29] to evaluate the increase of calcite dissolution resulting from 1,1,1-TCA degradation reactions in the aquifer.

3.1. Sampling and Analysis

On the existing monitoring network, 22 wells were sampled to analyze chlorinated solvents and among those wells, 13 were also sampled for a more detailed hydrochemical characterization of groundwater. Wells are screened in the upper part of the chalk except for the well 'S' equipped with a double casing allowing two sampling levels (25 m and 40 m below surface).

The sampling procedure was carried out by pumping at least 3 times the volume of water in the sampled well and monitoring of temperature, pH, electrical conductivity, redox potential, and dissolved oxygen. All these parameters were monitored using a multi probe handheld meter (WTW multi 350i, Weilheim, Germany) except dissolved oxygen that was measured by a luminescent DO probe (Hach Lange, Düsseldorf, Germany).

Samples for CAHs analyses were collected in 40 mL glass vials filled and acidified with sulphuric acid (H_2SO_4) at pH ≈ 2. 180 mL polypropylene bottles were used for the analysis in the lab of a standard package of major elements (Ca^{2+}, K^+, Mg^{2+}, Na^+, Cl^-, SO_4^{2-}, HCO_3^-) analyses and several minor chemical compounds (NH_4^+, Li^+, Sr^{2+}, NO_3^-, PO_4^{3-}, Br^-, F^- and SiO_2). 100 mL bottles were filled using 0.45 µm filter and acidified (37% HCl) to analyze dissolved iron and manganese. Before being analyzed, all samples were kept refrigerated at 4 °C and protected from light.

At the laboratory, the chemical analyses were carried out using the following methods:

- Ion chromatography for K^+, Mg^{2+}, Na^+, NH_4^+, Li^+, Sr^{2+}, PO_4^{3-}, Br^- Cl^-, F^-, NO_3^- and SO_4^{2-};
- Titrimetric method for Ca^{2+};
- Flame atomic absorption for Fe^{3+}, Mn^{2+} and SiO_2;
- The Carbonate speciation between CO_2, HCO_3^-, CO_3^{2-} is obtained from pH and total alkanity; according to Rodier's formula [30];
- Gas chromatography coupled with mass spectrometry (GC/MS) for the CAHs.

For the organic CAHs compounds, results show that three CAHs are dominant: 1,1,1-TCA, 1,1-DCE and TCE, with maximum concentrations of 1100 µg/L, 820 µg/L and 550 µg/L respectively. Concentrations have decreased of 85.13%, 72.5% and 80.47% respectively compared March 2013, before the beginning of remediation operations of the CAHs source. Other aliphatic hydrocarbons such as 1,1,2-TCA, tetrachloromethane, 1,2-dichloroethane and cis-1,2-DCE are detected with lower concentrations. Results showed also a decrease of concentrations of all CAHs in groundwater compared to the concentrations observed in March 2013.

Along the plume centerline, the sum of molar concentrations of the three dominant CAHs are presented in a same graph with the molar percentage of each (Figure 2) Despite the general decrease in CAHs concentrations, their spatial distribution is quite similar to that from Palau et al. (2016) obtained in March 2013 on this site. Along the centerline, concentrations (of the sum of 1,1,1-TCA + 1,1-DCE + TCE) show a decreasing trend from 20.66 µmol/L at well (E) to 1.28 µmol/L at well (Q). For almost all wells along the plume centerline, 1,1-DCE molar fraction is closer to the one of 1,1,1-TCA, with a slightly lower fraction for TCE.

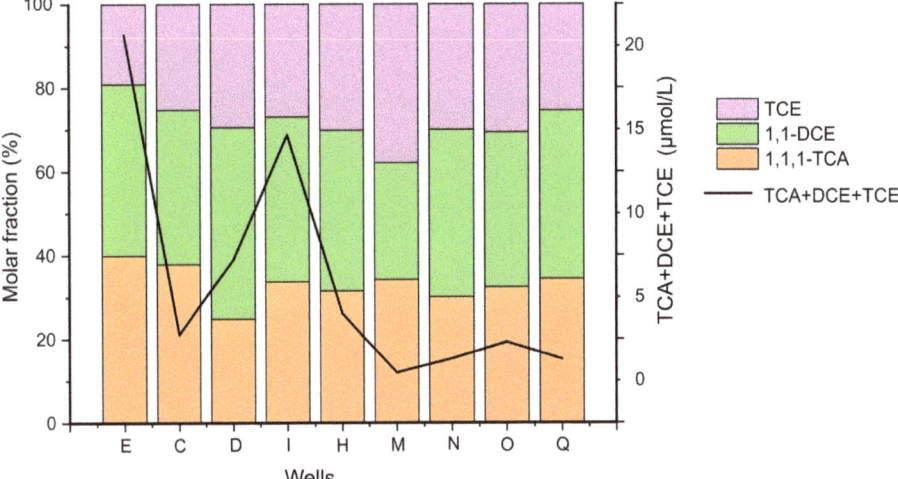

Figure 2. Total concentration (sum of 1,1,1-TCA, 1,1-DCE and TCE) (right y-axis, line) and concentration of 1,1,1-TCA, 1,1-DCE and TCE normalized by the total concentration (left y-axis, bars).

Results of chemical analyses of the inorganic elements in March 2017 are given in Table 2. Wells are listed according to the increasing distance from the source of the CAHs. Well labels are shown in Figure 1.

Table 2. Data from groundwater chemical analyses in the different wells of the monitoring network in March 2017.

Sample	Temperature (In Situ) (°C)	Dissolved Oxygen mg/L	pH (In Situ)	EC (In Situ) µS/cm	Ca^{2+} mg/L	K^+ mg/L	Mg^{2+} mg/L	Na^+ mg/L	Cl^- mg/L	NO_3^- mg/L	SO_4^{2-} mg/L	HCO_3^- mg/L	SiO_2 mg/L
G	10.6	2.95	6.68	848	310.09	1.28	31.13	59.40	160.97	90.36	343.63	423.99	4.22
P	12.5	1.04	6.12	1525	255.46	1.15	21.07	65.69	94.73	57.99	301.15	426.27	10.25
E	11.8	2.82	6.27	1120	241.67	1.02	23.09	45.88	144.55	58.81	273.00	298.28	6.29
A	12.1	1.4	6.47	1847	304.43	0.93	27.20	86.56	125.40	47.13	370.35	528.95	11.42
S-25	14.6	3.25	6.95	858	252.12	0.79	25.62	50.38	74.66	64.94	298.01	423.96	4.95
S-40	14.9	3.19	7.11	914	203.86	0.99	18.36	41.98	65.64	76.83	144.96	423.90	5.13
T	15.1	4.71	7.25	1056	242.09	2.02	22.68	53.70	102.63	95.20	235.28	394.66	1.59
D	10.8	2.08	6.75	1110	227.43	0.77	20.43	37.23	60.44	50.61	208.97	444.67	4.56
D_2	10.8	2.08	6.75	1110	226.86	0.80	20.26	36.58	59.99	51.27	207.79	449.26	n.d.
I	12.1	2.13	6.86	1190	189.58	0.81	17.83	24.93	61.73	47.06	133.61	393.23	n.d.
H	10.1	1.24	6.72	1293	208.41	1.00	20.01	37.11	68.40	75.68	161.21	415.20	12.54
M	9.2	4.08	6.9	787	173.14	1.07	18.16	22.36	49.41	68.03	99.50	387.72	13.75
N	11.3	3.1	6.83	1124	184.81	0.98	19.16	25.75	52.92	71.59	116.95	401.77	n.d.
O	10.5	3.75	6.6	1286	179.59	1.14	18.49	24.25	53.42	64.51	107.73	394.41	3.21
Q	12.2	3.47	6.45	1054	168.39	0.87	14.84	19.93	48.30	63.70	93.06	360.32	7.63

velocity of 15 ± 2 cm/day. Equilibrium time (i.e., the time at rest between the column saturation and the beginning of the test) was 96 h to reach the equilibrium between the solid grains and the water.

The collected 'liquid to solid fractions' (L/S) were: 0.1 ± 0.02, 0.2 ± 0.04, 0.5 ± 0.08, 1 ± 0.15, 2 ± 0.3, 5 ± 0.4, and 10 ± 1 L/kg. Eluates were filtered off-line with 0.45 µm membrane filters and analyzed for the same hydrochemical elements and using methods described in Section 3.1. The pH and the electrical conductivity were measured twice for each L/S fraction, first during the test and second during the chemical analyses.

For each analyzed element, the results are expressed as an average (on the 4 columns) with standard deviation (Figure 6). After a 96-h period, equilibrium conditions (pH difference < 0.5 unit) were reached without the need for water recirculation. pH and electrical conductivity measured during the test are shown in Figure 6a,b. The pH values are ranging between 7.38 and 8.7 with a slight increase from the beginning to the end of the test. Electrical conductivity decreases significantly from 1585 µS/cm to 92.35 µS/cm due to the depletion of leachable elements of the backfill during the test.

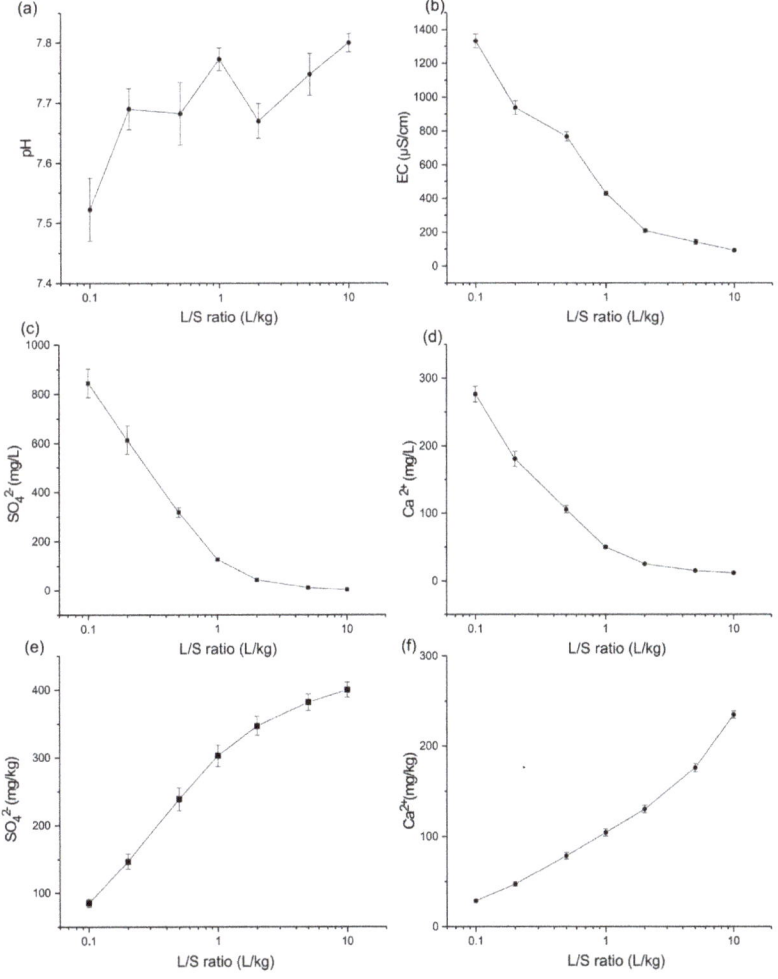

Figure 6. Parameters: (**a**) pH and (**b**) EC with concentrations of (**c**) SO_4^{2-} and (**d**) Ca^{2+} as a function of the liquid-to-solid ratio of the backfill soil. Cumulative releases of: (**e**) SO_4^{2-} and (**f**) Ca^{2+} as a function of liquid-to solid ratio for the backfill soil.

For each analyzed element, a decrease in concentrations was observed during the test with very low concentrations in the last L/S fractions.

Figure 6c,d shows the average concentrations of SO_4^{2-} and Ca^{2+} in mg/L. The maximum concentration of SO_4^{2-} was 984.62 mg/L, it was obtained for the first fraction of one of the columns. Although this concentration is below the inert waste acceptance limit of the European Directive 2003/33/EC (1500 mg/L for the 0.1 L/kg fraction) [41], it is significant in the present case because the tested samples were taken on site where the backfill was exposed to a partial leaching by the natural infiltration water for years (before being excavated).

In the beginning of the test, between 0.1 and 1 L/kg fractions, the relative chemical composition of eluates shows that (in meq/L) sulphate dominates the anions while calcium dominates the cations. Sulphate concentration declines from 844.26 mg/L to 126.80 mg/L. After 1 L/S fraction, it decreases following a different slope to reach 3.68 mg/L in the final test fraction. Calcium concentrations decrease from 276.48 mg/L at the beginning of the test to a concentration of 11.71 mg/L at the final test fraction. The other elements show lower concentrations compared to sulphate and calcium. HCO_3^- concentrations decrease from 96.61 mg/L for the first fraction to 38.60 mg/L at the end of the test. Whereas Na^+ concentrations decrease from 76.68 to 0.35 mg/L. The other elements, not presented in this section, show lower concentrations that have limited influence on the global ionic balance.

Ca-SO_4 is the dominant hydrochemical facies in eluates between fractions 0.1 and 1 L/kg. Further, eluates are less charged with SO_4^{2-} and the facies changed into a calcium bicarbonate Ca-HCO_3.

The average cumulative releases in mg/kg are shown in Figure 6e,f. Values are obtained using the equation [39]:

$$U_i = (V_i \times C_i)/m_0 \qquad (4)$$

i: index of the eluate fraction, U_i: average cumulative released quantity of a component per mass of the sample at the fraction i, expressed in milligram per kilogram of solid mass (mg/kg solid mass); V_i: volume of the eluate fraction i expressed in liters (L), C_i: concentration of the component concerned in the eluate fraction i (mg/L), m_0: solid mass of the test sample expressed in kilogram (kg).

The most important quantities of sulphate are released during the first four fractions. The cumulative quantities gradually increase from 85 mg/kg for the 0.1 L/S fraction to 302.5 mg/kg for the 1L/kg fraction. Further, the increase is less important, and the cumulative value reaches 400.06 mg/kg at the end of the test. This value is lower than the inert waste acceptance limit for sulphate waste according to European Directive 2003/33/EC (1000 mg/kg for L/S = 10 L/kg) [41] but it remains considerable for a backfill layer that was placed 40 years ago.

The calcium evolution is different from that of sulphate. After the fraction 1 L/kg, the slope is steeper than before, meaning that more calcium is dissolved at the end of the test with a final released quantity of 235.01 mg/kg for the last fraction.

Application at the Site Scale

To contextualize the results of the test, we estimated the actual time required to reach the L/S ratios, the total mass released from the entire backfill and the concentration in the recharge water for SO_4^{2-} and Ca^{2+} (Table 5). Cumulative release estimated from the test were used for extrapolation at the scale of the entire backfill using the following data: an area of 1034 m^2, an average thickness of 1.5m, a bulk density of 1500 kg/m^3, an average infiltration of 260 mm/year [26].

Considering an average recharge of 260 mm/year, the 0.1 L/kg ratio is reached after a period of around 0.86 year. Based on the calculation assumptions made for this specific site, for this ratio, the total mass of SO_4^{2-} leaching from the backfill is approximately 198 kg and the concentration in the infiltrated water is about 849.79 mg/L. Ca^{2+} quantity is about 66.45 kg and its concentration in the groundwater is about 285.62 mg/L. The 2 L/kg ratio is reached after 17.18 years and the cumulative total leaching mass is 806.83 kg for SO_4^{2-} and 303.05 kg for Ca^{2+}. At this time, the corresponding average concentrations in infiltration water is decreased to 173.40 mg/L and 65.13 mg/L respectively. The last fraction of the test (10 L/kg) corresponds to a period of 85.88 years. The cumulative leached mass is then 930.71 kg for sulphate and 546.76 kg for calcium. Concentrations in infiltration water are low for this fraction because it is the one with the lowest dissolved solutes, with 40 mg/L of SO_4^{2-} and 23.5 mg/L of Ca^{2+}.

Table 5. Estimation of the total released quantities of SO_4^{2-} and Ca^{2+} with average concentrations in recharge water (the different liquid to solid ratios are considered).

L/S Ratio (L/kg)	Estimated Time (Year)	SO_4^{2-} Released Quantity (mg/kg) from the Test	SO_4^{2-} Total Mass (kg)	SO_4^{2-} Concentration in Water Recharge (mg/L)	Ca^{2+} Released Quantity (mg/kg) from the Test	Ca^{2+} Total Mass (kg)	Ca^{2+} Concentration in Water Recharge (mg/L)
0.1	0.86	84.98	197.70	849.79	28.56	66.45	285.62
0.2	1.72	146.67	341.23	733.36	47.20	109.81	236.01
0.5	4.29	238.71	555.35	477.41	78.58	182.81	157.15
1	8.59	302.52	703.82	302.52	104.39	242.87	104.39
2	17.18	346.80	806.83	173.40	130.26	303.05	65.13
5	42.94	381.85	888.37	76.37	175.91	409.25	35.18
10	85.88	400.05	930.71	40.00	235.01	546.76	23.50

During the four first fractions (from 0.1 to 1 L/kg) which correspond to a duration of about 8.6 years, practically 76% of the total sulphate mass is leached and the rest is leached over a period of about 77 years.

The results of this test, mainly for the four first fractions, have confirmed that the backfill is potentially an important source of calcium and sulphate that can be released in the infiltrating water to reach the saturated zone. Transport time in the unsaturated loess layer, between the backfill layer and the saturated zone (1 m/year, see above), is not considered in this assessment. This means that, on the one hand, pollutants has got enough time to be transferred by infiltration through the whole thickness of the unsaturated zone (about 18 m), and on the other hand, the released quantities were most probably higher in the past than what we estimated from the leaching tests (i.e., performed on recent and somehow depleted backfill samples).

4.2. Mineralogical Analysis

A mineralogical analysis was carried out to identify the main mineral phases that are present in the backfill. First, XRD analyses were carried out. After quartering, 4 sub-samples of few grams, crushed and sieved to less than 150 μm, were placed in a sample holder by simple pressure to limit any preferential orientation of the minerals according to the Moore and Reynolds method [42]. The diffraction spectrum is recorded for diffraction angles between 2 and 70° 2 theta on the Bruker D8-Advance diffractometer (copper Kα1 radiance, λ = 1.5418 Å). Mineral identification was done by using the EVA 3.2 software and is then quantified via Topas the Bruker's software using Rietveld's refinement method [43].

Second, to complete the XRD analysis, SEM analysis were performed on one sample obtained after quartering and crushing to less than 1000 μm. A sample of few grams was used to obtain a polished section impregnated in an epoxy resin. Initially, an optical microscope analysis was carried out with the Zeiss Axio Imager.M2m microscope to pre-select areas of interest for advanced analysis of minerals under the electronic microscope (SEM). The individual stoichiometry of minerals was determined with a SEM (Zeiss Sigma 300) equipped with two energy dispersive spectrometers (EDS, Silicon Drift Detector XFlash by Bruker, 30 mm^2) with operating conditions: 20 keV, ~200 μA and a work distance of 8.5 mm.

For the four backfill samples, XRD results show almost identical diffractograms. The detected minerals are quartz, calcite, silicates (micas, plagioclase, chlorite, orthoclase, kaolinite, amphibole) and hematite. No sulphate nor sulphide minerals have been detected. Table 6 summarizes mineral quantification for the four analyzed samples.

Table 6. Relative quantification of minerals (in %) based on XRD analysis of backfill soil samples.

Samples	Quartz	Micas	Calcite	Plagioclase	Chlorite	Orthoclase	Kaolinite	Hematite	Amphibole
	%	%	%	%	%	%	%	%	%
1	53.1	12.7	7.3	10.3	4.6	5.3	3.7	1.2	1.9
2	49.7	15.4	10.8	8.2	5.6	4.3	4.4	1.6	0.0
3	49.7	14.4	11.6	8.9	5.0	5.1	4.0	1.3	0.0
4	48.1	14.8	12.3	8.3	5.2	5.6	4.3	1.4	0.0
Average	50.2	14.3	10.5	8.9	5.1	5.1	4.1	1.4	0.5
Standard error	0.9	0.5	1.0	0.4	0.2	0.2	0.1	0.1	0.4

Minerals proportions are very similar in all samples; the standard error is less than 1% for all detected minerals. Quartz is the dominant mineral with an average quantity of 50.2%, followed by 38% of silicates (including 14.3% of micas). Detected calcite quantity is 10.5%, and hematite corresponds to 1.4%. The presence of calcite can be explained by the recycled construction materials in the backfill layer and by the nature of the underlying loess that may also contain calcite. Micaceous shales are present in local Carboniferous formations locally called 'Houiller' formations [44]. Those formations were intensively exploited in the past for coal production, producing a lot of shaly waste at that time (i.e., often used as backfill materials in different places). In addition, micaceous sandstones have been widely used in construction [45]. The presence of mica in the backfill is therefore most likely from those 'Houiller' shales wastes and from recycled construction materials.

SEM results showed traces of pyrite with an advanced oxidized state surrounded by iron oxide (Figure 7). This explains the detection of hematite (Fe_2O_3) in backfill samples by XRD. The presence of Hematite in cases of pyrite oxidation under room temperature conditions were reported in different studies [46].

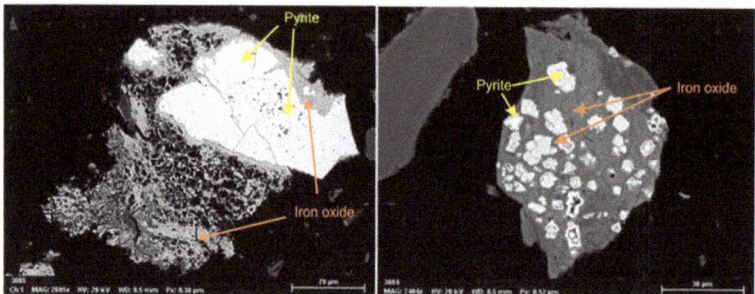

Figure 7. Traces of oxidized pyrite observed by SEM in the backfill analyzed sample.

Our results confirm that the backfill shales contained initially pyrites that oxidized in the presence of oxygen and infiltrating water, releasing sulphate in the environment (Reaction (5)):

$$2FeS_2(s) + 7O_2 + 2H_2O \rightarrow 2FeSO_4 + 2H_2SO_4 \quad (5)$$

Gypsum is often associated with recycled construction materials in Belgium [47]. In our case, even if gypsum was not detected, it could be initially present in the backfill within the recycled construction materials. On the other hand, it can be formed in the presence of SO_4^{2-} ions (Reactions (7) and (8)) and calcite in acidic conditions (Reaction (6)) created by pyrite oxidation and by HY/DH of 1,1,1-TCA (Reactions (1),(2) and (5)). In addition, the presence of a clear input of calcium and sulphate in groundwater (see Section 3.1) also supports this interpretation.

Acidity neutralization by calcite dissolution:

$$CaCO_3(s) + H^+ \rightarrow Ca^{2+} + HCO_3^- \quad (6)$$

Gypsum resulting from reaction between sulphuric acid and calcite:

$$H_2SO_4 + CaCO_3(s) + 2H_2O \rightarrow CaSO_4.2H_2O + CO_2 \tag{7}$$

Gypsum formation from sulphate and calcium ions:

$$Ca^{2+} + SO_4^{2-} + 2H_2O \rightarrow CaSO_4.2H_2O(s) \tag{8}$$

The dominance of sulphate in the ionic balance of the leaching test eluates and the significant released quantities, combined with XRD results, allow to conclude that it is very likely that some sulphate in the backfill existed under an amorphous form that is not identifiable correctly by mineralogical analyses.

5. $\delta^{34}S$ and $\delta^{18}O$ of SO_4^{2-} in Backfill Eluates and in Groundwater

The isotopic composition of dissolved sulphate ($\delta^{34}S$ et $\delta^{18}O$ de SO_4^{2-}) was studied to identify the origin of sulphate concentration increases in the study area. This was done by performing analyses on groundwater samples as well as on backfill eluates.

For groundwater, 8 points were sampled in December 2016 near the CAHs source zone. In March 2017, during the measurement and sampling campaign presented in Section 3.1, 11 samples were collected from the monitoring network. For backfill eluates, analyses were performed on the first four fractions (0.1, 0.2, 0.5, and 1 L/kg), taken from the same column. Polyethylene vials of 500 mL were used with a pinch of zinc acetate to stabilize the solution. Isotopic analyses were carried out in the laboratory of the Centre for Environmental Research Leipzig (UFZ). The method is based on the precipitation of $BaSO_4$ first, and then analyzing each isotopic ratio in this $BaSO_4$ precipitate. For the sulphur isotope, $BaSO_4$ is converted to SO_2 by a continuous flow combustion technique coupled with isotope ratio mass spectrometry (delta S, Thermo Finnigan). The result is expressed in ‰ of $\delta^{34}S$ of the deviation from Cañon Diablo Troilite (CDT). For the oxygen isotope, the used technique consists in a pyrolysis at high temperature (1450 °C) in a TC/EA connected to a delta plus XL mass spectrometer (Thermo Finnigan, Bremen, Germany). The result is expressed in ‰ of $\delta^{18}O$ of the deviation from the Vienna Standard Mean Ocean Water (VSMOW) standard. The analytical error is ±0.3‰ for $\delta^{34}S$ and ±0.5‰ for $\delta^{18}O$.

The results showed that for the eluate, $\delta^{34}S$ is quantified between 2.1 and 2.6‰ while $\delta^{18}O$ is the same for the 4 samples at 6.5‰. For groundwater in December 2016, $\delta^{34}S$ ranged between 1.2 and 2.6‰ while $\delta^{18}O$ ranged between 3.6 and 5.2‰. For March 2017 campaign, $\delta^{34}S$ was between 0.9 and 2.9 ‰ while $\delta^{18}O$ varied between 2.4 and 4.4‰.

$\delta^{34}S$ results versus $\delta^{18}O$ for all analyzed samples are shown in Figure 8a where they can be compared to typical domains of known sulphate sources reported by Mayer [48]. Results are also shown with regards to sulphate concentrations in Figure 8b (where the point size represents the relative sulphate concentration). The isotopic signature of leachate sulphate from the backfill eluates is close to that observed in groundwater. Compared to Mayer's sulphate sources [48], groundwater samples match with the 'soil SO_4' zone, which includes organic (ester and CS-mineralization) and inorganic sulphides. The eluate samples from the backfill are within the anthropogenic source zone of atmospheric deposition near the boundary of the 'soil SO_4' zone.

As described previously, the studied backfill soil may be highly heterogeneous: a mixture of sandy loamy soil with recycled construction and mining wastes that contain shale and coal in small quantities. The isotopic signature of sulphate in this backfill therefore represents the signature of different anthropogenic sources.

In an urban context, Bottrell et al. [17] studied, using isotopes, sulphate sources around the city of Birmingham. Sulphate isotopic signature, for a group of wells in the city center and some industrial sites, was quite similar to our study results with $\delta^{34}S$ between 0.2‰ and 3.2‰ and $\delta^{18}O$ between 2.8‰

and 7.73‰. This isotopic signature was interpreted as from SO_4^{2-} urban pollution resulting from artificial (made ground) soils and/or from wastewater.

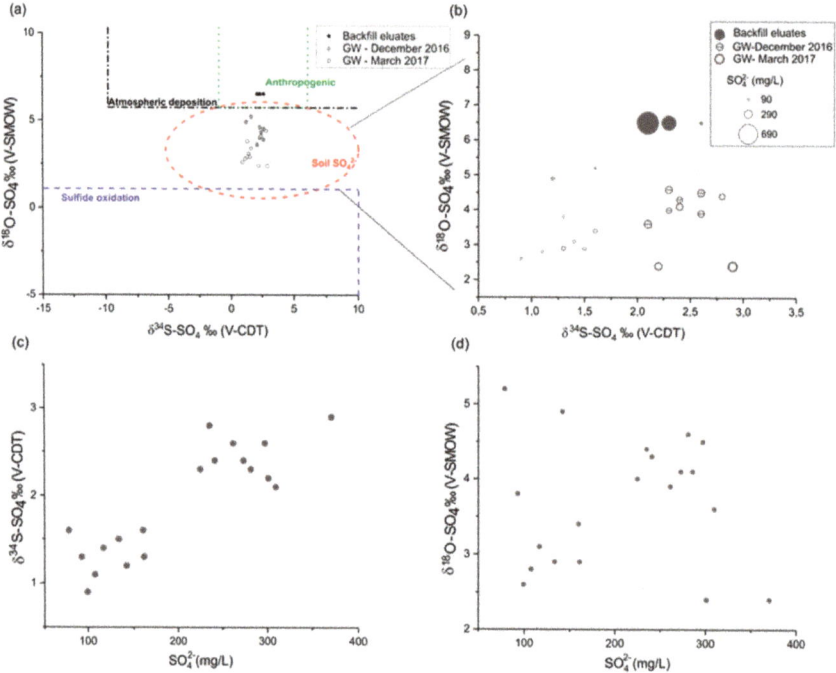

Figure 8. $\delta^{34}S$ versus $\delta^{18}O$ values of sulphate from groundwater and eluates samples compared to typical values depending on: (**a**) Sulphate sources and (**b**) with regards to SO_4^{2-} concentrations. The isotopic composition for the sulphate sources are taken from Mayer [48]. Concentration of SO_4^{2-} in groundwater versus: (**c**) $\delta^{34}S$ and (**d**) $\delta^{18}O$.

The isotopic signature obtained in our study reflects a mixture between background sulphate in the aquifer and additional sulphate released from the backfill layer. In the study by Jurado et al. [49], the isotopic signature of SO_4^{2-} in the same aquifer for points with low sulphate concentrations (between 34.54 and 71.49 mg/L) was between −2.08‰ and −0.13‰ for $\delta^{34}S$, and between 2.1‰ and 4.3‰ for $\delta^{18}O$. These results are similar to our sampling points with low SO_4^{2-} concentrations, and therefore we can consider these values as the natural background signature of sulphate in the chalk aquifer.

Locally, in the study area, the higher the sulphate concentrations increase, the more we have an enrichment of SO_4^{2-} in groundwater at $\delta^{34}S$ and $\delta^{18}O$, and a trend towards the isotopic signature of backfill eluates Figure 8c,d. Sampling wells showing the highest SO_4^{2-} concentrations are located below the backfill of the industrial site.

The increase in sulphate concentration is clearly accompanied by an increase of $\delta^{34}S$-SO_4. For $\delta^{18}O$-SO_4, this trend is not so evident because of the background isotopic signature of sulphate in groundwater comes mainly from the mineralization of organic matter which causes a depletion in $\delta^{18}O$-SO_4 but not of $\delta^{34}S$-SO_4 which does not show significant fractionation through this process [50].

6. Synthesis of Process Leading to Groundwater Mineralization Changes

The main processes explaining groundwater mineralization changes that occurs at the industrial site are conceptually summarized in Figure 9:

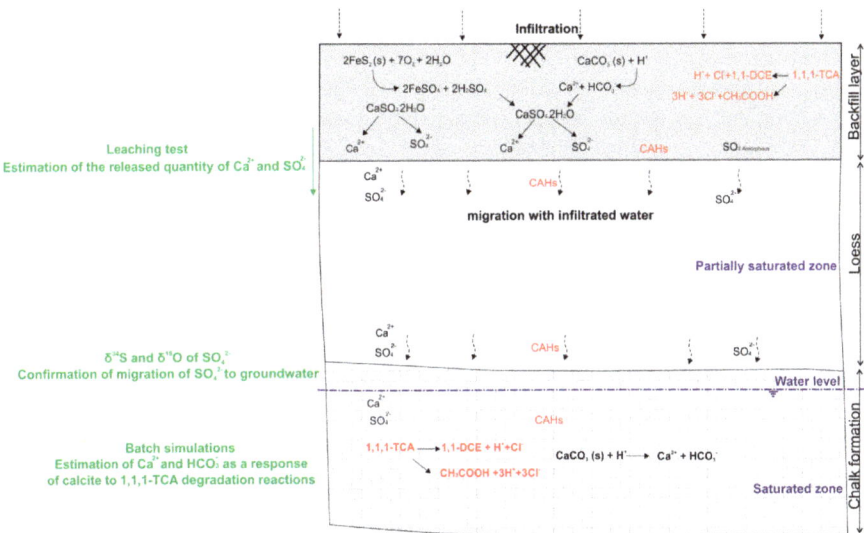

Figure 9. Conceptual scheme summarizing the obtained results with the main process controlling the groundwater mineralization under the industrial site (Scale not respected).

The investigated backfill layer had been in place for more than 40 years prior the present study. It is made of a mixture of materials with considerable calcium sulphate release capacity. The detected traces of oxidized pyrite with hematite indicate that the backfill soil contained pyrite in the past. Gypsum was also present, either with recycled building materials or formed as a result of calcite reaction to acidity. Some sulphate in the backfill soil were in amorphous form. In addition, the improper handling of products containing CAHs has produced a contamination of soil and groundwater. The mixture of pollutants had enough time to percolate across the unsaturated zone with the infiltrated water to reach the saturated zone.

The current groundwater hydrochemistry changes have resulted from the combination of an increase of calcite dissolution along with the migration of calcium and sulphate from the backfill soil, as verified by the correlation of sulphate concentrations with non-carbonated calcium (Ca-HCO$_3$). Furthermore, isotopic analyses results confirm the mixing between the sulphate released from the backfill soil with the background sulphate in the aquifer.

Considering the leaching test results for the first 4 fractions, where easily leachable components are leached quite extensively, calcium and sulphate are the dominant compounds in the backfill eluates. Results extrapolation at the field scale shows that for a period between 0.86 to 8.59 years, the infiltrated water passing through the backfill layer produces leachate with sulphate concentrations between 9.2 and 3.3 times the background concentrations of sulphate in the aquifer. While for calcium, for the same duration, concentrations in infiltrated water correspond to values between 1.9 and 0.7 times the background concentration of calcium in the aquifer.

On the other hand, geochemical simulations provided an assessment of calcite dissolution in groundwater in the presence 1,1,1-TCA degradation reactions. Results showed that for the maximum concentration of 1,1,1-TCA (1100 µg/L) observed in groundwater, HY/DH degradation reactions decrease the pH by a value of 1.01 units for a 1-year degradation time. Consequently, the calcite dissolution produces an additional amount of 0.64 times the background calcium concentration, and 0.93 times the background bicarbonates concentration.

7. Conclusions

Through this study, the understanding of a pollution problem in the considered chalky aquifer was improved. Using a combination of tests on the backfill material extracted from the site and groundwater quality analyses, the mechanisms that lead to changes in groundwater mineralization at this site was clarified. These changes are mainly due to two sources. The first is the increase in calcite dissolution as a buffer reaction to the acidity released by the degradation of 1,1,1-TCA by HY/DH. Geochemical simulation results showed that calcite can release up to 95.28 mg/L of calcium and 289.93 mg/L of bicarbonate during one year of 1,1,1-TCA degradation.

The backfill layer at the industrial site represents a second source that affected the hydrochemistry by releasing calcium sulphate that migrated to the saturated zone with recharge water. Leaching test results showed that for a period between 0.86 and 8.59 years, the average concentration in backfill leachate is 302.52 mg/L for sulphate and 104.39 mg/L calcium.

Thus, in the present case study, the improper backfill materials quality caused an additional pollution source influencing groundwater mineralization. A better control of backfilling materials would have prevented a part of the groundwater contamination. Furthermore, the advanced characterization of backfill soil and groundwater hydrochemistry provided an appropriate estimation of the effect of 1,1,1-TCA degradation reactions on the hydrochemistry compared to backfill leachates effect, leading to a more correct conceptual site model that will improve the remediation plan.

The current research demonstrated also the benefits of analyzing major inorganic chemical elements in cases of local pollution with organic pollutants such as CAHs. These analyses not only make it possible to identify other potential pollutions, but also to improve the understanding of CAHs degradation reactions in groundwater.

Author Contributions: Conceptualization, Y.B., S.B. and A.D.; methodology, Y.B., S.B., P.O. and A.D.; investigation, Y.B. and D.G.; resources, P.J., D.G. and A.D.; writing—original draft preparation, Y.B.; writing—review and editing, S.B. and A.D.; supervision, S.B. and A.D.

Funding: This research was partly funded by the research unit Hydrogeology and Environmental Geology at the University of Liège.

Acknowledgments: Many thanks to Joël Otten for XRD analysis on soil samples and for the analysis of inorganic chemical elements performed on water samples at the University of Liège. The authors are grateful to Dr. Kay Knöller from UFZ for the isotopic analysis. Many thanks to Dr. Mohammed Tayeb Sadani and to Youcef Hakimi for their enriching discussions.

Conflicts of Interest: The authors declare no conflict of interest.

References

1. Westrick, J.J.; Mello, J.W.; Thomas, R.F. The Groundwater Supply Survey. *J. Am. Water Work. Assoc.* **1984**, *76*, 52–59. [CrossRef]
2. Lawrence, S.J. Description, Properties, and Degradation of Selected Volatile Organic Compounds Detected in Ground Water—A Review of Selected Literature. *U.S. Geol. Surv. Open-File Rep.* **2006**, *2006–2133*, 65. [CrossRef]
3. Higgo, J.J.W.; Nielsen, P.H.; Bannon, M.P.; Harrison, I.; Christensen, T.H. Effect of geochemical conditions on fate of organic compounds in groundwater. *Environ. Geol.* **1996**, *27*, 335–346. [CrossRef]
4. Wiedemeier, T.H.; Rifai, H.S.; Newell, C.J.; Wilson, J.T. *Natural Attenuation of Fuels and Chlorinated Solvents in the Subsurface*; John Wiley & Sons, Inc.: Hoboken, NJ, USA, 1999; ISBN 9780470172964.
5. Stroo, H.F.; Ward, C.H. (Eds.) *In Situ Remediation of Chlorinated Solvent Plumes*; Springer: New York, NY, USA, 2010; ISBN 1441914013.
6. Chapelle, F.H. The significance of microbial processes in hydrogeology and geochemistry. *Hydrogeol. J.* **2000**, *8*, 41–46. [CrossRef]
7. Azadpour-Keeley, A.; Keeley, J.W.; Russell, H.H.; Sewell, G.W. Monitored Natural Attenuation of Contaminants in the Subsurface: Processes. *Groundw. Monit. Remediat.* **2001**, *21*, 97–107. [CrossRef]

8. Wiedemeier, T.H.; Swanson, M.A.; Moutoux, D.E.; Gordon, E.K.; Wilson, J.T.; Wilson, B.H.; Kampbell, D.H.; Haas, P.E.; Miller, R.N.; Hansen, J.E. Technical protocol for evaluating natural attenuation of chlorinated solvents in groundwater. *Cincinnatiohiousa: Usepa Natl. Risk Manag. Res. Lab. Off. Res. Dev.* **1998**. Available online: https://clu-in.org/download/remed/protocol.pdf (accessed on 21 August 2019).
9. Field, J.; Sierra-Alvarez, R. Biodegradability of chlorinated solvents and related chlorinated aliphatic compounds. *Rev. Environ. Sci. Bio/Technol.* **2004**, *3*, 185–254. [CrossRef]
10. Doherty, R.E. A History of the Production and Use of Carbon Tetrachloride, Tetrachloroethylene, Trichloroethylene and 1,1,1-Trichloroethane in the United States: Part 2—Trichloroethylene and 1,1,1-Trichloroethane. *Environ. Forensics* **2000**, *1*, 83–93. [CrossRef]
11. McCarty, P.L. Biotic and abiotic transformations of chlorinated solvents in ground water. In Proceedings of the Symposium on Natural Attenuation of Chlorinated Organics in Ground Water, Dallas, TX, USA, 11–13 September 1996; pp. 5–9.
12. Tobiszewski, M.; Namieśnik, J. Abiotic degradation of chlorinated ethanes and ethenes in water. *Environ. Sci. Pollut. Res.* **2012**, *19*, 1994–2006. [CrossRef]
13. Scheutz, C.; Durant, N.D.; Hansen, M.H.; Bjerg, P.L. Natural and enhanced anaerobic degradation of 1,1,1-trichloroethane and its degradation products in the subsurface—A critical review. *Water Res.* **2011**, *45*, 2701–2723. [CrossRef]
14. Foster, S.S.D.; Morris, B.L.; Chilton, P.J. Groundwater in urban development-a review of linkages and concerns. *Iahs Publ.* **1999**, *259*, 3–12.
15. Foster, S.S.D. The interdependence of groundwater and urbanisation in rapidly developing cities. *Urban Water* **2001**, *3*, 185–192. [CrossRef]
16. Chambel, A.; Duque, J.; Madeira, M.M. HYDROCHEMICAL QUALITY OF GROUNDWATER IN URBAN AREAS OF SOUTH PORTUGAL. In *Urban Groundwater Management and Sustainability*; Springer: Dordrecht, The Netherlands, 2006; pp. 241–250.
17. Bottrell, S.; Tellam, J.; Bartlett, R.; Hughes, A. Isotopic composition of sulfate as a tracer of natural and anthropogenic influences on groundwater geochemistry in an urban sandstone aquifer, Birmingham, UK. *Appl. Geochem.* **2008**, *23*, 2382–2394. [CrossRef]
18. Howard, K.W. Urban Groundwater Issues—An Introduction. In *Current Problems of Hydrogeology in Urban Areas, Urban Agglomerates and Industrial Centres*; Howard, K.W.F., Israfilov, R.G., Eds.; Springer: Dordrecht, The Netherlands, 2002; pp. 1–15. ISBN 978-94-010-0409-1.
19. Maes, E. Methodological Manual: Management of Local Soil Pollution-Notice Méthodologique: Gestion de la Pollution Locale des sols (in French). Direction de l'Etat Environnemental (DEE)-Service Public de Wallonie (SPW). Available online: http://etat.environnement.wallonie.be/files/indicateurs/SOLS/SOLS5/Notice_methodologique_Gestionpollutionlocaledessols.pdf (accessed on 4 March 2019).
20. Orban, P.; Brouyère, S.; Compère, J.; Six, S.; Hallet, V.; Goderniaux, P.; Dassargues, A. Aquifère crayeux de Hesbaye. In *Watervoerende Lagen en Grondwater in België-Aquifères et eaux Souterraines en Belgique*; Dassargues, A., Walraevens, K., Eds.; Academia Press: Gent, Belgium, 2014; partie 1-Chapitre 12; pp. 143–159.
21. Palau, J.; Jamin, P.; Badin, A.; Vanhecke, N.; Haerens, B.; Brouyère, S.; Hunkeler, D. Use of dual carbon-chlorine isotope analysis to assess the degradation pathways of 1,1,1-trichloroethane in groundwater. *Water Res.* **2016**, *92*, 235–243. [CrossRef] [PubMed]
22. Commission, E. Council Directive 98/83/EC of 3 November 1998 on the quality of water intended for human consumption. *Off. J. Eur. Communities* **1998**, *41*, 32–54.
23. Brouyère, S.; Dassargues, A.; Hallet, V. Migration of contaminants through the unsaturated zone overlying the Hesbaye chalky aquifer in Belgium: A field investigation. *J. Contam. Hydrol.* **2004**, *72*, 135–164. [CrossRef]
24. Dassargues, A.; Monjoie, A. The chalk in Belgium. In *The Hydrogeology of the Chalk of North-West Europe*; Downing, R.A., Price, M., Jones, G.P., Eds.; Clarendon Press: Oxford, UK, 1993; pp. 153–269. ISBN 0198542852.
25. Hallet, V. Etude de la Contamination de la Nappe Aquifere de Hesbaye par les Nitrates: Hydrogéologie, Hydrochimie et Modélisation (Contamination of the Hesbaye Aquifer by Nitrates: Hydrogeology, Hydrochemistry and Mathematical Modeling). Ph.D. Thesis, University of Liege, Liege, Belgium, 1998.
26. Orban, P. Solute Transport Modelling at the Groundwater Body Scale: Nitrate Trends Assessment in the Geer Basin (Belgium). Ph.D. Thesis, University of Liege, Liege, Belgium, 2009.
27. Dassargues, A. *Hydrogeology: Goundwater Science and Engineering*; Taylo & Francis CRC Press: Boca Raton, FL, USA, 2018; ISBN 9781498744003.

28. Brouyère, S.; Hallet, V.; Dassargues, A. Effets de retard et de piégeage des polluants dus à la présence d'eau immobile dans le milieu souterrain: Importance de ces effets et modélisation. In Proceedings of the Nat. Colloquium van de BCIG/CBGI; KULeuven: Leuven, Belgium, 1997; pp. 21–27. Available online: https://orbi.uliege.be/handle/2268/2363 (accessed on 21 August 2019).
29. Parkhurst, D.L.; Appelo, C.A.J. *Description of input and Examples for PHREEQC Version 3: A Computer Program for Speciation, Batch-Reaction, One-Dimensional Transport, and Inverse Geochemical Calculations*; US Geological Survey: Reston, VA, USA, 2013.
30. Rodier, J.; Legube, B. *L'analyse de l'eau*; Dunod: Paris, France, 2009; ISBN 2100072463.
31. Orban, P.; Brouyère, S.; Batlle-Aguilar, J.; Couturier, J.; Goderniaux, P.; Leroy, M.; Maloszewski, P.; Dassargues, A. Regional transport modelling for nitrate trend assessment and forecasting in a chalk aquifer. *J. Contam. Hydrol.* **2010**, *118*, 79–93. [CrossRef]
32. Szczucińska, A.; Dłużewski, M.; Kozłowski, R.; Niedzielski, P. Hydrochemical Diversity of a Large Alluvial Aquifer in an Arid Zone (Draa River, S Morocco). *Ecol. Chem. Eng. S* **2019**, *26*, 81–100. [CrossRef]
33. Wiejaczka, Ł.; Prokop, P.; Kozłowski, R.; Sarkar, S. Reservoir's Impact on the Water Chemistry of the Teesta River Mountain Course (Darjeeling Himalaya). *Ecol. Chem. Eng. S* **2018**, *25*, 73–88. [CrossRef]
34. Kimblin, R.T. The chemistry and origin of groundwater in Triassic sandstone and Quaternary deposits, northwest England and some UK comparisons. *J. Hydrol.* **1995**, *172*, 293–311. [CrossRef]
35. Gerkens, R.R.; Franklin, J.A. The rate of degradation of 1,1,1-trichloroethane in water by hydrolysis and dehydrochlorination. *Chemosphere* **1989**, *19*, 1929–1937. [CrossRef]
36. Palau, J.; Shouakar-Stash, O.; Hunkeler, D. Carbon and Chlorine Isotope Analysis to Identify Abiotic Degradation Pathways of 1,1,1-Trichloroethane. *Environ. Sci. Technol.* **2014**, *48*, 14400–14408. [CrossRef] [PubMed]
37. Cline, P.V.; Delfino, J.J. Transformation kinetics of 1, 1, 1-trichloroethane to the stable product 1, 1-dichloroethene. In *Biohazards of drinking water treatment*; Larson, R.A., Ed.; Lewis Publishers, Inc.: Chelsea, MI, USA, 1989; pp. 47–56.
38. Gauthier, T.D.; Murphy, B.L. Age Dating Groundwater Plumes Based on the Ratio of 1,1-Dichloroethylene to 1,1,1-Trichloroethane: An Uncertainty Analysis. *Environ. Forensics* **2003**, *4*, 205–213. [CrossRef]
39. European Committee for Standardization. *CEN-TS 14405: Characterization of waste-Leaching behaviour tests-Up-flow percolation test (under specified conditions)*; CEN: Brussels, Belgium, 2004.
40. International Organization for Standardization. *ISO/TS 21268-3 Soil quality—Leaching procedures for subsequent chemical and ecotoxicological testing of soil and soil materials—Part 3: Up-flow percolation test*; ISO: Geneva, Switzerland, 2007.
41. Council, E.U. Council Decision 2003/33/EC of 19 December 2002 establishing criteria and procedures for the acceptance of waste at landfills persuant to Article 16 of and Annex II to Directive 1999/31/EC. *Off. J. Eur. Commun.* **2003**, *16*, L11.
42. Moore, D.M.; Reynolds, R.C. *X-ray Diffraction and the Identification and Analysis of Clay Minerals*; Oxford University Press: Oxford, UK, 1989; Volume 322.
43. Coelho, A.A. *TOPAS User's Manual, Version 3.0*; Bruker AXS GmbH: Karlsruhe, Germany, 2003.
44. Stainier, X. Matériaux pour la faune du Houiller de Belgique (note 3). Available online: http://biblio.naturalsciences.be/rbins-publications/bulletin-de-la-societe-belge-de-geologie/007-1893/bsbg_07_1893_mem_p135-160.pdf (accessed on 9 April 2019).
45. Ruthy, I.; Dassargues, A. Carte Hydrogéologique de Wallonie, Jehay-Bodegnée-Saint-Georges-sur-Meuse 41/7-8. Notice explicative: Première édition: Mai 2003-Actualisation partielle: Décembre 2010; Namur, Belgium, 2010. Available online: http://environnement.wallonie.be/cartosig/cartehydrogeo/document/Notice_4178.pdf (accessed on 20 August 2019).
46. Evangelou, V.P.; Zhang, Y.L. A review: Pyrite oxidation mechanisms and acid mine drainage prevention. *Crit. Rev. Environ. Sci. Technol.* **1995**, *25*, 141–199. [CrossRef]
47. Vrancken, K.C.; Laethem, B. Recycling options for gypsum from construction and demolition waste. *Waste Manag. Ser.* **2000**, *1*, 325–331. [CrossRef]
48. Mayer, B. Assessing sources and transformations of sulphate and nitrate in the hydrosphere using isotope techniques. In *Isotopes in the Water Cycle*; Springer: Dordrecht, The Netherlands, 2005; pp. 67–89.

49. Jurado, A.; Borges, A.V.; Pujades, E.; Hakoun, V.; Otten, J.; Knöller, K.; Brouyère, S. Occurrence of greenhouse gases in the aquifers of the Walloon Region (Belgium). *Sci. Total Environ.* **2018**, *619–620*, 1579–1588. [CrossRef]
50. Krouse, H.R.; Mayer, B. Sulphur and Oxygen Isotopes in Sulphate. In *Environmental Tracers in Subsurface Hydrology*; Springer: Boston, MA, USA, 2000; pp. 195–231.

© 2019 by the authors. Licensee MDPI, Basel, Switzerland. This article is an open access article distributed under the terms and conditions of the Creative Commons Attribution (CC BY) license (http://creativecommons.org/licenses/by/4.0/).

Article

Groundwater Contribution to Sewer Network Baseflow in an Urban Catchment-Case Study of Pin Sec Catchment, Nantes, France

Fabrice Rodriguez [1,2,*], **Amélie-Laure Le Delliou** [1,3], **Hervé Andrieu** [1,2] **and Jorge Gironás** [4,5,6,7]

1. Gers Laboratoire Eau et Environnement, Université Gustave Eiffel, Ifsttar, 44343 Bouguenais, France; al.le.delliou@groupeginger.com (A.-L.L.D.); andrieuherve@orange.fr (H.A.)
2. Irstv Fr Cnrs 2488, rue de la Noé, 44321 Nantes, France
3. Ginger Burgeap Nantes, 44806 St Herblain, France
4. Departamento de Ingeniería Hidráulica y Ambiental, Pontificia Universidad Católica de Chile, Santiago 7820436, Chile; jgironas@ing.puc.cl
5. Centro de Investigación para la Gestión Integrada de Riesgo de Desastres Anid/Fondap/15110017, Santiago 7820436, Chile
6. Centro de Desarrollo Urbano Sustentable Anid/Fondap/15110020, Santiago 7530092, Chile
7. Centro Interdisciplinario de Cambio Global, Pontificia Universidad Católica de Chile, Santiago 7820436, Chile
* Correspondence: fabrice.rodriguez@ifsttar.fr; Tel.: +33-(0)24084588

Received: 21 December 2019; Accepted: 21 February 2020; Published: 3 March 2020

Abstract: Sewer systems affect urban soil characteristics and subsoil water flow. The direct connection observed between baseflow in sewer systems under drainage infiltrations and piezometric levels influences the hydrological behavior of urban catchments, and must consequently be considered in the hydrologic modeling of urban areas. This research studies the groundwater contribution to sewer networks by first characterizing the phenomenon using experimental data recorded on a small urban catchment in Nantes (France). Then, the model MODFLOW was used to simulate the infiltration of groundwater into a sewer network and model dry weather flows at an urban catchment scale. This application of MODFLOW requires representing, in a simplified way, the interactions between the soil and the sewer trench, which acts as a drain. Observed average groundwater levels were satisfactorily simulated by the model while the baseflow dynamics is well reproduced. Nonetheless, soil parameters resulted to be very sensitive, and achieving good results for joint groundwater levels and baseflow was not possible.

Keywords: groundwater; urban hydrology; drainage; modeling; sewer; baseflow

1. Introduction

Urbanization modifies land use and affects soil, sub-soil and subsurface processes in different ways. Urban features as well as surface and underground infrastructure can have a strong impact on groundwater levels. Some causes explaining a reduction in groundwater include an infiltration decrease due to additional imperviousness, groundwater pumping for various urban water uses, and groundwater flow into drainage trenches [1]. On the other hand, leakage from water supply and waste water networks becomes a source of recharge for urban groundwater [2–5]. Furthermore, leakage from waste water systems is also a possible source of groundwater contamination [6]. Despite aquifer levels and stream flow having proved to be related in various rural contexts [7,8], few studies have focused on examining this relationship in urban catchments [5,9–11]. Soil water infiltration in sewers, however, is a phenomenon affecting urban hydrology, sewerage and waste water treatment plant management [12,13]. Initially, the presence of soil water in wastewater sewerage was revealed by [14,15], when flow variations in separate wastewater sewerage during wet weather periods were commonly attributed to inappropriate or irregular stormwater runoff connections. These authors showed that groundwater seep into the sewers through defective cracks once the water table level reaches the depth at which sewers are buried. This phenomenon has been recently reinforced by [13] for several Flemish catchments; some interesting stable isotopes methods have been used to detect and estimate groundwater intrusions in sewers in Nancy, France [16] and in Brussels, Belgium [17] and temperature-based methods in Trondheim, Norway [18]. Similarly, [9] showed that soil water may explain the variations of runoff coefficients observed on a small urban catchment in the city of Nantes, France, as groundwater drainage flow becomes significant once the water table exceeds a given threshold. The flowrate in sewers is the sum of runoff during rain events, wastewaters from housing (for separative wastewater or combined sewers), and a groundwater infiltration component, representing the baseflow in sewers; this component is often part of "extraneous" water in sewers. Interestingly, [19] suggest the appropriate terminology of urban "karst" to denote the multiple water soil and pipe interactions below the surface. Despite the relevance of improving our understanding of the impact of groundwater drained by urban sewer systems, only a few efforts are currently devoted to the problem. Groundwater drainage by sewers is often noted in literature [20], although is rarely quantified when assessing the urban water budget. Only a few papers deal with the determination of the urban water budget, likely due to the lack of reliable data of all the water budget components (i.e., rainfall, evapotranspiration, surface and groundwater flow rates). This lack of knowledge must be addressed as urban stormwater practices increasingly consider innovative infiltration-based technologies and approaches to mitigate the hydrological impacts of urbanization [21]. Thus, a better understanding and quantification of the fate of urban soil water becomes essential [22,23].

The first attempts to represent the interaction between surface water and the aquifer in urban integrated hydrological models were simple and conceptual. Aquacycle [24] and the Storm Water Management Model (SWMM, [25]) are examples of such models. In a comprehensive review of ten stormwater models, [26] identified the models able to simulate the groundwater baseflow component, mostly through a conceptual linear reservoir. Some of the more recent physically-based modeling efforts consider soil drainage by sewers. An integrated sewer-aquifer model was developed [27] to test the effects of future buried drains on both groundwater flow and sewer infiltration phenomena. This integration proved to be relevant to describe the groundwater-sewer interactions. The Network Exfiltration and Infiltration Model (NEIMO) model [28] was developed in close connection with already available groundwater models such as MODFLOW [29] or FEFLOW [30], to simulate infiltration and exfiltration processes. This modeling approach requires developing an integrated modeling framework like Urban Volume and Quality (UVQ) developed by CSIRO, Australia [31,32]. MODFLOW was used [33] to analyze the relevant infiltration parameters at a local scale, and compared a 1D-infiltration approach and a MODFLOW modeling approach to model groundwater infiltration at a larger scale. Overall, coupling groundwater models with the simulation of infiltration/exfiltration processes is a major challenge, especially due to the high uncertainty associated with these models [34]. The coupling of a hydraulic model was realized by [35] with both a groundwater model and a sewer failure estimation approach to identify the sections potentially affected by infiltration in coastal urban areas. Satisfying results were obtained by [11] with regard to the groundwater infiltration by coupling two commercial models MIKE URBAN for simulation of sewer flow and MIKE SHE for simulation of groundwater transport. The Urban Runoff Branching Structure Model (URBS-MO) [36] was developed to represent surface and subsurface water flows while focusing on the impact of sewer networks. Although the model simulates groundwater drainage by sewer, real observations at a local scale to test this component of the model were not available. Finally, [5] coupled the hydrological/water management WEAP model and MODFLOW to verify a strong stream-aquifer interaction in areas with shallow groundwater, as well as quantify the local recharge associated with pipe leaks and inefficient urban irrigation. Overall, the literature review shows that much more experimental data and modeling using comprehensive 3D modeling tools such as MODFLOW, are needed to better understand and quantify the impact of groundwater infiltration and groundwater levels on low flows in urban environments.

This study investigates the interactions between urban groundwater and baseflow in wastewater and stormwater sewer systems, and its first goal is to better understand their role on the hydrological budget of an urban catchment. These interactions are analyzed and quantified using a field study conducted in the "Pin Sec" catchment, located in the city of Nantes (France) [37]. The second objective consists of testing the ability of a groundwater modeling tool to mimic these interactions. The results are used to build a numerical model focused on simulating the hydrological behavior of the urban soil and low flows in the sewer system. The outline of this paper is as follows: the next section presents the case study and the methods adopted in this work, which focus on using MODFLOW to model baseflow and interactions between the sewer systems and groundwater at catchment scale. The results section presents and compares observed and simulated groundwater levels and baseflow discharges. Finally, the last section summarizes our results and highlights the main conclusions.

2. Materials and Methods

2.1. In-Situ Interactions Between Baseflow and Groundwater Levels

2.1.1. Case Study

The experimental area is part of the Observatoire Nantais des Environnements Urbains (ONEVU) initiative, devoted to long-term monitoring of water and pollutant fluxes and soil-atmosphere exchanges in urban settings [37]. The Pin Sec catchment is located in the East of Nantes (France) between the Loire and the Erdre rivers. The area is under an oceanic climate, with frequent but not very intense rainfalls. The Pin Sec neighborhood developed between 1930 and 1970 is mainly composed of single and multi-family housing (Figure 1). This 31 ha catchment (called urban catchment) belongs to a larger hydrogeological catchment of 120 ha, both of them are represented on Figure 1. This catchment has a gentle slope, the highest altitude of the catchment is 28 m asl and the lowest is 13 m. The Pin sec catchment has an imperviousness of 45% and is equipped with a 50 years old separate sewer network. The wastewater and stormwater sewer systems have total lengths of 7.3 and 4 km, and mean depths of 2 and 2.9 m respectively. Although the wastewater sewer network is denser than the stormwater system, both are mainly superimposed, and both present cracks or faulty sealing joints. Both system outlets, while collecting water from the same geographic area, are not located in the same place. The stormwater system is connected to the Gohards river, whereas the wastewater sewer system drains into a downstream combined sewer system. Gohards river stream is an old river which was buried between 1945 and 2012 and which has been re-opened through a renovation urban planning of this area in 2012. It is a perennial stream.

Recorded data include (1) 5 min rainfall records from 1999 to 2010, deduced from the average of 3 rain gauges covering the study area, (2) 5-min flow discharge records at the outlet of both the stormwater and the wastewater systems from 1 September 2006 to 31 August 2008, (3) 20-min groundwater levels measured with pressure sensors in eleven piezometers located throughout the catchment from 1 September 2006 to 31 August 2008 (one piezometer, PZGN being only used to map the groundwater level contours), and (4) other meteorological records as well as evapotranspiration estimated with the Penman method, measured 12 km away from the catchment outlet, from 1999 to 2010. We focus our analysis on the main period between 1 September 2006 and 31 August 2008 (i.e., two hydrological years), but we also used year 2002 for some modeling tests as rainfall characteristics were representative of the average conditions in the area.

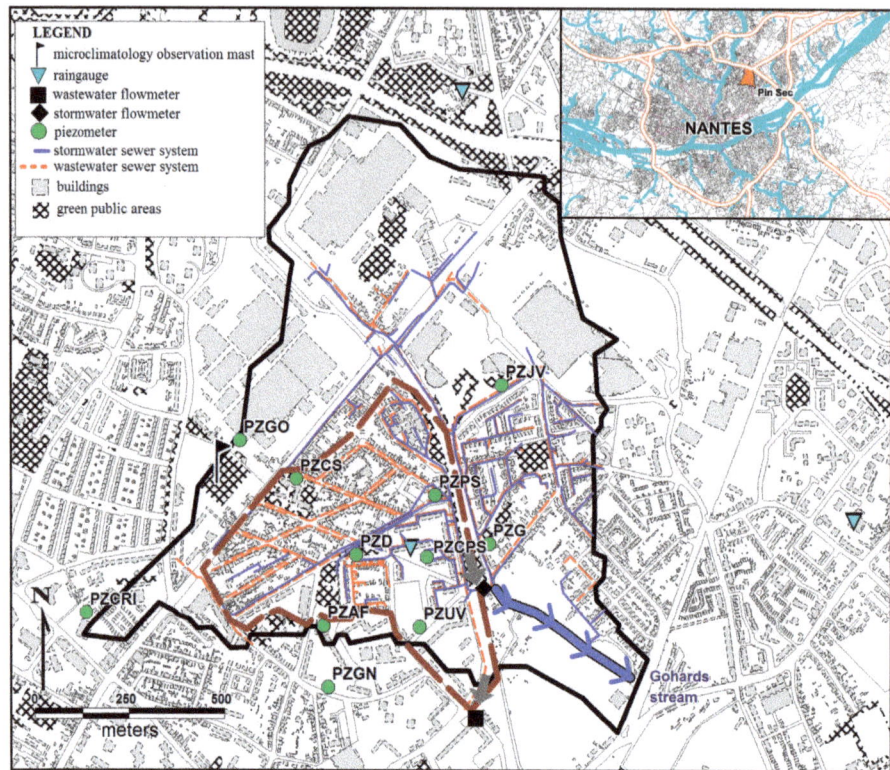

Figure 1. Pin Sec catchment within the Observatoire Nantais des Environnements Urbains (ONEVU) territory and location of the piezometers (green circles). The dotted brown and continuous black lines represent the boundary of the urban catchment and the hydrogeological catchment respectively. The large arrows represent the main sewer flow directions. By convention, north is placed at the top of the following maps. PZPS is located 47°14′43.0″ N 1°31′09.1″ W.

2.1.2. Geology and Hydrogeology

The city of Nantes has developed on the Armorican massif, and especially in the south Armorican shear zone. The geology is mainly composed of a mica-schists bedrock and a stack of alluvial deposits along the Loire River. On the Pin sec catchment, the mica-schists bedrock is covered with altered mica-schists, silty eolian deposits and alluvial deposits along the Gohards stream (Figure 2a) [38,39]. The piezometers are mainly located in the top layers and especially within altered mica-schists, with a drilling depth varying from 4 to 8 m (Figure 2b); for two of them, the depth was limited due to hard non altered micaschists. The hydraulic conductivities estimated from in-situ measurements by water bail tests [40] using the Hvorslev method [41] varies from 8.5×10^{-8} to 1.8×10^{-5} m s^{-1} [42]. Piezometer PZGN is out of the hydrogeological catchment and was not considered for further results. From now on groundwater level will designate the level of the saturated zone, which is equivalent to the local groundwater level in our case study.

The piezometers were used for mapping the behavior of the water table on the catchment. Groundwater level contours were interpolated using inverse distance weighting (IDW) techniques with Vertical Mapper© in Mapinfo platform, and used to determine groundwater main flow directions. The groundwater levels of the high water table periods are presented in Figure 3. Note that the upstream condition of the water table is given by two piezometers (PZCRI and PZGO) located at the upper limit capacity, using the drain package (DRAIN); and (3) a representation of the sewer network using a river with the river package (RIV).

A preliminary sensitivity analysis was carried out at a street scale for a 100 × 100 m unit element, both in steady state and transient conditions (not presented here). This analysis helped to understand the impact of the modeling configurations (grid resolution and the sewer trench representation option) on both groundwater levels and flows simulation. Moreover, the influence of the main parameters used in the model (i.e., sewer trench conductance, hydraulic conductivities, specific yield and specific storage) was analyzed. The choices about the parameters or configurations described below were not adopted to replicate experimental results, but to retain realistic characteristics of the subsurface processes to be represented.

The grid resolution in our analysis varied between 1 and 10 m and affected mainly the water flows. Finally, a 4 m grid cell resolution was chosen. This size is quite large to represent the sewer trenches encompassing sewer pipes, but is a good compromise given the area of the study catchment (120 ha). Along the three tested options, the option using the DRAIN package to represent the sewer trench proved best to represent the saturated zone drainage by a sewer, because it can better reproduce the decreasing groundwater near the trench. Specific yields and conductivities are the main sensitive parameters, influencing both the groundwater levels and the flows. Due to the low impact of anisotropy in our sensitivity analysis, conductivities are considered homogenous in the three directions. The water flux increases with the hydraulic conductivities, while the groundwater level decreases. The initial hydraulic conductivities are estimated from in-situ measurements described above (§2.1.2). The influence of the specific yield of the topsoil layer is significant for the sewer trench outflow: the downstream flow reduces by 7% when specific yield decreases from 20% to 5%, which are typical values assessed by [50].

The field drainage system representing the sewer trench is characterized by a drain hydraulic conductance (L^2/T) (conductance thereafter), which depends a priori on the material and characteristics of the drain. Since the conductance is usually unknown, it must be estimated by calibration [49,51]. We first assessed the range of variation of the conductance from the sewer baseflow observations. As discussed by [51], the discharge rate to drain cells may be calculated as the product of the conductance and the head gradient (h-d), where h is the hydraulic head and d is the elevation of the drain. We integrated this relation at the whole sewer network scale, which allows estimating a global conductance associated with the total groundwater flowrate in sewers. Assuming that the baseflow in sewers is only due to groundwater infiltrations, the application of this relation for the high water level period during year 2007 and for the total length of flooded sewer pipes generates a conductance of ~0.9 m^2/day. This parameter only changes the water flows and its effect on groundwater levels is less significant. In the end, the specific storage slightly affects the modeling results, as it ranges theoretically from 3.3×10^{-6} m^{-1} (rock) to 2×10^2 m^{-1} (plastic clay) [51]. A medium value of 1×10^{-5} m^{-1} was adopted initially.

2.2.2. Urban Groundwater Modeling at the Catchment Scale and Application to the Case Study

The MODFLOW model is used on the Pin Sec catchment to assess soil water-sewer interactions at the urban catchment scale. Furthermore, certain simplifications not considered in the sensitivity analysis for the unit element were adopted. The two soil layers are represented in this way: (1) the basement layer is located in the entire catchment at a mean depth of 18 m, and (2) the topsoil layer is located in the valley area around the old stream bed, with a mean depth of 2 m. Because both the stormwater and wastewater systems are often placed within the same trench below the street surface, both networks have been combined to represent the draining trench in the soil. Thus, the field drain represents both stormwater and wastewater networks.

Boundary conditions are stipulated as follows and summarized on Figure 5: (1) a zero-flux condition is defined in the upstream boundary of the hydrogeological catchment by considering no-flow cells and (2) a river boundary condition representing the Gohards stream is defined as the downstream condition by using the River package (RIV) on a 200 m long stream; it is a head-dependant boundary condition. The combined 'field drain', representing the sewer systems, drains into this stream. A one hydrological year simulation in transient conditions is performed on the catchment using a daily time step. The transient simulation is initialized using the steady state case calibrated to the field data observed the first day of simulation. The recharge is calculated on a daily basis using the urban surface hydrological model URBS-MO, already successfully applied to urban catchments in the area to simulate the various components of the water budget [36]. The recharge is assumed to be uniform throughout the catchment, and equal to the infiltration calculated by URBS-MO (i.e., rainfall intensity minus both the actual evapotranspiration and surface runoff). Rainfall and potential evapotranspiration data available for the Pin Sec catchment were used with URBS-MO. The initial set of parameters for the simulation is deduced from the observations and a first steady state simulation conducted on the catchment (Table 1). Specific yield and specific storage values are the same as those used for the sensitivity analysis; they vary in space along with the two geological layers defined in this catchment. Due to the lack of information about the spatial distribution of cracks and faulty sealing joints in sewers and to an homogeneous age of the pipes, the drain conductance is assumed to be uniform and the initial value is deduced from the groundwater baseflow observed in the catchment sewers presented in next section.

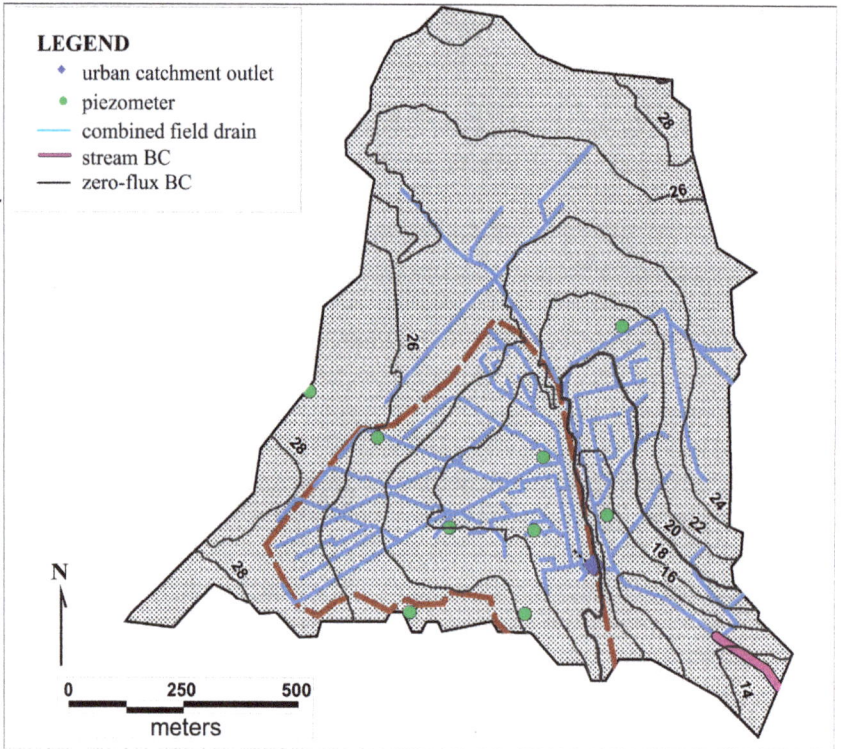

Figure 5. Main boundary conditions applied to the catchment scale model. The 4 × 4 m grid is not perceptible at this scale; the relief is represented with gray elevation contours (altitude asl in m).

The model simulates (1) the hydraulic head and groundwater flux at any grid cell of the domain, and (2) the drain flux at any point of the field drain system. Special attention is paid to the hydraulic head at field piezometers and the flow at the outlet of the catchment, where a flow gauge is located

(Figure 1). The period going from September 2006 to September 2007 is used for calibration, while the period between September 2007 and September 2008 is used for validation.

Table 1. Parameters used within the catchment modelling for the different layers.

Parameter	Loess and Altered Mica-Schists	Mica-Schists	Alluvial Deposits
		Initial values	
Hydraulic conductivity (m/s)	1×10^{-6}	1×10^{-7}	1×10^{-6}
Specific yield (%)	20	20	20
Specific storage (1/m)	1×10^{-5}	1×10^{-5}	1×10^{-5}
Drain conductance (m²/day)	-	0.9	-
		Calibrated values	
Hydraulic conductivity (m/s)	1×10^{-4}	1×10^{-6}	1×10^{-3}
Specific yield (%)	5	10	10
Specific storage (1/m)	1×10^{-2}	5×10^{-4}	1×10^{-2}
Drain conductance (m²/day)	-	5	-

3. Results

3.1. Groundwater Dynamics and Experimental Relationship Between Groundwater Level and Baseflow

The temporal dynamics of groundwater levels are presented in Figure 6 for two of the piezometers; more details may be found in [39]. The mean difference between the high and low water table was approximately 1.7 m during the period 2006–2007, and 1.5 m during the period 2007–2008. This difference varied from ~3 m upstream to 1 m downstream (Table 2).

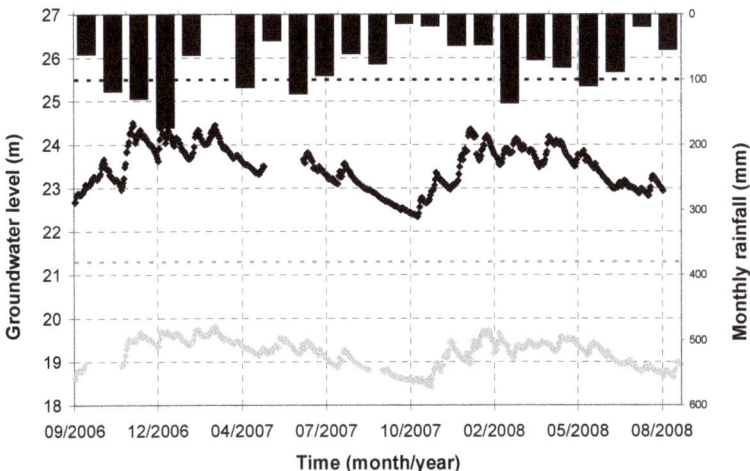

Figure 6. Daily dynamics of the groundwater levels in piezometer PZCS (black) and PZD (grey) from 2006 to 2008 on the Pin Sec catchment. The corresponding surface elevations are plotted with black and grey dotted lines. Bars correspond to monthly rainfall.

Groundwater infiltration in sewer networks depended on several factors such as the location and shape of the water table, the depth of the sewers, and their conditions (e.g., presence of defects). [9] collected information in a small urban catchment in the metropolitan area where the Pin Sec catchment is located, and noticed that groundwater drainage flow became significant once the water table reached a depth of 1.5 m below the surface, and increased as the water table rose. In fact, the mean sewer depth in the area is ~1.2 m below the surface.

Table 2. Main features of the groundwater levels during the studied period (September, 2006 to September 2008). The piezometers are presented from upstream (left) to downstream (right). Z_{soil} is the altitude of the corresponding piezometer; 'gw depth' is the groundwater depth from the surface level; R^2(SW) and R^2(WW) are the determination coefficient of the polynomial regression relationship for stormwater and wastewater baseflow, respectively (2006–2007 period) (See Figure 7 for piezometer PZPS).

	PZCRI	PZGO	PZCS	PZAF	PZUV	PZJV	PZD	PZCPS	PZPS	PZG
Z_{soil} (m)	28.70	26.70	25.50	24.50	23.80	22.14	21.30	21.05	20.00	17.20
Average gw depth (m)	4.14	2.73	2.04	2.24	1.82	2.77	2.09	2.17	3.35	3.35
Minimum gw depth (m)	5.18	4.14	3.16	3.01	3.60	3.52	2.94	2.75	3.72	4.01
Maximum gw depth (m)	2.85	1.65	1.00	1.53	0.10	0.13	1.48	1.66	2.76	2.73
R^2(SW)	0.380	0.649	0.640	0.533	0.711	0.609	0.713	0.729	0.817	0.660
R^2(WW)	0.250	0.766	0.828	0.646	0.753	0.729	0.802	0.762	0.822	0.640

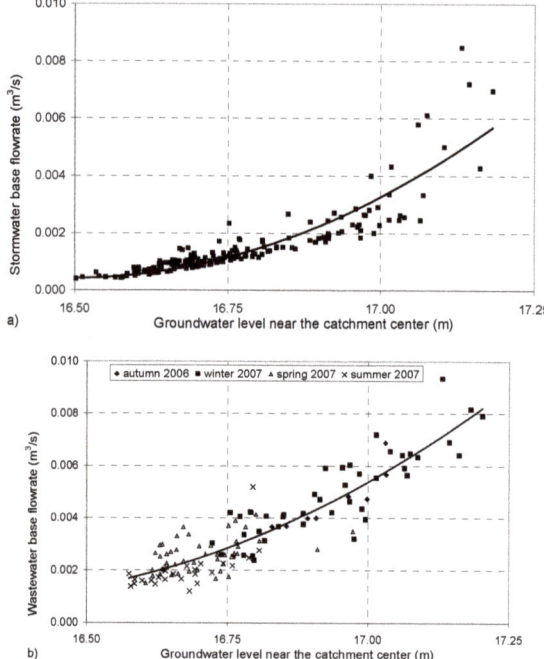

Figure 7. Relationship between daily groundwater drainage flow Q in (**a**) stormwater and (**b**) wastewater sewers, and the groundwater level H for the 2006–2007 period. The groundwater level is observed on piezometer PZPS, located on the main sewer intersection. Polynomial regression relationships are y = $0.0112x^2 - 0.3693x + 3.0460$ (R^2(SW) = 0.817) for (**a**) stormwater and y = $0.0083x^2 - 0.2711x + 2.2056$ (R^2(WW) = 0.822) for (**b**) wastewater.

The connection between the average groundwater level of the catchment H and daily baseflow Q in the sewer systems is evident from Figure 7. A specific piezometer (PZPS) located near the main sewer intersection and representative of the average behaviour of the groundwater level dynamics has been chosen; it is considered to be in the catchment center according to groundwater main flow directions. Figure 7a shows a Q~H^2 relationship for the baseflow in the storm sewer, which is typical of ideal field drains used in rural hydrology [52]. Groundwater drainage is observed at the outlet of the stormwater network during the entire observation period, which demonstrates that sewer infiltrations take place all year round. The behavior is similar for the wastewater sewers (Figure 7b). The wastewater baseflow barely varies during summer when the water table is low (between July and October). Baseflow becomes substantial again when the groundwater level for this piezometer PZPS exceeds a threshold value of ~16.75 m (i.e., 3.2 m below the surface). Waste- and storm-baseflows differ because the drainage density of the wastewater system is higher than that of the stormwater system. This is the reason why the Q~H^2 relationships are more consistent for the wastewater baseflow that the stormwater baseflow, as revealed by the determination coefficient variation R^2(SW) or R^2(WW) (Table 2). Additionally, these relationships are generally more pronounced for the downstream piezometers, closer to the baseflow monitoring locations.

This observation is confirmed by the map showing the sewer receiving groundwater contributions during the high water table period of 2007 (Figure 8). This map was realized through the superimposition of groundwater level contours (Figure 3) and the sewer depth levels deduced from GIS data. The total sewer length soaked during winter is 2.6 km for the wastewater sewers and 1.1 km for the stormwater sewers. Finally, the annual subsurface water volume drained into both sewer systems deduced from the sum of daily base-flows between September 2006 and August 2007 is 28% (wastewater sewer) and 14% (stormwater sewer) of the total rainfall respectively. This result proves that the base flow discharges in artificial sewer systems can be a major component of the urban water budget, as 42% of the total annual rainfall is drained to the sewer systems.

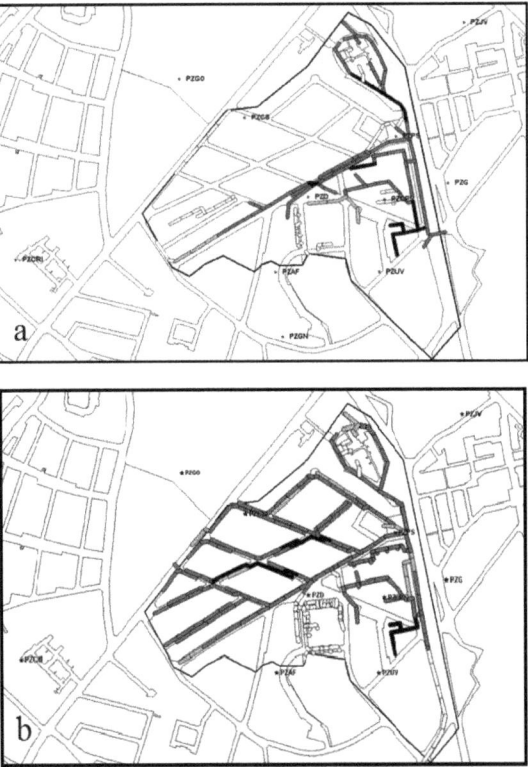

Figure 8. Sewer pipe network affected by groundwater during winter (2007/03/07): (**a**) stormwater network and (**b**) wastewater network. Black and light gray sewer pipes are located 1 m below and above the groundwater table respectively. Dark gray sewer pipes are between them.

3.2. Comparison between Modeling Results and Observed Data at the Catchment Scale

Three metrics were adopted for the evaluation of the model: the root mean square error (RMSE), a Bias error, and the determination coefficient R^2:

$$RMSE = \sqrt{\frac{1}{n}\sum_{i=1}^{n}[V_2(t) - V_1(t)]^2} \quad (1)$$

$$Bias\ error = \frac{\sum_{i=1}^{n} V_2(t) - V_1(t)}{\sum_{i=1}^{n} V_1(t)} \quad (2)$$

$$R^2 = \frac{\left(\sum_{i=1}^{n}[V_2(t) - \overline{V_2}][V_2(t) - \overline{V_1}]\right)^2}{\sum_{i=1}^{n}\left(V_1(t) - \overline{V_2}\right)^2 \sum_{i=1}^{n}\left(V_1(t) - \overline{V_1}\right)^2} \quad (3)$$

where n is the number of time steps involved, V_1 and V_2 are the observed and simulated values, $\overline{V_1}$ and $\overline{V_2}$ the temporal average of V_1 and V_2.

Absolute error (m) was used as well to estimate the difference between simulated and observed groundwater levels in specific piezometers.

3.2.1. Implementation and Calibration of MODFLOW at the Catchment Scale

In the MODFLOW model of the catchment, we first calibrated the hydraulic conductivity, specific yield and specific storage of each soil type by minimizing the global bias error between observed and simulated groundwater levels. Then we calibrated the drain conductance by maximizing the coefficient of determination R^2 between observed and simulated combined baseflows (i.e., stormwater + wastewater sewers baseflows) at the outlet of both sewer systems. The final calibration presented in Table 1 is achieved through a trial and error method. The obtained hydraulic conductivities are two orders of magnitude higher than the observed hydraulic conductivities using the piezometers. This result is consistent with previous statements following the application of other models to ditch-groundwater interactions [53], and confirms the difficulty to simulate groundwater flows and groundwater levels using a few specific point observations of the hydraulic conductivities. High soil heterogeneity, particularly in urban areas, cannot be well characterized using soil properties measurements sampled in a few locations, which leads to a poor simulation of the general groundwater behavior. Despite the differences between observed and calibrated conductivities, the observed differences between the different types of soil are consistent with the theory. Hence, the highest hydraulic conductivity is obtained for alluvial deposits, and the smallest one for mica-schist. Finally, the calibrated drain conductance is higher than initially estimated, which allows a better simulation of high values and the dynamics of the baseflow.

In order to assess the influence of the sewer system on the evolution of the groundwater levels, a complementary simulation of the catchment was done on the same period and with the same set of parameters by removing the sewer system and the corresponding field drain. This simulation exercise shows that the groundwater reaches the soil surface during the winter 2007–2008 on a significant part of the basin; that result clearly confirmed the importance of the drainage ensured by the sewer system [39].

3.2.2. Groundwater Level Distribution Assessment

The temporal and spatial simulation of groundwater levels is assessed using available piezometric records. Figure 9 illustrates the simulation of the isopiezometric contours for the high water table period (7th March, 2007), and can be compared to the observed isopiezometric contours at the same date shown in Figure 3. At this period, the absolute errors between simulated and observed values may be larger than 2 m in some piezometers, especially on the south catchment boundary. They are however better in the middle of the catchment and near the outlet. In addition, absolute errors are higher around the piezometers distant from the sewer trenches. Note that the simulated contours in Figure 9 are plotted on a surface area larger than the experimental one, as the simulated catchment exceeds the piezometric set boundary. Consequently, the simulated flow directions differ significantly from observed ones near the urban catchment boundary, but are rather similar within the urban catchment. Nonetheless, at this time the simulated saturation level is often higher than the observed one. The simulated vs. observed groundwater level dynamics is assessed by computing the RMSE, bias error and R^2 for all the piezometer locations, both for the calibration and validation periods (Table 3). R^2 values are often smaller than 0.5 for the calibration stage, but increase for the validation stage. In general, the model underestimates the groundwater level for half of the piezometers (PZCPS, PZCS, PZGO, PZUV, PZJV) and overestimates it for the others. The boxplot analysis of the bias error (Figure 10) shows that the underestimation is mainly concentrated on the upstream part of the catchment (PZCRI, PZGO, PZCS, PZJV, PZUV), and might be explained by either the upstream boundary condition or the uniform recharge assumption. Indeed, the spatial analysis of the imperviousness coefficient of the Pin Sec catchment reveals that underestimated piezometric observations are located in more pervious areas within the catchment (i.e., green public parks). Hence, using a spatially distributed recharge could allow a more accurate modeling of the groundwater levels dynamics in these areas. Overall, the simulated groundwater levels compared properly with observed ones both for the calibration and validation periods (Figure 11), despite an underestimation of the model compared to the observed data at the beginning of the validation period, when a particularly dry autumn took place. In the absence of rain, the groundwater model simulates a groundwater level decrease higher than observed, which to some extent questions both the upstream boundary condition assumption and the recharge estimation. Moreover, the quite stable groundwater level observed during this period could be explained by a groundwater influx coming from upstream, which would disprove the assumed upstream boundary condition. Nonetheless, no observations outside the catchment are available to clarify this issue. Finally, drinking water leakage in the soil can occur, which would increase the recharge throughout the year and prevent the high decrease of groundwater levels observed in the simulation; this phenomenon has been discussed in several cities like Bucharest, Romania [4,49], St Louis, Missouri(US) [54] or Santiago, Chile [5].

Table 3. Metrics of simulated groundwater levels simulated in each piezometer for the calibration and validation stages. Data availability indicates the percentage of valid data available during the simulation period. Columns *in italics* denote piezometers with poor data availability.

Criterion	PZCRI	PZGO	PZCS	PZAF	PZUV	PZJV	PZD	PZCPS	PZPS	PZG
	Calibration (2006–2007)									
Data availability (%)	100	96	87	96	100	100	88	94	100	96
R^2	0.66	0.32	0.48	0.50	0.44	0.48	0.24	0.36	0.45	0.44
Bias error (%)	6.87	−0.38	−3.47	3.36	−9.16	−3.12	7.02	−3.15	0.90	7.93
RMSE (m)	1.80	0.56	0.87	0.87	2.10	0.64	1.39	0.70	0.23	1.11
	Validation (2007–2008)									
Data availability (%)	96	*56*	98	*63*	95	99	98	*38*	100	*51*
R^2	0.81	*0.45*	0.64	*0.60*	0.26	0.05	0.63	*0.76*	0.78	*0.70*
Bias Error (%)	−0.23	*−3.68*	−6.19	*1.72*	−2.07	−9.79	5.52	*−3.07*	3.72	*5.45*
RMSE (m)	0.61	*0.40*	1.42	*0.31*	4.91	1.78	1.03	*0.28*	0.58	*0.55*

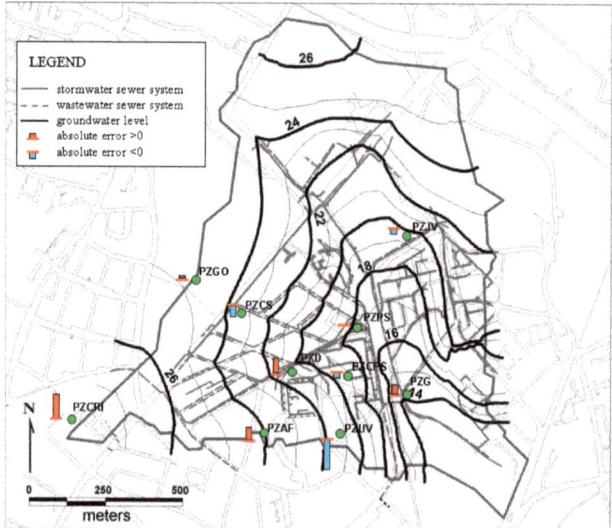

Figure 9. Groundwater level contours (hydraulic heads in m) simulated in transient state during the low groundwater table period (7th March, 2007) in the Pin Sec catchment. Absolute errors are represented by red and blue bars at each piezometer. Bar sizes show a variation from −2.66 m (PZUV) to 2.11 m (PZCRI).

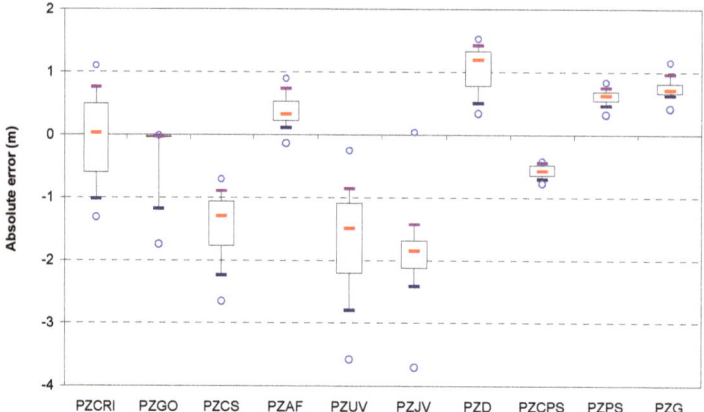

Figure 10. Box-plot characterizing the absolute error distribution between simulated and observed groundwater levels for the validation period (2007–2008). The piezometers are presented from upstream (left) to downstream (right). Small circles are minimum and maximum values red mark are the median, whereas the ends of the whiskers represent the 25th and 75th percentiles.

Figure 9 shows a small curvature of the simulated groundwater level contours near the sewer trenches caused by the local groundwater drawdown. Groundwater decrease is quite significant in places where sewer density is high (i.e., central and north portions of the catchment). This phenomenon is enhanced when the water table rises above the sewer system, and becomes less significant in summer. For simulated data samples, groundwater drawdown is higher during the wet year 2006–2007 with a decrease of about 0.3 m near the sewer trench, and a zone of influence ranging up to 120 m away from the trench axis.

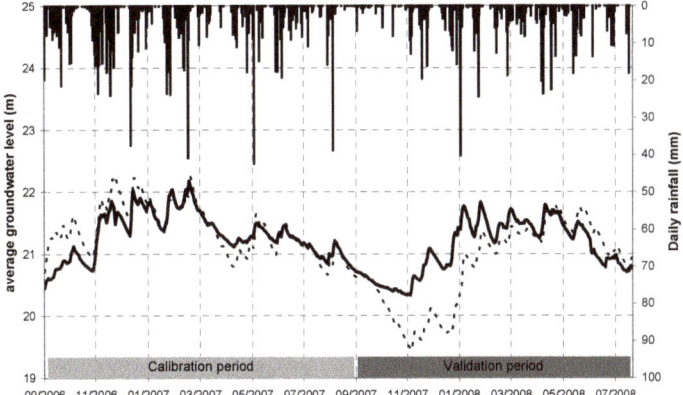

Figure 11. Comparison between simulated (dotted line) and observed (bold line) average hydraulic heads in the catchment from September 2006 to August 2008.

3.2.3. Baseflow Rate Assessment

The simulated soil water flux combines two components simulated in MODFLOW: groundwater flux within the soil and groundwater drainage inside the sewer trench. Both components flow in the same general direction forced by the catchment topography and the downstream condition imposed by the Gohards stream. Groundwater drainage may be compared with the baseflow discharge observed in both waste- and storm- sewer systems during the 2006–2007 simulation (note that the period 2007–2008 is not presented due to the lack of data). Figure 12 shows that the daily groundwater drainage component varies significantly, ranging from 50 m^3/day during very dry periods (i.e., the end of summer 2007) to 2600 m^3/day during the particularly humid 2006–2007 winter. Indeed, ~630 mm of rain (i.e., ~80% of the catchment mean annual rainfall) fell between November 2006 and April 2007 in the area. Unfortunately, a direct comparison between the simulated groundwater drainage flow and observed baseflow is difficult because of the lack of flow rate data. Considering the data actually available, the comparison focuses on the dynamic evolution of the baseflow. Simulated groundwater drainage and in-situ observations over the simulated period vary in a quite similar manner. Nonetheless the overestimation of the observed flows could not be reduced by further calibration, as drain conductance variations do not affect groundwater drainage significantly. As seen in the sensitivity analysis, soil hydraulic conductivities have more impact on the simulation than drain conductance. Because soil hydraulic conductivities affect not only groundwater drainage but also groundwater level, a simultaneous enhancement of both is not possible.

Special attention must be paid to the relationship between baseflow rates and groundwater levels. As discussed above and shown in Figure 7, both are strongly linked through a power-law function. Figure 13 overlaps simulated and observed values when considering the combined baseflow, and shows that the model can satisfactorily reproduce this groundwater level-baseflow relationship despite of the baseflow overestimation during high water table periods. Although only 32% of the experimental results during the 2006–2007 period are available simultaneously for both sewers, and apart from the low groundwater levels occurring in summer and discussed above, the validity of the relationship holds quite well.

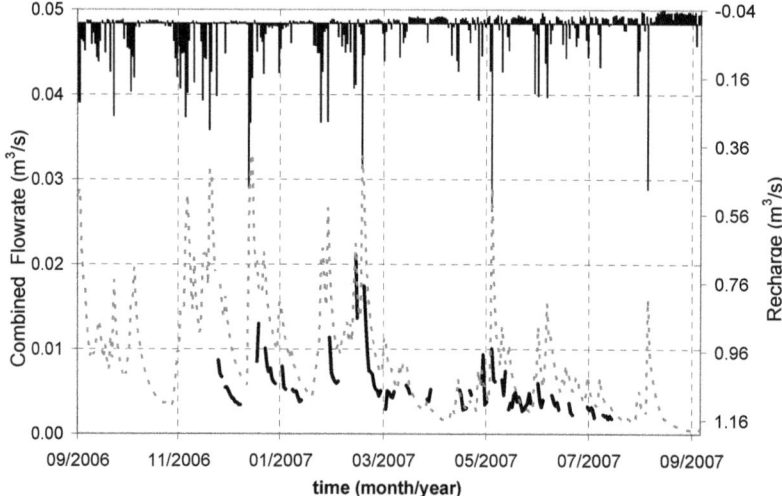

Figure 12. Comparison between simulated (dotted line) and observed (bold line) daily combined baseflow at the catchment outlet from September 2006 to September 2007. The recharge evolution shows the infiltration and evaporation periods during this simulation.

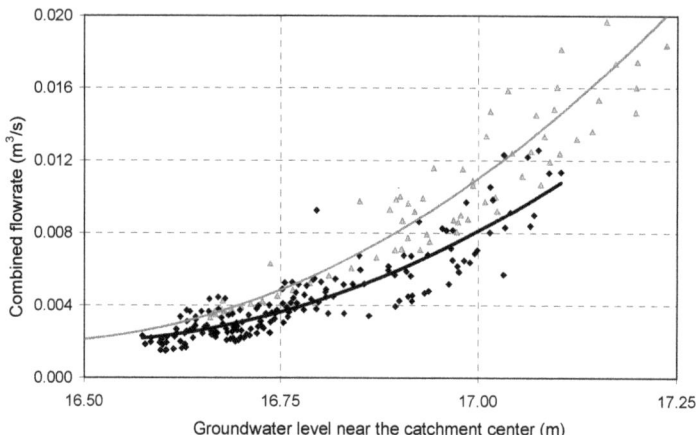

Figure 13. Combined daily baseflow in sewers vs. piezometric levels between September 2006 and September 2007. Both observed values (black points) and simulated values (grey points) are shown in the (0–0.02 m^3/s) range of flowrates, along with their respective polynomial regression relationships: y = 0.0225x^2 − 0.7404x + 6.1016 (R^2 = 0.803) for the observed values and y = 0.0276x^2 − 0.9060x + 7.4436 (R^2 = 0.908) for the simulated values.

4. Conclusions

This study presents an experimental analysis and modeling of the interactions between groundwater and urban sewer systems in a 31 ha urban catchment located near Nantes, France. Groundwater contributions to flow in stormwater and wastewater systems due to sewers' defects were deduced from observed flow rates, whereas piezometric records were used to characterize the dynamics of groundwater levels in the catchment.

Overall, the experimental analysis reveals a strong co-fluctuation of groundwater levels and sewer baseflow, which is more pronounced in downstream piezometers. This study highlights that two or three piezometers located near the catchment outlet are likely adequate to estimate the groundwater-baseflow relationship in an urban catchment. The total baseflow volume in wastewater sewers is larger than in stormwater sewers because of its larger coverage. On an annual basis, the total volume of soil water drained by both sewers is 42% of the total rainfall.

A MODFLOW model was implemented in the study catchment to simulate the groundwater dynamics and contribution to the sewer systems on a daily basis. This application proves the model ability to represent satisfactorily the spatial and temporal evolutions of groundwater levels, although low groundwater levels were underestimated. Spatial differences between modeled and observed groundwater levels in some locations could be explained by the spatial variability of the land uses and the urban soil configuration around each piezometer. The simulation of combined waste and storm water sewer baseflows is not entirely relevant because the model fails to reproduce the high values of baseflow. The calibration of the model lead to hydraulic conductivities values higher than those observed in the piezometers. In fact, the calibrated values are similar to those reported in the literature, and they are attributed to possible secondary porosities and hydraulic conductivity distributions comparable to those of a karstic system [55]. Hydraulic conductivity varies significantly in urban soils due to differences in their characteristics, the existence of different land-uses, and the presence of previous buried constructions.

More work is needed to improve our knowledge of urban soil water flow paths. Baseflow must be better measured, as flow meter sensors are not always reliable to be used with low flows. In addition, more observations at a local scale are needed to better understand groundwater-sewer interactions and improve large scale modeling efforts. Such data will come from either (1) physical models relying on better estimations of soil parameters, which can represent the influence area of a sewer trench, or (2) sensors located near the sewers. Groundwater models applied to urban catchments will require using a graded grid size with mesh refinements near sewers and more urbanized areas. Furthermore, better spatially distributed recharge estimations can also improve the performance of these models. This could be done by coupling the groundwater and surface water models, adopting better modeling frameworks like the modeling chain proposed by [56], or implementing other specific coupling modeling approaches. Finally, a better understanding of the occurrence and spatial distribution of leaks from drinking water pipes is also needed to improve the characterization of the recharge.

Author Contributions: This study was designed by F.R., A.-L.L.D., and H.A. The manuscript was prepared by F.R. and revised by H.A. and J.G. All authors have read and agreed to the published version of the manuscript.

Funding: This study received financial support from the French research programs ANR and INSU-EC2CO within the project Hy^{2}ville, and from the Région Pays de la Loire. Funding from CONICYT/FONDAP grants 15110017 and 15110020 were also available.

Acknowledgments: We thank Nantes Métropole (Direction de la Géographie et de l'Observation) for providing the GIS data.

Conflicts of Interest: The authors declare no conflict of interest.

References

1. Lee, J.Y.; Choi, M.J.; Kim, Y.Y.; Lee, K.K. Evaluation of hydrologic data obtained from a local groundwater monitoring network in a metropolitan city, Korea. *Hydrol. Process.* **2005**, *19*, 2525–2537. [CrossRef]
2. Barrett, M.H.; Hiscock, K.M.; Pedley, S.; Lerner, D.N.; Tellam, J.H.; French, M.J. Marker species for identifying urban groundwater recharge sources: A review and case study in Nottingham, UK. *Water Res.* **1999**, *33*, 3083–3097. [CrossRef]
3. Lerner, D.N. Identifying and quantifying urban recharge: A review. *Hydrogeol. J.* **2002**, *10*, 134–152. [CrossRef]
4. Gogu, C.R.; Gaitanaru, D.; Boukhemacha, M.A.; Serpescu, I.; Litescu, L.; Zaharia, V.; Moldovan, A.; Mihailovici, M.J. Urban hydrogeology studies in Bucharest City, Romania. *Procedia Eng.* **2017**, *209*, 135–142. [CrossRef]

5. Sanzana, P.; Gironás, J.; Braud, I.; Muñoz, J.F.; Vicuña, S.; Reyes-Paecke, S.; de la Barrera, F.; Branger, F.; Rodriguez, F.; Vargas, X.; et al. Impact of Urban Growth and High Residential Irrigation on Streamflow and Groundwater Levels in a Peri-Urban Semiarid Catchment. *JAWRA J. Am. Water Resour. Assoc.* **2019**, *55*, 720–739. [CrossRef]
6. Reynolds, J.H.; Barrett, M.H. A review of the effects of sewer leakage on groundwater quality. *Water Environ. J.* **2003**, *17*, 34–39. [CrossRef]
7. Evans, M.G.; Burt, T.P.; Holden, J.; Adamson, J.K. Runoff generation and water table fluctuations in blanket peat: Evidence from UK data spanning the dry summer of 1995. *J. Hydrol.* **1999**, *221*, 141–160. [CrossRef]
8. Scibek, J.; Allen, D.M.; Cannon, A.J.; Whitfield, P.H. Groundwater–surface water interaction under scenarios of climate change using a high-resolution transient groundwater model. *J. Hydrol.* **2007**, *333*, 165–181. [CrossRef]
9. Berthier, E.; Andrieu, H.; Creutin, J.D. The role of soil in the generation of urban runoff: Development and evaluation of a 2D model. *J. Hydrol.* **2004**, *299*, 252–266. [CrossRef]
10. Wittenberg, H. Effects of season and man-made changes on baseflow and flow recession: Case studies. *Hydrol. Process.* **2003**, *17*, 2113–2123. [CrossRef]
11. Thorndahl, S.; Balling, J.D.; Larsen, U.B.B. Analysis and integrated modelling of groundwater infiltration to sewer networks. *Hydrol. Process.* **2016**, *30*, 3228–3238. [CrossRef]
12. Semadeni-Davies, A.; Hernebring, C.; Svensson, G.; Gustafsson, L.-G. The impacts of climate change and urbanisation on drainage in Helsingborg, Sweden: Combined sewer system. *J. Hydrol.* **2008**, *350*, 100–113. [CrossRef]
13. Dirckx, G.; Van Daele, S.; Hellinck, N. Groundwater Infiltration Potential (GWIP) as an aid to determining the cause of dilution of waste water. *J. Hydrol.* **2016**, *542*, 474–486. [CrossRef]
14. Belhadj, N.; Joannis, C.; Raimbault, G. Modelling of rainfall induced infiltration into separate sewerage. *Water Sci. Technol.* **1995**, *32*, 161–168.
15. Dupasquier, B. Modélisation Hydrologique et Hydraulique des Infiltrations D'eaux Parasites dans les Réseaux Séparatifs D'eaux Usées [Hydrological and Hydraulic Modelling of Infiltration into Separate Sanitary Sewers]. Ph.D. Thesis, ENGREF, Paris, France, 26 February 1999.
16. Houhou, J.; Lartiges, B.S.; France-Lanord, C.; Guilmette, C.; Poix, S.; Mustin, C. Isotopic tracing of clear water sources in an urban sewer: A combined water and dissolved sulfate stable isotope approach. *Water Res.* **2010**, *44*, 256–266. [CrossRef] [PubMed]
17. De Bondt, K.; Seveno, F.; Petrucci, G.; Rodriguez, F.; Joannis, C.; Claeys, P. Potential and limits of stable isotopes ($\delta 18O$ and δD) to detect parasitic water in sewers of oceanic climate cities. *J. Hydrol Reg. Stud.* **2018**, *18*, 119–142. [CrossRef]
18. Beheshti, M.; Saegrov, S. Quantification assessment of extraneous water infiltration and inflow by analysis of the thermal behavior of the sewer network. *Water* **2018**, *10*, 1070. [CrossRef]
19. Kaushal, S.S.; Belt, K.T. The urban watershed continuum: Evolving spatial and temporal dimensions. *Urban Ecosyst.* **2012**, *15*, 409–435. [CrossRef]
20. Gessner, M.O.; Hinkelmann, R.; Nützmann, G.; Jekel, M.; Singer, G.; Lewandowski, J.; Nehls, T.; Barjenbruch, M. Urban water interfaces. *J. Hydrol.* **2014**, *514*, 226–232. [CrossRef]
21. Schlea, D.; Martin, J.F.; Ward, A.D.; Brown, L.C.; Suter, S.A. Performance and water table responses of retrofit rain gardens. *J. Hydrol. Eng.* **2013**, *19*. [CrossRef]
22. Furumai, H. Rainwater and reclaimed wastewater for sustainable urban water use. *Phys. Chem. Earth* **2008**, *33*, 340–346. [CrossRef]
23. Roldin, M.; Fryd, O.; Jeppesen, J.; Mark, O.; Binning, P.J.; Mikkelsen, P.S.; Jensen, M.B. Modelling the impact of soakaway retrofits on combined sewage overflows in a 3km^2 urban catchment in Copenhagen, Denmark. *J. Hydrol.* **2012**, *452*, 64–75. [CrossRef]
24. Mitchell, V.G.; Mein, R.G.; McMahon, T.A. Modelling the urban water cycle. *Environ. Modell. Softw.* **2001**, *16*, 615–629. [CrossRef]
25. Rossman, L.A. Storm Water Management Model User's Manual Version 5.0. U.S. Environmental Protection Agency, EPA/600/R-05/040. 2009. Available online: https://csdms.colorado.edu/w/images/Epaswmm5_user_manual.pdf (accessed on 24 February 2020).
26. Elliott, A.H.; Trowsdale, S.A. A review of models for low impact urban stormwater drainage. *Environ. Modell. Softw.* **2007**, *22*, 394–405. [CrossRef]

27. Gustafsson, L.-G. Alternative Drainage Schemes for Reduction of Inflow/Infiltration—Prediction and Follow-Up of Effects with the Aid of an Integrated Sewer/Aquifer Model. 1st International Conference on Urban Drainage via Internet. 2000. Available online: http://www.dhigroup.com/upload/publications/mouse/Gustafsson_Alternative_Drainage.pdf (accessed on 24 February 2020).
28. Wolf, L.; Klinger, J.; Hoetzl, H.; Mohrlok, U. Quantifying mass fluxes from urban drainage systems to the urban soil-aquifer system. *J. Soils Sediments* **2007**, *7*, 85–95. [CrossRef]
29. Harbaugh, A.W. MODFLOW–2005, the U.S. Geological Survey Modular Ground-Water Model-the Ground-Water Flow Process, Techniques and Methods 6–A16, U.S. Geol. Surv., Reston, Va. 2005. Available online: https://pubs.er.usgs.gov/publication/tm6A16 (accessed on 24 February 2020).
30. Trefry, M.G.; Muffels, C. FEFLOW: A Finite-Element Ground Water Flow and Transport Modeling Tool. *Ground Water* **2007**, *45*, 525–528. [CrossRef]
31. Mitchell, V.; Diaper, C.; Gray, S.R.; Rahilly, M. UVQ: Modelling the movement of water and contaminants through the total urban water cycle. In Proceedings of the 28th International Hydrology and Water Resources Symposium: About Water, Wollongong, Australia, 10–13 November 2003. Symposium Proceedings 3.131–3.138.
32. Burn, S.; DeSilva, D.; Ambrose, M.; Meddings, S.; Diaper, C.; Correll, R.; Miller, R.; Wolf, L. A decision support system for urban groundwater resource sustainability. *Water Pract. Technol.* **2006**, *1*, wpt2006010. [CrossRef]
33. Karpf, C.; Krebs, P. Modelling of groundwater infiltration into sewer systems. *Urban Water J.* **2013**, *10*, 221–229. [CrossRef]
34. Schirmer, M.; Leschik, S.; Musolff, A. Current research in urban hydrogeology–A review. *Adv. Water Resour.* **2013**, *51*, 280–291. [CrossRef]
35. Liu, T.; Su, X.; Prigiobbe, V. Groundwater-sewer interaction in urban coastal areas. *Water* **2018**, *10*, 1774. [CrossRef]
36. Rodriguez, F.; Andrieu, H.; Morena, F. A distributed hydrological model for urbanized areas—Model development and application to case studies. *J. Hydrol.* **2008**, *351*, 268–287. [CrossRef]
37. Ruban, V.; Rodriguez, F.; Lamprea-Maldonado, K.; Mosini, M.L.; Lebouc, L.; Pichon, P.; Letellier, L.; Rouaud, J.M.; Martinet, L.; Dormal, G.; et al. Le secteur atelier pluridisciplinaire (SAP), un observatoire de l'environnement urbain: Présentation des premiers résultats du suivi hydrologique et microclimatique du bassin du Pin Sec (Nantes). [Interdisciplinary Experimental Observatory dedicated to the urban environment: Presentation of initial hydrological and microclimatic monitoring results from the Pin Sec catchment (Nantes)]. *Bulletin de liaison des Lab. des Ponts et Chaussées* **2010**, *277*, 5–18.
38. BRGM. Notice de la Carte Géologique de Nantes. [Geological Map of Nantes—Instruction] Carte Géologique au 1/50000, Nantes (481). 1970. Available online: http://infoterre.brgm.fr/page/cartes-geologiques (accessed on 24 February 2020).
39. Le Delliou, A.-L. Rôle des Interactions Entre les Réseaux D'assainissement et les Eaux Souterraines dans le Fonctionnement Hydrologique D'un Bassin Versant en Milieu Urbanisé—Approche Expérimentale et Modélisations [Sewer Network and Underground Water Interactions Role on the Hydrological Behaviour of an Urban Catchment—Experimental and Modelling Approaches]. Ph.D. Thesis, Ecole Centrale de Nantes (SPIGA), Nantes, France, 7 December 2009. (In French).
40. Hiscock, K. *Hydrogeology: Principles and Practice*; Blackwell Science Ltd.: London UK, 2005.
41. Hvorslev, M.J. Time Lag and Soil Permeability in Ground-Water Observations. Bull. No. 36 1951, Waterways Exper. Sta. Corps of Engrs, U.S. Army, Vicksburg, Mississippi, 1–50. Available online: https://books.google.com (accessed on 24 February 2020).
42. Le Delliou, A.L.; Rodriguez, F.; Andrieu, H. Hydrological modelling of sewer network impacts on urban groundwater. *La Houille Blanche* **2009**, *5*, 152–158. [CrossRef]
43. De Bénédittis, J.; Bertrand-Krajewski, J.-L. Infiltration in sewer systems: Comparison of measurement methods. *Water Sci. Technol.* **2005**, *52*, 219–227. [CrossRef]
44. Wittenberg, H.; Aksoy, H. Groundwater intrusion into leaky sewer systems. *Water Sci. Technol.* **2010**, *62*, 92–98. [CrossRef] [PubMed]
45. Zhang, M.; Liu, Y.; Cheng, X.; Zhu, D.Z.; Shi, H.; Yuan, Z. Quantifying rainfall-derived inflow and infiltration in sanitary sewer systems based on conductivity monitoring. *J. Hydrol.* **2018**, *558*, 174–183. [CrossRef]

46. Carrera-Hernandez, J.J.; Gaskin, S.J. The groundwater modeling tool for GRASS (GTMG): Open source groundwater flow modeling. *Comput. Geosci.* **2006**, *32*, 339–351. [CrossRef]
47. Yang, Y.; Lerner, D.N.; Barrett, M.H.; Tellam, J.H. Quantification of groundwater recharge in the city of Nottingham, UK. *Environ. Geol.* **1999**, *38*, 183–198. [CrossRef]
48. Fatta, D.; Naoum, D.; Loizidou, M. Integrated environmental monitoring and simulation system for use as a management decision support tool in urban areas. *J. Env. Manag.* **2002**, *64*, 333–343. [CrossRef]
49. Boukhemacha, M.A.; Gogu, C.R.; Serpescu, I.; Gaitanaru, D.; Bica, I. A hydrogeological conceptual approach to study urban groundwater flow in Bucharest city, Romania. *Hydrogeol. J.* **2015**, *23*, 437–450. [CrossRef]
50. Johnson, A.I. *Specific Yield—Compilation of Specific Yields for Various Materials*; US Geological Survey Water-Supply Paper 1662-D; United States Government Printing Office: Washington, DC, USA, 1967; p. 74.
51. Chiang, W.-H. *3D-Groundwater Modeling with PMWIN: A Simulation System for Modeling Groundwater Flow and Transport Processes*; Springer: Berlin/Heidelberg, Germany, 2005.
52. Cassan, M. *Aide-Mémoire D'hydraulique Souterraine [Groundwater Hydraulic Aide-Memoire]*; Presses de l'Ecole Nationale des Ponts et Chaussées: Paris, France, 1993.
53. Adamiade, V. Influence d'un Fossé sur les Ecoulements Rapides au Sein d'un Versant [The Hydrological Influence of the Ditches on Hillslope Flow: Application to the Pesticides Transfer]. Ph.D. Thesis, Université Pierre et Marie Curie, Géosciences et Ressources Naturelles, Paris, France, 15 March 2004. (In French)
54. Lockmiller, K.A.; Wang, K.; Fike, D.A.; Shaughnessy, A.R.; Hasenmueller, E.A. Using multiple tracers (F−, B, δ11B, and optical brighteners) to distinguish between municipal drinking water and wastewater inputs to urban streams. *Sci Total Environ* **2019**, *671*, 1245–1256. [CrossRef]
55. Garcia-Fresca, B.; Sharp, J.-M. Hydrogeologic considerations of urban development: Urban-induced recharge. *Rev. Eng. Geol.* **2005**, *16*, 123–136.
56. Vizintin, G.; Souvent, P.; Veselic, M.; Cencur Curk, B. Determination of urban groundwater pollution in alluvial aquifer using linked process models considering urban water cycle. *J. Hydrol.* **2009**, *377*, 261–273. [CrossRef]

© 2020 by the authors. Licensee MDPI, Basel, Switzerland. This article is an open access article distributed under the terms and conditions of the Creative Commons Attribution (CC BY) license (http://creativecommons.org/licenses/by/4.0/).

Article

Nitrogen Mass Balance and Pressure Impact Model Applied to an Urban Aquifer

Mitja Janža [1,*], Joerg Prestor [1], Simona Pestotnik [1] and Brigita Jamnik [2]

1. Geological Survey of Slovenia, Dimičeva ulica 14, SI–1000 Ljubljana, Slovenia; joerg.prestor@geo-zs.si (J.P.); simona.pestotnik@geo-zs.si (S.P.)
2. Javno Podjetje Vodovod Kanalizacija Snaga d.o.o; Vodovodna cesta 90, SI-1000 Ljubljana, Slovenia; brigita.jamnik@vokasnaga.si
* Correspondence: mitja.janza@geo-zs.si

Received: 13 February 2020; Accepted: 16 April 2020; Published: 19 April 2020

Abstract: The assurance of drinking water supply is one of the biggest emerging global challenges, especially in urban areas. In this respect, groundwater and its management in the urban environment are gaining importance. This paper presents the modeling of nitrogen load from the leaky sewer system and from agriculture and the impact of this pressure on the groundwater quality (nitrate concentration) in the urban aquifer located beneath the City of Ljubljana. The estimated total nitrogen load in the model area of 58 km^2 is 334 ton/year, 38% arising from the leaky sewer system and 62% from agriculture. This load was used as input into the groundwater solute transport model to simulate the distribution of nitrate concentration in the aquifer. The modeled nitrate concentrations at the observation locations were found to be on average slightly lower (2.7 mg/L) than observed, and in general reflected the observed contamination pattern. The ability of the presented model to relate and quantify the impact of pressures from different contamination sources on groundwater quality can be beneficially used for the planning and optimization of groundwater management measures for the improvement of groundwater quality.

Keywords: urban hydrogeology; groundwater quality; sewer system; agriculture; groundwater modeling

1. Introduction

In 2018, 55% of the world's population lived in urban areas, and this proportion is expected to increase to 68% by 2050 [1]. Europe, with 74% of its population living in urban areas in 2018, is expected to reach an urban population of 80% in 2040 and approaching 85% by 2050 [1]. Considering this population growth in light of climate change projections, the estimates show that with current practices, the world will face a 40% shortfall between the forecasted demand and available supply of water by 2030. Water security is thus emerging as one of the biggest global risks in terms of developmental impact [2].

Groundwater will have an important role, as it represents over 95% of the world's available freshwater reserves [3]. Compared to surface water, it is better protected and a more stable water resource. Better quality, proximity of the resource, and the relatively low cost of pumping wells make groundwater an increasingly important source for city populations in the future.

Aside from socio-economic benefits, urbanization has also created environmental problems, manifested in the deterioration of air and water quality as well as a significant change in water balance [4–7]. Urban subsurface infrastructures strongly influence the chemical as well as the quantitative status of groundwater. The subsurface infrastructure of Bucharest (sewer system and water supply network) contributes more than 71% of the groundwater recharge. In addition, more than 77% of the groundwater discharge is associated with non-natural sinks (mainly the sewer system and the drain of the main sewer collector) [4]. In the Baltimore metropolitan area, the infiltration of groundwater into wastewater pipelines was found to be the most important factor of urban development affecting the groundwater storage, which can cause its decrease by 11.1% [8].

The assurance of good quality groundwater in the urban environment is a complex task, due to numerous human activities and the number of potential sources of contamination [9,10]. Additionally, the limited ability for observation and the delay between the contamination and its detection make it complicated to define the causal relationship between human activities that exert pressures and their impacts on groundwater quality.

Wastewater exfiltration from the sewer system is an important factor influencing groundwater quality, with its impact being strongly variable both in space and time [11–14]. A review of exfiltration modeling approaches and measurement techniques in terms of both volume and pollutant load showed a wide range of exfiltration rates estimated on the catchment scale, ranging from 1% to 56% of dry-weather flow (DWF) [15].

Nitrate is one of the key groundwater contaminants, and measurements of its concentration are regularly included in groundwater quality monitoring. In Slovenia, about 37% of the groundwater in alluvial aquifers has a poor chemical status, mainly due to high concentrations of nitrate (above 50 mg/L) [16]. Background concentrations of nitrate largely depend on the geological properties of aquifers; in Slovenia, they are below 10 mg/L [17]. Increased concentrations are related to human activities, in particular to agriculture. The main source of diffuse nitrogen input on agricultural land in Slovenia are mineral fertilizers and livestock manure, which account for 84% of the total input of nitrogen [18]. Among the important non-agricultural sources of nitrate in groundwater, wastewater has been identified as one of the major sources of nitrogen in urban aquifers [19].

The impact of nitrogen sources on groundwater have been investigated in a number of studies. Different sources of nitrogen and their contribution to the nitrate content in the groundwater have been analyzed [20–24]. The research on the role of land use [25], vulnerability of groundwater [26,27], processes of leaching [28], and the migration and transformation of nitrogen in groundwater [29–31] have furthered the knowledge on factors controlling groundwater nitrate contamination and its mitigation. The spatial and temporal variation of nitrate in groundwater has been commonly simulated with numerical groundwater flow and transport models [24,29,32,33]. A fundamental step in the development of such models is the estimation of nitrogen loading. The broad range of nitrogen sources in urban areas makes reliable estimation of nitrogen load a difficult task in these areas [19].

The drinking water supply of the City of Ljubljana relies on groundwater, primarily on its abstraction from the Ljubljansko polje aquifer, situated beneath the city. The well fields Kleče (55%), Hrastje (12%), and Šentvid (9%) (Figure 1) assure around three-quarters of all the water needs in Ljubljana's drinking water supply system [34]. To determine the urban contamination in the Ljubljansko polje alluvial aquifer, an urban water balance modeling approach was applied [35]. The study focused on an area (approximately 0.76 km^2) located within the City of Ljubljana, and indicated that residential land use in an urban area with a high unsaturated thickness may have a significantly smaller impact on groundwater quality than agriculture or industry.

The first catchment scale pressure impact model for the Ljubljansko polje aquifer was developed in the frame of the INCOME project [36]. The aim of the model was to simulate the concentrations in groundwater for a number of pollutants. The model was able to explain practically the entire mass of nitrogen in the sewer system by the known sources. On the basis of these results and an encouraging decrease of observed nitrate concentration in groundwater, more optimistic groundwater management objectives for the year 2017 were set. The targeted average nitrate concentration in the aquifer was set below 15 mg/L and in all observation wells below 25 mg/L [37]. Nitrogen input from agriculture was identified as the most important source of uncertainty in the model. The explanation of high nitrate concentration at the local level, that is, individual observation wells, was pointed out as an important groundwater management activity.

In the present study, a refined and updated groundwater flow and transport model [38] was applied in the catchment scale approach. The aim of the modeling was to analyze the causal relationship between pressure (nitrogen load) arising from the leaky sewer system and from agriculture, and its impact on the quality of groundwater (nitrate concentration) in the Ljubljansko polje aquifer [39]. Nitrogen load from the leaky sewer system was estimated with the empirical model of spatially distributed sewer exfiltration [40] and together with estimated nitrogen surpluses from agriculture [41] used as input into the groundwater transport model. Modeled and observed nitrate concentrations in groundwater were analyzed to assess the uncertainty of the model and to identify priority areas for remediation.

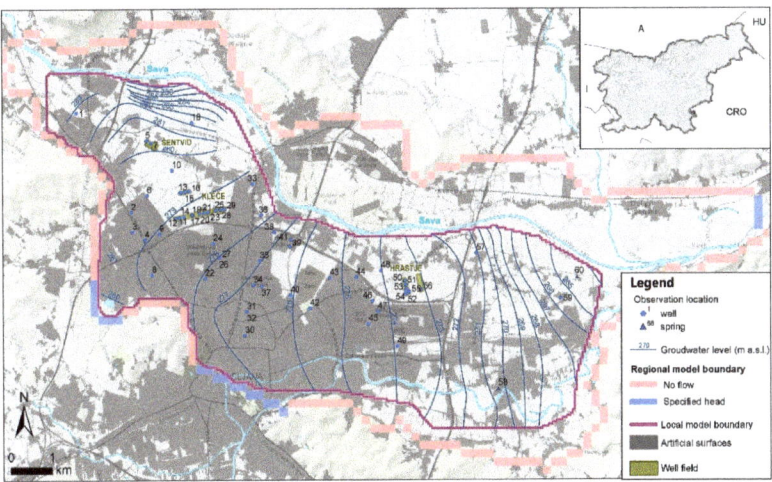

Figure 1. Map of the study area showing observation locations, groundwater level isolines, model boundaries, artificial surfaces, and well fields.

2. Materials and Methods

2.1. Study Area

The study area (57.91 km^2) is located in the flat part of the catchment of the alluvial Ljubljansko polje aquifer forming the subsurface of the City of Ljubljana, the capital of Slovenia. It corresponds to the area of the local groundwater model (Figure 1). The mean altitude of the area is 292 m above sea level (m a.s.l.) and prevailing land cover types are artificial surfaces (59%) and agricultural areas (39%) [42].

The Ljubljansko polje alluvial aquifer is developed in a tectonic basin, filled with highly permeable Quaternary gravel and sand beds, which are partly conglomerated (cemented with calcite cement) [43]. The thickness of the sedimentary fill is up to 100 m down to its very low permeable pre-Quaternary basement, composed of Carboniferous–Permian siliciclastic rocks. The groundwater table is on average 25 m below the surface. The unconfined intergranular aquifer with high transmissivity (Figure 2) is mainly recharged from the Sava River in the northwestern part of the study area. It drains into the river in the northwestern part, which largely predisposes the groundwater flow direction (Figure 1). The main water balance components of the regional model area (Figure 1) for the 3 year modeling period (2013–2015) are presented in Table 1.

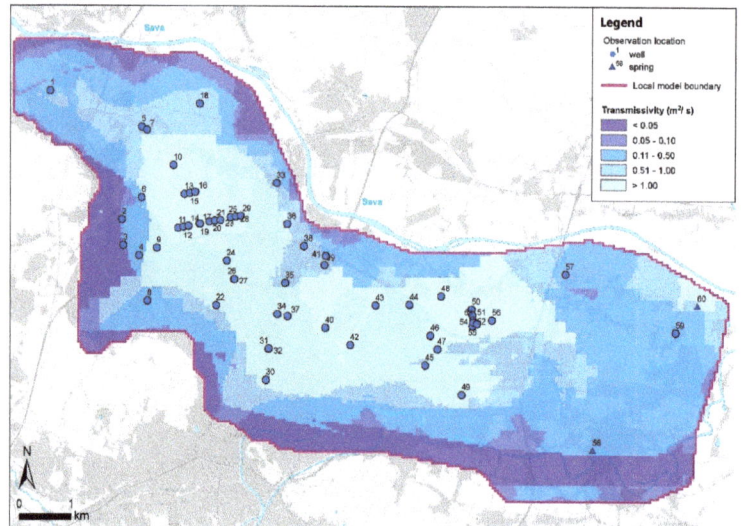

Figure 2. The calibrated transmissivity of the aquifer.

Table 1. Water balance components (modeling period 2013–2015).

Water Balance Component	(m³/s)
Recharge from Sava River	+3.80
Recharge from precipitation	+1.51
Subsurface inflow	+0.27
Exfiltration from water supply system	+0.16
Abstraction from wells	−1.13
Discharge to Sava River	−4.63

2.2. Nitrogen Mass Balance and Pressure Impact Model

The applied methodology [36,37,40] comprises the following main steps: (1) estimation of the spatial distribution of nitrogen load from the leaky sewer system and agriculture (nitrogen mass balance model), (2) use of the nitrogen load as input into the groundwater transport model, and (3) analysis of the modeling results with observed nitrate concentrations in the observation wells.

2.2.1. Nitrogen Load from the Leaky Sewer System

Assessment of nitrogen load from the leaky sewer system was based on the model of spatially distributed sewer exfiltration developed by Prestor et al. (2011) [36] for the sewer system of the City of Ljubljana. The main steps of workflow for the estimation of nitrogen load from the sewer system are presented in Figure 3. Firstly, the exfiltration rate (losses) from the sewer system was determined by the mass balance of wastewater released to sewer system and outflow from the sewer system. Then, the total mass of pollutant in the sewer system was calculated. In the next step, the exfiltration rate from the tested section of the sewer system was assessed and then upscaled to the spatial distribution of exfiltration of the entire sewer system.

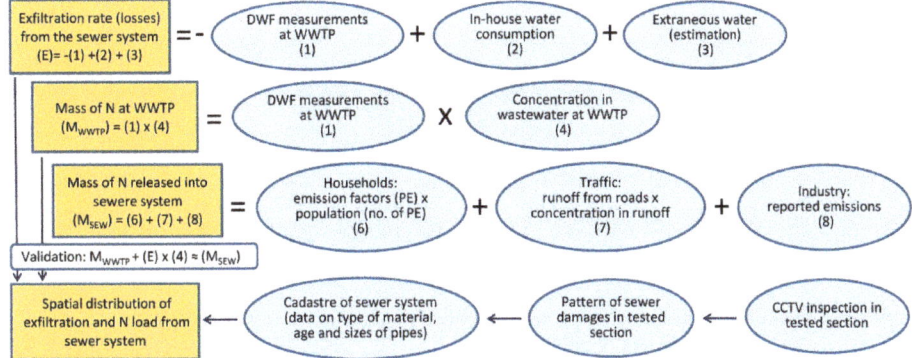

Figure 3. The flowchart of the workflow for the estimation of nitrogen (N) load from the sewer system (adapted after Pestotnik et al., 2019 [40]).

Exfiltration Rate (Losses) from the Sewer System

Daily flow rate measurements at the wastewater treatment plant (WWTP) for the period 2000–2010 were compared with in-house water consumption for that period, provided by the drinking water supply and the wastewater system operator. In the analysis, 13 DWF measurements were used, mainly in the winter period, when rainfall infiltration and outdoor water use have a low influence on the sewer system water balance. Nevertheless, the measured flow rate at the WWTP exceeded the quantity of in-house water consumption, which indicated the presence of extraneous water. Estimation of the extraneous water was based on the assumption that 0.01 km^2 of surface in the catchment of WWTP contributes 0.05 L/s of inflow of water [44], which accounted for the entire catchment 276.6 L/s. On the basis of this, the exfiltration rate from the entire sewer system in the catchment of WWTP was estimated to be 168 L/s (on average 0.21 L/s/km) [40].

Mass of Nitrogen in Wastewater

The estimation of the mass of nitrogen in water is based on the mass balance model [36,37]. The aim of this model was to investigate the shares of known and unknown sources of different pollutants (e.g., nitrogen, phosphorus, chloride, sulphate, copper, chromium, lead, organic carbon) in the sewer system. The mass of pollutants released into sewer system was calculated in the model as a sum of inputs from households, from roads and other traffic surfaces, and from industry. The model results were compared with two extensive analyses of chemical composition of wastewater at WWTP, made in 2008. More frequent measurements were performed only for nitrogen. The concentration of total nitrogen was correlated with discharge and both varied over time depending on the amount of rainfall (Figure 4). The yearly mass of other pollutants discharged at WWTP was a product of discharge and concentrations defined in the two mentioned analyses. Data from the first analysis was used as representative for the dry-weather periods (121 days), and from the second measurement as representative for the wet periods (121 days).

Good agreement between the modeled and observed content of phosphorous (−2.1%), nitrogen (+5.9%), and lead (13.1%) was found [36,37,40]. The content of other pollutants could not be well explained by the known sources.

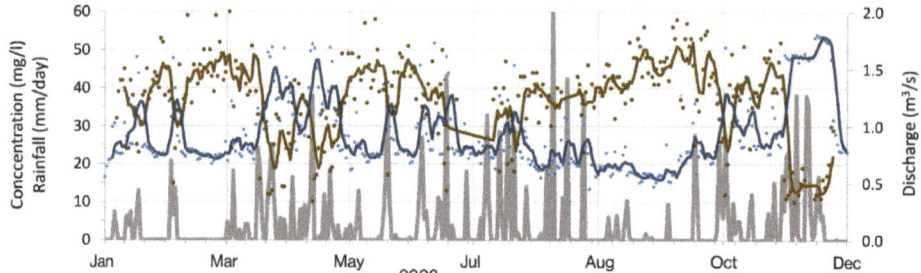

Figure 4. Concentration of total nitrogen at wastewater treatment plant (WWTP) presented with brown dots and 5 days moving average (brown line), discharge presented with blue dots and 5 days moving average (blue line), and rainfall high (gray line) measured at Ljubljana Bežigrad weather station (year 2008).

Spatial Distribution of Sewer Exfiltration

The model of spatial distribution of sewer exfiltration [36] was based on a detailed inspection of a 14.82 km section of the sewer system, which represents 2% of the entire sewer system. Closed-circuit television (CCTV) provided by the sewer system operator was used to detect the pattern of sewer damages in this section. Among 1246 detected damages, 104 were selected as significant and characterized in more detail to quantify exfiltration. This pattern and information on type of material, age, and size of pipes [45] were used to estimate the spatial distribution of the exfiltration and nitrogen load from the entire sewer system (Table 2).

Table 2. Estimated exfiltration and nitrogen load from pipelines of the entire sewer system [40].

Material	Time of Construction	Exfiltration (L/s/km)	Length (km)	Exfiltration Sum (L/s)	Nitrogen Load (ton/year)
Concrete	<1975	0.34	277.24	95.20	121.07
	1975–1995	0.23	255.00	58.80	74.78
	>1995	0.12	11.20	1.29	1.64
Plastic	<1975	0.11	22.37	2.39	3.04
	1975–1995	0.07	48.92	3.49	4.44
	>1995	0.04	186.22	6.65	8.45
	Total		800.93	167.82	213.42

For the calculation of the exfiltration flow rate from the leaky areas, an approach based on Darcy's law was used. It assumes linear dependency between the hydraulic gradient (I), the hydraulic conductivity of the clogging layer (k_f), the leakage area of the pipe (A_{leak}), and the exfiltration flow rate (Q_{ex}) [46]:

$$Q_{ex} = k_f \times A_{leak} \times I \quad (1)$$

where the hydraulic gradient (I) is the ratio between the sum of the water level inside the sewer (h_s) and the pipe wall thickness (h_w), and the pipe wall thickness (h_w):

$$I = (h_s + h_w)/h_w \quad (2)$$

In the calculation, three different hydraulic conditions in the pipelines were considered: (1) dry-weather conditions, days with mainly waste water discharge from water consumers (DWF); (2) days with significant rainfall inflow; and (3) extreme storms events with very high rainfall inflow. An analysis of rainfall data of a 10 year period (2000–2010) was performed. It showed that on average, in 1 year, DWF conditions last for 146 days, significant rainfall inflow occurs in 218 days, and once a year extreme storms are expected. It was assumed that sewer pipes are 10% full in DWF period, 50% in significant inflow period, and 100% in storm events (Figure 5). These data were used together with data on the diameter of pipes and a representative sample of damages for different materials and ages of pipes to calculate the spatial distribution of yearly exfiltration (Figure 6).

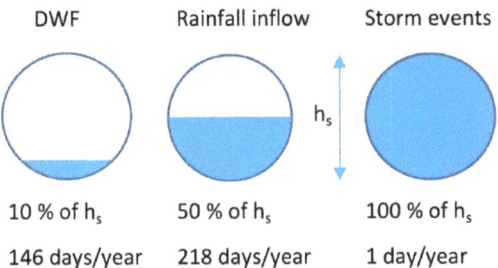

Figure 5. Scheme showing the different degrees (10%, 50%, 100%) of fullness in sewer pipes.

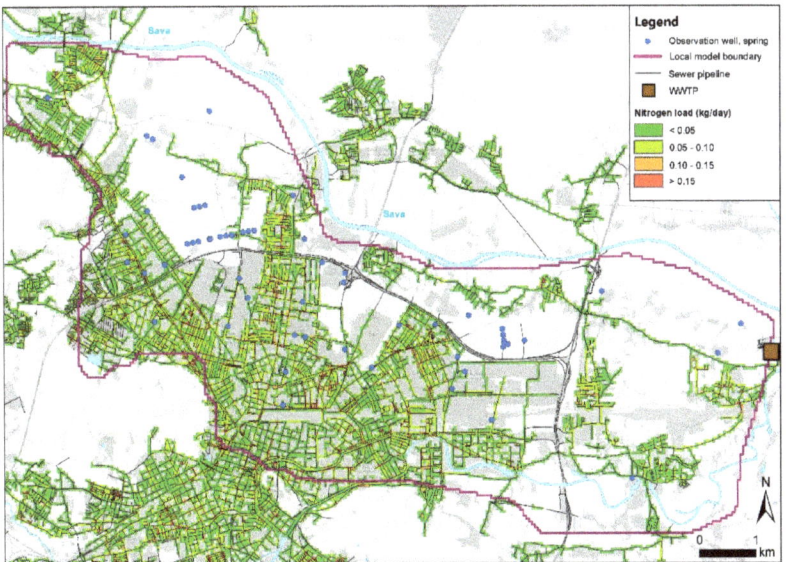

Figure 6. The spatial distribution of the estimated nitrogen load from the leaky sewer system (in kg/day per 50 m grid cell).

2.2.2. Nitrogen Load from Agriculture

Nitrogen load from agriculture used in the model was the nitrogen surplus calculated as the difference between nitrogen added to an agricultural land (mineral fertilizers, livestock manure, atmospheric deposition, and mineralization of organic matter) and nitrogen removed from the agricultural land (removed field crops and denitrification in the soil) [41]. On the basis of literature and statistical data, the nitrogen mass balance for different agricultural land use areas (arable fields, vineyards, orchards, and meadows and pastures) was calculated, and the results were provided in the polygon information layer. These data were used to generate a model input grid (Figure 7).

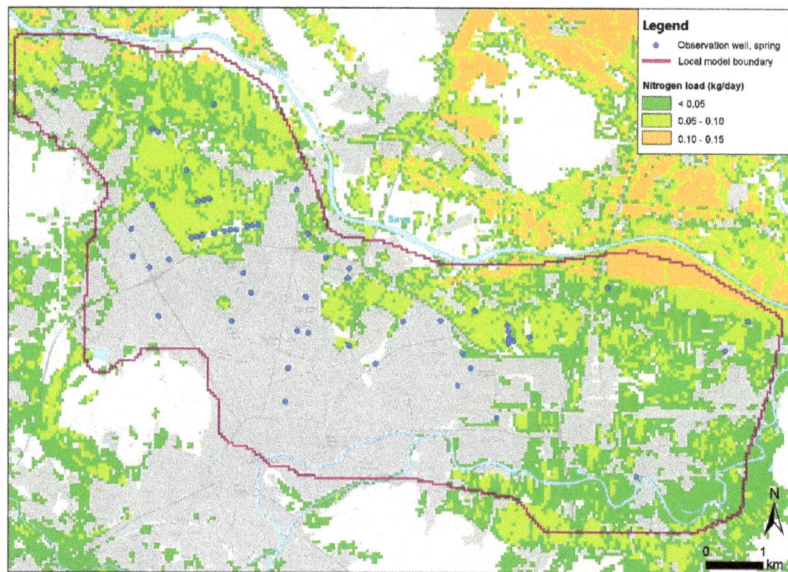

Figure 7. The spatial distribution of the estimated nitrogen load from agriculture (in kg/day per 50 m grid cell).

2.2.3. Groundwater Transport Model

Groundwater flow and transport simulations were carried out using the MIKE SHE/MIKE 11 modeling framework [47], which enabled transient simulation of the most important processes of the hydrological cycle in the study area. The solute transport model used in the study [38] is a refined and local version of the calibrated regional hydrological model [48]. The regional model (Figure 1) is a spatially distributed (200 m grid cells) and transient model with implemented daily values of time-dependent variables (temperature, precipitation, river levels and discharges at boundaries, groundwater level observations, abstractions, and leakage from the water supply system). The calibration and validation of the model was based on a comparison of the observed and simulated groundwater heads at the locations of 12 observation wells distributed over the entire study area, and the Sava River levels at one gauging station [48]. Additionally, the solute transport model was validated against the observed concentrations of the trichloroethylene pollution plume discovered in 2004 [49].

The local model (Figure 1) has one calculation layer and at boundaries uses hydraulic conditions from the regional model. Its model domain is discretized into horizontal grid cells of the size 50 × 50 m. The estimated nitrogen load was used as a spatially distributed, subsurface and constant mass input into the saturated zone of the aquifer. Neither retardation nor degradation processes were considered in the simulations. For the nitrate background concentration, 6 mg/L was used. This concentration is comparable to the lowest observed nitrate concentration (7.2 mg/L in observation location 18, Table S1 and is slightly higher than the observed concentration (4.9 mg/L) in the Sava River [50].

Scenarios for a 6 year simulation period (2010–2015) were performed. The last 3 years (2013–2015) of simulations were considered for the analysis. The first half of the simulation period was used for the model warm-up, which assured the avoidance of biased results affected by initial conditions.

3. Results

The estimated sources of nitrogen from the sewer system and from agriculture are distributed over an area of 19.84 km^2 (34%) and 27.81 km^2 (48% of model area), respectively. The mean nitrogen load in these model cells from the sewer system is 45 g/day and from agriculture is 52 g/day (Table 3).

In the local model area of 58 km^2, this accounts for a total nitrogen of 334.22 ton/year, 38% arising from the leaky sewer system (Figure 6) and 62% from agriculture (Figure 7).

Table 3. Statistics for the estimated nitrogen load expressed in g/day per 50 m cell.

N Source	No. of Cells	Mean (g/day)	Median (g/day)	Min (g/day)	Max (g/day)	SD (g/day)
Sewer system	11,125	45.26	41.00	0.10	223.10	33.36
Agriculture	7936	51.75	43.59	0.07	118.63	29.43

Nitrate concentration in groundwater was available from 58 observation wells and 2 springs. The observed locations were numbered according to their geographical positions, in accelerating order from the west towards the east (Figure 1). The modeled total nitrate concentration (Figure 8) is the sum of the modeled impacts of nitrogen load from the sewer system and agriculture (Figure S1, Figure S2) with an assumed natural background concentration of 6 mg/L. Median values of observed and modeled concentrations in a 3 year period (2013–2015) at observation locations (obs) are presented in Table S1 and summary statistics in Table 4.

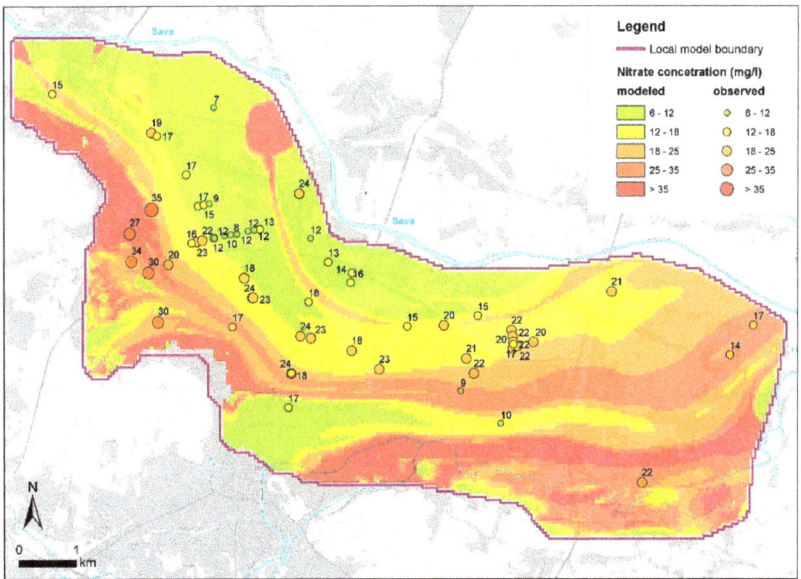

Figure 8. The spatial distribution of modeled compared to observed nitrate concentrations in groundwater.

Table 4. Summary statistics for modeled and observed nitrate concentrations at observation locations (Table S1).

Statistics	Modeled Concentration (mg/L)			Observed Concentration (mg/L)	Difference ** (mg/L)
	Agriculture	Sewer System	Total *		
Mean	4.15	5.48	15.63	18.36	2.73
SD	2.85	5.54	6.63	6.07	6.32
Max	14.79	26.09	41.49	35.10	17.83
Min	0.00	0.15	6.37	7.20	−14.76

* Sum of the modeled impacts from the sewer system and agriculture with an assumed natural background of 6 mg/L. ** Difference between observed and modeled total concentrations.

The average difference between the observed and modeled concentrations at the observation locations was found to be 2.7 mg/L (Table 4, Figure 9). The distribution of differences was left-skewed (Sk = −0.73, Figure 10). The correlation coefficient between the observed and modeled concentrations was 0.51 (Figure 11).

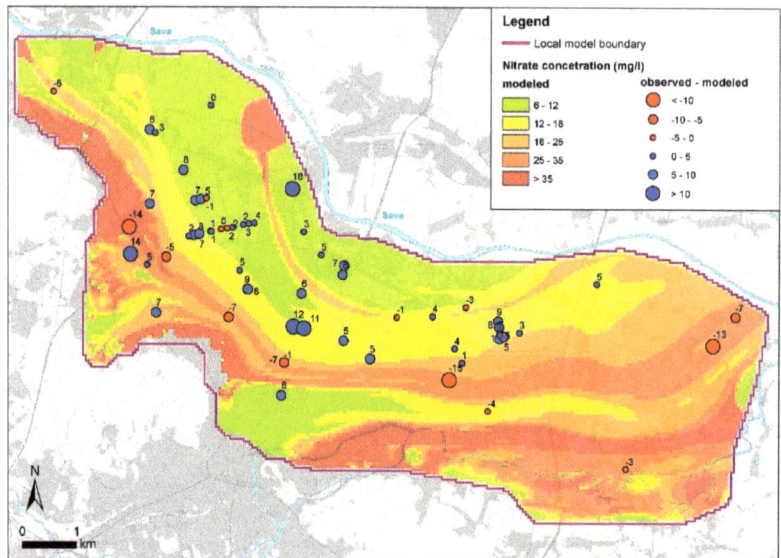

Figure 9. Spatial distribution of differences between the observed and the modeled nitrate concentrations in the groundwater in mg/L with red showing and overestimation by the model.

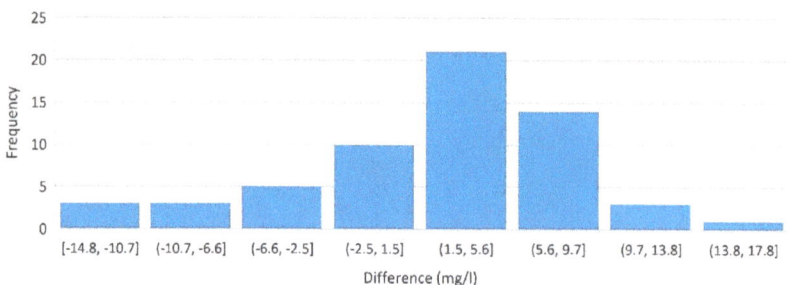

Figure 10. Histogram of differences between the observed and modeled nitrate concentrations at the observation locations.

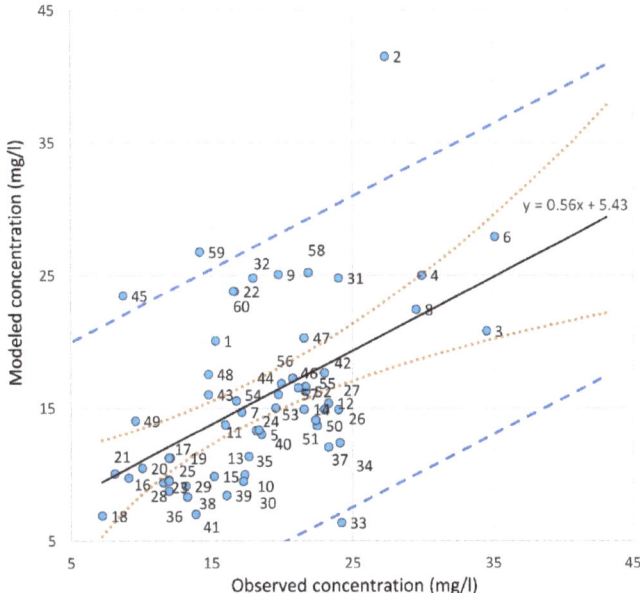

Figure 11. Scatterplot of observed and modeled nitrate concentrations at the observation locations, with fitted line (black line), 95% confidence band (light-brown dotted line), and 95% prediction band (blue dashed line).

4. Discussion

The modeled spatial distribution of nitrate followed the groundwater flow pattern. It reflected the general characteristics of the observed concentrations, which are higher in the southeastern and central parts of the model area (Figure 8). In spite of higher total nitrogen load from agriculture, the highest nitrate concentrations (>25 mg/L) were observed in areas (obs 2, 3, 4, 6 and 8) with predominant impact from sewer system (Figure 8).

The model gives insight into the impacts of two contamination sources in the capture zones of abstraction wells that are used for drinking water supply. Nitrate concentrations in observation wells in the Šentvid (obs 5 and 7) and Kleče well fields (obs 11–17, 19, 20, 21, 23, 25, 28 and 29) were found to be relatively low (on average below 15 mg/L) and, according to the model, arise predominantly from agriculture (72% of anthropogenic impact). Nitrate concentrations in observation wells in the Hrastje well field (obs 50–56) were higher (on average above 20 mg/L), and the shares of nitrogen load from agriculture (47%) and the sewer system (53%) are comparable. These findings give a bigger picture of the impact of different nitrogen sources on groundwater quality in comparison to the local study, focused on a small part of the urban area [35].

On average, the modeled concentrations were found to be 2.7 mg/L lower than the observed concentrations (Table 4, Figure 9). The discrepancy is relatively low and gives a good estimate of the nitrogen mass balance in the study area. Overestimation of observed nitrate concentrations by a factor of 4.5 was reported in the case study solely by using the nitrogen mass balance estimate for infiltration water [22]. This emphasizes the important contribution of the integration of groundwater dynamics in modeling in order to reduce the uncertainty in the estimate of the nitrate concentration in groundwater.

The modeled concentrations were most evidently underestimated in the area of the highest transmissivity and groundwater discharge (Figure 2). This zone stretches from the northeastern part of the model area, in which the most intensive recharge from the Sava River occurs and further follows the groundwater flow path. This discrepancy could be attributed either to the overestimated groundwater discharge rates or to underestimated nitrogen load, which in this area arises mainly from agriculture. In the urbanized part of the study area, underestimated concentrations could be related to unknown non-agricultural sources of nitrogen [19]. On the contrary, the overestimated nitrate concentrations are mainly in the southern part of the model area. This zone stretches from the southeastern part of the model area, in which the highest concentrations were observed (obs 2, 3 and 4) and follows the groundwater flow path towards the southeast.

The differences between the modeled and observed nitrate concentrations are related to the uncertainties arising from different sources. The level of complexity of used models was adjusted to available data and knowledge. The selection of the models followed the principle as simple as possible, as complex as necessary [51,52]. The used estimates of nitrogen load were based on the limited available data and a simplified nitrogen mass balance model. The wastewater exfiltration model used characteristics of 2% of the sewer system and scales up these findings to the entire sewer system. Due to this generalization and the lack of observations and the complexity of exfiltration process [14], the quantification of impact of sewer leaking on groundwater quality is uncertain. The available quantification of nitrogen load (surpluses) from agriculture was based on literature and statistical data [41], and thus it did not consider site-specific agricultural practices and their changes over time.

The groundwater model is a mathematical representation of an aquifer that is highly productive, characterized by a saturated zone up to 70 m deep and a fast groundwater flow with up to 20 m/day. The simulated pollution plumes are narrow and fast progressing, thus difficult to observe [49]. The modeling of the nitrate concentration is based on simulated groundwater discharge and flow direction. These two processes in the aquifer are difficult to measure and the lack of available data limits the possibility of validating the groundwater transport model. Especially in the areas close to the boundaries of the model, in which the transmissivity of the aquifer (Figure 2) is low, modeled concentrations were found to be very high (Figure 8) and sensitive to the interpreted shape of the aquifer basement. It is assumed that in these areas, the uncertainty of the model is the highest.

Data on nitrate concentration in groundwater was acquired from three monitoring networks managed by the Slovenian Environmental Agency, the City of Ljubljana, and the water supply company JP VOKA SNAGA. The samples were taken from observation wells using withdrawal method. No multilevel sampling technique for collecting depth-specific groundwater samples was applied. Representative depth-specific groundwater sampling in unconsolidated and coarse-grained aquifers is difficult to obtain. In such aquifers, even the application of packers within a screened well does not assure representative depth-specific sampling, due to the high risk of vertical circulations in the annular space around the well screen or casing [53]. The available data on groundwater sampling were insufficient for characterization of vertical distribution of nitrate in the aquifer. Due to this, the median value of measurements (2013–2015) in each well was considered as a representative concentration of the whole groundwater column at the observation location. This fact led to the simplification of the model and the use of a single calculation layer, which limited the simulation of vertical stratification and interpretation of the site-specific vertical concentration distribution.

5. Conclusions

The presented modeling approach enabled relation and quantification of the impact of pressure from contamination sources on groundwater quality. Modeled spatial distribution of nitrate content in groundwater pointed out the leaky sewer system as the prevailing source of nitrogen load inducing locally high nitrate concentration (above 25 mg/L). Reconstruction of the part of the sewer system in the catchments of observation wells with high nitrate content was recognized as a priority measure for the improvement of groundwater quality. The developed model can be used as a supportive tool in the optimization of mitigation measures, providing the simulation of effects of different remediation scenarios in a cost–benefit analysis. This approach could be applied also to assess the efficiency of groundwater management plans and to analyze the contribution of the stakeholders' activities to the improvement of groundwater quality.

Nitrate concentration in groundwater could be well explained by the known sources of nitrogen that approved the applied integrated mass balance and groundwater modeling concept. Application of this concept to other pollutants revealed a high share of unknown sources for several pollutants; among them, chromium (CrIV) was pointed out as the most threatening for the water supply [40]. Identification and remediation of unknown sources of pollutants whose content in groundwater is high, considering drinking water quality standards, or increases in time, is a priority for groundwater management and the assurance of safe drinking water supply for the City of Ljubljana.

Supplementary Materials: The following are available online at http://www.mdpi.com/2073-4441/12/4/1171/s1, Table S1: Median values of modeled and observed nitrate concentrations (2013–2015), Figure S1: The spatial distribution of the modeled nitrate concentration arising from the sewer system; Figure S2: The spatial distribution of the modeled nitrate concentration arising from agriculture.

Author Contributions: Conceptualization, M.J. and J.P.; methodology, J.P., S.P. and M.J.; investigation, J.P., S.P. and M.J.; data curation S.P., J.P. and M.J.; writing—original draft preparation, M.J.; writing—review and editing, M.J., J.P. and B.J.; project administration, J.P., B.J. and S.P.; funding acquisition, J.P., B.J. and M.J. All authors have read and agreed to the published version of the manuscript.

Funding: This research was funded by projects AMIIGA (Interreg Central Europe, No. CE32), INCOME (LIFE07, ENV/SLO/000725), IGCP 684 (UNESCO-IUGS), and by the Slovenian Research Agency (research core funding Groundwaters and Geochemistry (P1-0020)).

Acknowledgments: The authors wish to thank Petra Meglič and Dejan Šram for the help in cartography. The authors are very grateful to reviewers for very constructive reviews that significantly improved the manuscript.

Conflicts of Interest: The authors declare no conflict of interest.

References

1. UN. *World Urbanization Prospects 2018: Highlights*; United Nations, Department of Economic and Social Affairs, Population Division (ST/ESA/SER.A/421); United Nations: New York, NY, USA, 2019.
2. WBG. *Climate Change Action Plan*; World Bank Group: Washington, DC, USA, 2016.
3. Howard, K.W.F. Sustainable cities and the groundwater governance challenge. *Environ. Earth Sci.* **2014**. [CrossRef]
4. Boukhemacha, M.A.; Gogu, C.R.; Serpescu, I.; Gaitanaru, D.; Bica, I. A hydrogeological conceptual approach to study urban groundwater flow in Bucharest city, Romania. *Hydrogeol. J.* **2015**, *23*, 437–450. [CrossRef]
5. Diamond, M.L.; Hodge, E. Urban contaminant dynamics: From source to effect. *Environ. Sci. Technol.* **2007**, *41*, 3796–3805. [CrossRef] [PubMed]
6. Gogu, C.R.; Campbell, D.; de Beer, J. PREFACE: The Urban Subsurface – from Geoscience and Engineering to Spatial Planning and Management. *Procedia Eng.* **2017**, *209*, 1–3. [CrossRef]
7. Jago-on, K.A.B.; Kaneko, S.; Fujikura, R.; Fujiwara, A.; Imai, T.; Matsumoto, T.; Zhang, J.; Tanikawa, H.; Tanaka, K.; Lee, B.; et al. Urbanization and subsurface environmental issues: An attempt at DPSIR model application in Asian cities. *Sci. Total Environ.* **2009**, *407*, 3089–3104. [CrossRef]
8. Bhaskar, A.S.; Welty, C.; Maxwell, R.M.; Miller, A.J. Untangling the effects of urban development on subsurface storage in Baltimore. *Water Resour. Res.* **2015**, *51*, 1158–1181. [CrossRef]
9. Colombo, L.; Alberti, L.; Mazzon, P.; Formentin, G. Transient Flow and Transport Modelling of an Historical CHC Source in North-West Milano. *Water* **2019**, *11*, 1745. [CrossRef]
10. Foster, S.S.D. The interdependence of groundwater and urbanisation in rapidly developing cities. *Urban Water* **2001**, *3*, 185–192. [CrossRef]
11. Schirmer, M.; Reinstorf, F.; Leschik, S.; Musolff, A.; Krieg, R.; Strauch, G.; Molson, J.W.; Martienssen, M.; Schirmer, K. Mass fluxes of xenobiotics below cities: Challenges in urban hydrogeology. *Environ. Earth Sci.* **2011**, *64*, 607–617. [CrossRef]
12. Klinger, J.; Wolf, L.; Hoetzl, H. Leaky sewers-measurements under operating conditions. In Proceedings of the 4th World Wide Workshop for Young Environmental Scientists (WWW-YES), Vitry sur Seine, France, 10–13 May 2005.
13. Peche, A.; Graf, T.; Fuchs, L.; Neuweiler, I. Physically based modeling of stormwater pipe leakage in an urban catchment. *J. Hydrol.* **2019**, *573*, 778–793. [CrossRef]

14. Wolf, L.; Klinger, J.; Held, I.; Hötzl, H. Integrating groundwater into urban water management. *Water Sci. Technol.* **2006**, *54*, 395–403. [CrossRef] [PubMed]
15. Rutsch, M.; Rieckermann, J.; Cullmann, J.; Ellis, J.B.; Vollertsen, J.; Krebs, P. Towards a better understanding of sewer exfiltration. *Water Res.* **2008**, *42*, 2385–2394. [CrossRef] [PubMed]
16. Uhan, J. Data-driven modelling of groundwater vulnerability to nitrate pollution in Slovenia (Podatkovno vodeno modeliranje ranljivosti podzemne vode na nitratno onesnaženje v Sloveniji). *RMZ—Mater. Geoenviron.* **2012**, *59*, 201–212.
17. Mihorko, P.; Gacin, M. *Environmental Indicators in Slovenia, Nitrates in Groundwater [VD05]*; Slovenian Environment Agency: Ljubljana, Slovenia, 2019; p. 14.
18. Andjelov, M.; Kunkel, R.; Uhan, J.; Wendland, F. Determination of nitrogen reduction levels necessary to reach groundwater quality targets in Slovenia. *J. Environ. Sci.* **2014**, *26*, 1806–1817. [CrossRef]
19. Wakida, F.T.; Lerner, D.N. Non-agricultural sources of groundwater nitrate: A review and case study. *Water Res.* **2005**, *39*, 3–16. [CrossRef]
20. Graham, J.P.; Polizzotto, M.L. Pit latrines and their impacts on groundwater quality: A systematic review. *Environ. Health Perspect.* **2013**, *121*, 521–530. [CrossRef]
21. Koda, E.; Sieczka, A.; Osinski, P. Ammonium Concentration and Migration in Groundwater in the Vicinity of Waste Management Site Located in the Neighborhood of Protected Areas of Warsaw, Poland. *Sustainability* **2016**, *8*, 1253. [CrossRef]
22. Kringel, R.; Rechenburg, A.; Kuitcha, D.; Fouepe, A.; Bellenberg, S.; Kengne, I.M.; Fomo, M.A. Mass balance of nitrogen and potassium in urban groundwater in Central Africa, Yaounde/Cameroon. *Sci. Total Environ.* **2016**, *547*, 382–395. [CrossRef]
23. Soldatova, E.; Guseva, N.; Sun, Z.; Bychinsky, V.; Boeckx, P.; Gao, B. Sources and behaviour of nitrogen compounds in the shallow groundwater of agricultural areas (Poyang Lake basin, China). *J. Contam. Hydrol.* **2017**, *202*, 59–69. [CrossRef]
24. Szabó, G.; Bessenyei, É.; Karancsi, G.; Balla, D.; Mester, T. Effects of nitrogen loading from domestic wastewater on groundwater quality. *Water SA* **2019**, *45*. [CrossRef]
25. Zhang, H.; Hiscock, K.M. Modelling the effect of forest cover in mitigating nitrate contamination of groundwater: A case study of the Sherwood Sandstone aquifer in the East Midlands, UK. *J. Hydrol.* **2011**, *399*, 212–225. [CrossRef]
26. Arauzo, M. Vulnerability of groundwater resources to nitrate pollution: A simple and effective procedure for delimiting Nitrate Vulnerable Zones. *Sci. Total Environ.* **2017**, *575*, 799–812. [CrossRef] [PubMed]
27. Nolan, B.T.; Hitt, K.J. Vulnerability of Shallow Groundwater and Drinking-Water Wells to Nitrate in the United States. *Environ. Sci. Technol.* **2006**, *40*, 7834–7840. [CrossRef] [PubMed]
28. Padilla, F.M.; Gallardo, M.; Manzano-Agugliaro, F. Global trends in nitrate leaching research in the 1960–2017 period. *Sci. Total Environ.* **2018**, *643*, 400–413. [CrossRef] [PubMed]
29. Lalehzari, R.; Tabatabaei, S.H.; Kholghi, M. Simulation of nitrate transport and wastewater seepage in groundwater flow system. *Int. J. Environ. Sci. Technol.* **2013**, *10*, 1367–1376. [CrossRef]
30. Podlasek, A.; Bujakowski, F.; Koda, E. The spread of nitrogen compounds in an active groundwater exchange zone within a valuable natural ecosystem. *Ecol. Eng.* **2020**, *146*, 105746. [CrossRef]
31. Zuo, R.; Jin, S.; Chen, M.; Guan, X.; Wang, J.; Zhai, Y.; Teng, Y.; Guo, X. In-situ study of migration and transformation of nitrogen in groundwater based on continuous observations at a contaminated desert site. *J. Contam. Hydrol.* **2018**, *211*, 39–48. [CrossRef]
32. Sieczka, A.; Bujakowski, F.; Koda, E. Modelling groundwater flow and nitrate transport: A case study of an area used for precision agriculture in the middle part of the Vistula River valley, Poland. *Geologos* **2018**, *24*, 225–235. [CrossRef]
33. Valivand, F.; Katibeh, H. Prediction of Nitrate Distribution Process in the Groundwater via 3D Modeling. *Environ. Model. Assess.* **2020**, *25*, 187–201. [CrossRef]
34. Bračič Železnik, B.; Jamnik, B. Javna oskrba s pitno vodo. In *Podtalnica Ljubljanskega Polja*; Rejec Brancelj, I., Smrekar, A., Kladnik, D., Eds.; Založba ZRC: Ljubljana, Slovenia, 2005; pp. 101–117.
35. Vižintin, G.; Souvent, P.; Veselič, M.; ČenČur Curk, B. Determination of urban groundwater pollution in alluvial aquifer using linked process models considering urban water cycle. *J. Hydrol.* **2009**, *377*, 261–273. [CrossRef]

36. Prestor, J.; Pestotnik, S.; Meglič, P.; Janža, M. *Model of Environmental Pressures and Impacts*; A.3.3 Final Report of INCOME Project (LIFE+ Programme); Geological Survey of Slovenia: Ljubljana, Slovenia, 2011.
37. Prestor, J.; Pestotnik, S.; Meglič, P.; Janža, M. Načrtovanje zaščitnih ukrepov na podlagi modela obremenitev in vplivov na podzemno vodo Ljubljanskega polja. In *Skrb za Pitno Vodo*; Jamnik, B., Janža, M., Smrekar, A., Eds.; Skrb za pitno vodo: Ljubljana, Slovenia, 2014; Volume GEOGRAFIJA SLOVENIJE 31, pp. 79–93.
38. Janža, M.; Meglič, P.; Prestor, J.; Jamnik, B.; Pestotnik, S.D. *T2.2.3—Report on the Improved Transport and Surface-Groundwater Interactions Model, Version 2. Project AMIIGA (Interreg, Central Europe)*; Geological Survey of Slovenia: Ljubljana, Slovenia, 2019; pp. 15–22.
39. Meglič, P.; Janža, M.; Prestor, J.; Pestotnik, S.; Jamnik, B.D. *T2.2.7—Report on the Results of the Most Probable Scenarios Threatening Groundwater, Version 2. Project AMIIGA (Interreg, Central Europe)*; Geological Survey of Slovenia: Ljubljana, Slovenia, 2019.
40. Pestotnik, S.; Prestor, J.; Meglič, P.; Janža, M.; Šram, D.D. *T2.2.4—Report on Actualized Contaminants Mass Balance and Pressures-Impacts Model, Version 2. Project AMIIGA (Interreg, Central Europe)*; Geological Survey of Slovenia: Ljubljana, Slovenia, 2019; pp. 15–22.
41. Pintar, M.; Sluga, G.; Bremec, U. *Določitev Obremenitev iz Kmetijstva za Izbrane Prostorske Enote*; Inštitut za vode Republike Slovenije: Ljubljana, Slovenia, 2005.
42. EEA. *Corine Land Cover 2012 Seamless Vector Data (Version 18)*; Agency, E.E., Ed.; EEA: Kopenhagen, Denmark, 2016.
43. Janža, M.; Lapanje, A.; Šram, D.; Rajver, D.; Novak, M. Research of the geological and geothermal conditions for the assessment of the shallow geothermal potential in the area of Ljubljana, Slovenia. *Geologija* **2017**, *60*, 309–327. [CrossRef]
44. Panjan, J. *Količinske in Kakovostne Lastnosti Voda: Študijsko Gradivo*; FGG, Inštitut za zdravstveno hidrotehniko: Ljubljana, Slovenia, 2004.
45. Smrekar, A.; Bole, D.; Breg Valjavec, M.; Gabrovec, M.; Gašperič, P.; Ciglič, R.; Pavšek, M.; Topole, M. *Register and Evaluation of the Active and the Potential Sources of Pollution, Project INCOME Action A.2.1 Report*; Anton Melik Geographical Institute, ZRC SAZU: Ljubljana, Slovenia, 2010.
46. Wolf, L.; Klinger, J.; Hoetzl, H.; Mohrlok, U. Quantifying Mass Fluxes from Urban Drainage Systems to the Urban Soil-Aquifer System (11 pp). *J. Soils Sediments* **2007**, *7*, 85–95. [CrossRef]
47. MIKE SHE, Integrated Catchment Modelling. Available online: https://www.mikepoweredbydhi.com/products/mike-she (accessed on 6 April 2020).
48. Janža, M. A decision support system for emergency response to groundwater resource pollution in an urban area (Ljubljana, Slovenia). *Environ. Earth Sci.* **2015**, *73*, 3763–3774. [CrossRef]
49. Janža, M.; Prestor, J.; Urbanc, J.; Jamnik, B. TCE contamination plume spreading in highly productive aquifer of Ljubljansko polje. In *Abstracts of the Contributions of the EGU General Assembly 2005*; EGU: Vienna, Austria, 2005; Volume 7.
50. Urbanc, J.; Jamnik, B. Distribution and origin of nitrates in the groundwater of Ljubljansko polje. *Geologija* **2007**, *50*, 467–475. [CrossRef]
51. Brunetti, G.; Šimůnek, J.; Glöckler, D.; Stumpp, C. Handling model complexity with parsimony: Numerical analysis of the nitrogen turnover in a controlled aquifer model setup. *J. Hydrol.* **2020**, *584*. [CrossRef]
52. Höge, M.; Wöhling, T.; Nowak, W. A Primer for Model Selection: The Decisive Role of Model Complexity. *Water Resour. Res.* **2018**, *54*, 1688–1715. [CrossRef]
53. Ducommun, P.; Boutsiadou, X.; Hunkeler, D. Direct-push multilevel sampling system for unconsolidated aquifers. *Hydrogeol. J.* **2013**, *21*, 1901–1908. [CrossRef]

© 2020 by the authors. Licensee MDPI, Basel, Switzerland. This article is an open access article distributed under the terms and conditions of the Creative Commons Attribution (CC BY) license (http://creativecommons.org/licenses/by/4.0/).

Article

Managing Stormwater by Accident: A Conceptual Study

Carly M. Maas [1,2,*], William P. Anderson, Jr. [3] and Kristan Cockerill [4]

1. Department of Geology, University of Maryland, College Park, MD 20742, USA
2. Earth System Science Interdisciplinary Center, University of Maryland, College Park, MD 20742, USA
3. Department of Geological and Environmental Sciences, Appalachian State University, Boone, NC 28608, USA; andersonwp@appstate.edu
4. Department of Interdisciplinary Studies, Appalachian State University, Boone, NC 28608, USA; cockerillkm@appstate.edu
* Correspondence: maascm@umd.edu

Abstract: Stormwater-driven road salt is a chronic and acute issue for streams in cold, urban environments. One promising approach for reducing the impact of road salt contamination in streams and adjacent aquifers is to allow "accidental wetlands" to flourish in urban areas. These wetlands form naturally as a byproduct of human activities. In this study, we quantified the ability of an accidental wetland in northwestern North Carolina, USA, to reduce the timing and peak concentration of road salt in a stream. Monitoring suggests that flow and transport processes through the wetland reduce peak concentrations and delay their arrival at the adjacent stream. We expand these findings with numerical simulations that model multiple meltwater and summer storm event scenarios. The model output demonstrates that small accidental wetland systems can reduce peak salinities by 94% and delay the arrival of saltwater pulses by 45 days. Our findings indicate that accidental wetlands improve stream water quality and they may also reduce peak temperatures during temperature surges in urban streams. Furthermore, because they find their own niche, accidental wetlands may be more effective than some intentionally constructed wetlands, and provide opportunities to explore managing stormwater by letting nature take its course.

Keywords: accidental wetland; road salt; urban hydrogeology; headwater stream

Citation: Maas, C.M.; Anderson, W.P., Jr.; Cockerill, K. Managing Stormwater by Accident: A Conceptual Study. *Water* **2021**, *13*, 1492. https://doi.org/10.3390/w13111492

Academic Editor: C. Radu Gogu

Received: 26 April 2021
Accepted: 24 May 2021
Published: 26 May 2021

Publisher's Note: MDPI stays neutral with regard to jurisdictional claims in published maps and institutional affiliations.

Copyright: © 2021 by the authors. Licensee MDPI, Basel, Switzerland. This article is an open access article distributed under the terms and conditions of the Creative Commons Attribution (CC BY) license (https://creativecommons.org/licenses/by/4.0/).

1. Introduction

The idea of "accidental" wetlands is gaining traction as a potentially valuable component of various ecosystems. These are water bodies that "result from human activities, but are not designed or managed for any specific outcome" [1]. As the moniker implies, these "accidental" systems arise on their own, often in vacant lots or various low spots where stormwater or irrigation runoff collect. In a study of the South Platte River basin, for example, 89% of the extant wetlands exist because of various irrigation conveyances [2]. Accidental wetlands typically form in low-lying abandoned or underutilized landscapes with poor drainage [3]. Flooding often occurs in these areas due to a lack of stormwater management and sediment transport processes, which contribute to repeated sediment deposition, allowing wetland plants to grow and establish a habitat [3]. Brooks et al. [4] documented that in arid, urban areas, effluent from human sources can account for up to 90% of streamflow during dry periods. Any wetland-like areas on those rivers exist because of human activity. Although they are not yet heavily studied, there is evidence that these accidental wetlands offer benefits to rival constructed versions [1,5,6]. In fact, Palta et al. [3] argue that "accidental wetlands may provide more services than designed environments because the latter are commonly over-designed for a limited set of specific functions." There is also evidence that small wetlands, which many accidental wetlands are, can be as or more effective than larger ones [7,8]. Although they are just beginning to be intentionally documented and studied, Palta et al. [3] offer evidence highlighting the ubiquity of accidental wetlands and their potential to offer diverse benefits. In the limited literature available, researchers have found that accidental wetlands can remove pollutants, mitigate heat, store groundwater, restore surface-mined lands, and provide the social benefits of additional green space in urban settings [3,9–11].

In these few known studies, accidental wetlands were shown to help with pollutant remediation, and a key pollutant in cold urban areas is road salt. Like accidental wetlands, in recent years researchers have paid increasing attention to de-icing salt in stormwater runoff as a key issue for stream quality, especially in urban areas. Although there are numerous salt sources in urban environments, including water-softening, septic field discharge, natural rock weathering, and wastewater treatment plants [12], the dominant source is road salt applied to the road surface in order to melt snow and ice [13,14]. The annual road salt application in the United States in 2014 was about 56.5 million metric tons, costing about $1.18 billion [15]. Elevated salt levels in waterways contribute to negative consequences, including contaminating drinking wells [16] and various ecosystem impacts. In their review of the literature, Hintz and Relyea [17] found that salt negatively affects all freshwater species, but that the level at which salt becomes a concern is highly variable across species.

There are complex relationships between surface and groundwater systems, such that salt remains available for long periods of time as it travels between surface and groundwater systems [18–20]. Furthermore, numerous studies have focused on how effective various "green infrastructure" (e.g., rain gardens, bioretention systems) are in cold climates (see Kratky et al. [21] for a review). Retention systems capture salt as it runs off and can slow its path to waterways [18,22]. Retention systems do not, however, eliminate salt or retain it for long periods. Studies have found that salinity remains a key stream stressor even in restored streams and in those with stormwater management efforts in place [23,24]. In addition to its own negative consequences, salt can mobilize various metals and lower plants' uptake of metals, thereby reducing the effectiveness of bioretention [21,25–27]. Cockerill et al. [20], however, did find that retention had the potential to reduce salinity spikes, which can reduce the long-term salt levels in a given hydrologic system. Flattening the curve, as it were, and alleviating those salt spikes, may over time lower the summer pulses of salt into that system. Given Hintz and Relyea's [17] findings on the variability in how much salt triggers negative consequences, reducing salinity spikes and reducing summer inputs may reduce the effects on some aquatic species.

Although researchers are studying accidental wetlands and the negative effects of road salt on urban waterways, as far as we know, these two subjects have not been assessed in concert. We offer a conceptual study addressing the potential for accidental wetlands to ameliorate salinity levels. Because constructed wetlands offer highly variable results depending on scale, lifespan, and local hydrologic conditions [28–31], paying more focused attention to sites where wetlands form themselves may offer a cheaper and equally or more effective alternative means of lessening road salt impacts. Though no wetland will completely remove the salt from an urban stream system, there is still potential to improve urban water quality by documenting and encouraging accidental wetlands.

The project documented here is a testimony to the value of paying attention to local environments and the role of serendipity in research. Copeland [32] describes serendipity in science as "the intersection of chance and wisdom" and notes that the value of the serendipitous process or event can only be assessed in hindsight. In our case, authors Anderson and Cockerill initially noticed a "wetland" forming in a concrete culvert on the Appalachian State University campus and joked about its potential value. As we continued to observe the accumulation of sediment and increased vegetation, we began thinking that the site actually warranted more focused attention. At that point, author Maas was seeking a project and began to intentionally explore the effect of that wetland on road salt contamination for a headwater stream in an urban setting. In searching for background information to guide this more intentional study, we encountered the premise of "accidental wetlands" and recognized that we had one.

Once intentionality was established, this study focused on these research questions:

1. Can flow through the shallow subsurface and chloride transport processes through accidental wetlands reduce the magnitude of saline peaks arriving at an adjacent stream?
2. Can flow through the shallow subsurface and chloride transport processes through accidental wetlands delay the arrival of a salt plume at an adjacent stream?

Site Description

Boone Creek and the accidental wetland investigated in this study are located in the small yet heavily urbanized town of Boone, North Carolina (Figure 1). Boone Creek is a headwater tributary of the South Fork New River, located within the Blue Ridge Mountains of northwestern North Carolina [20]. The town of Boone has high runoff ratios that result in flashy, flood-prone streams due to heavy urbanization near the stream. Although the total catchment area has 24.3% impervious surface cover, much of the more natural land use occurs on the mountain slopes near the catchment boundaries (Figure 2). Within 25 m of Boone Creek, the impervious surface cover ranges from 1% to 75% [33]. The Town of Boone has about 19,500 residents (U.S. Census Bureau, Quick Facts, https://www.census.gov/quickfacts/fact/table/boonetownnorthcarolina,US/PST045219, accessed 25 May 2021) and Appalachian State University (ASU) adds an additional 20,000 students to the local population (Appalachian State University Facts, https://www.appstate.edu/about/ accessed on 25 May 2021). The stream is in a mountainous area, with a total catchment relief of nearly 500 m, although the main channel gradient is fairly low, at 2%. The stream has previously had trout populations and retains a state classification as a "trout stream" [34]. Other tributaries to Boone Creek have gradients of greater than 10%. The catchment of the basin considered in this study has an area of 5.2 km^2 [20].

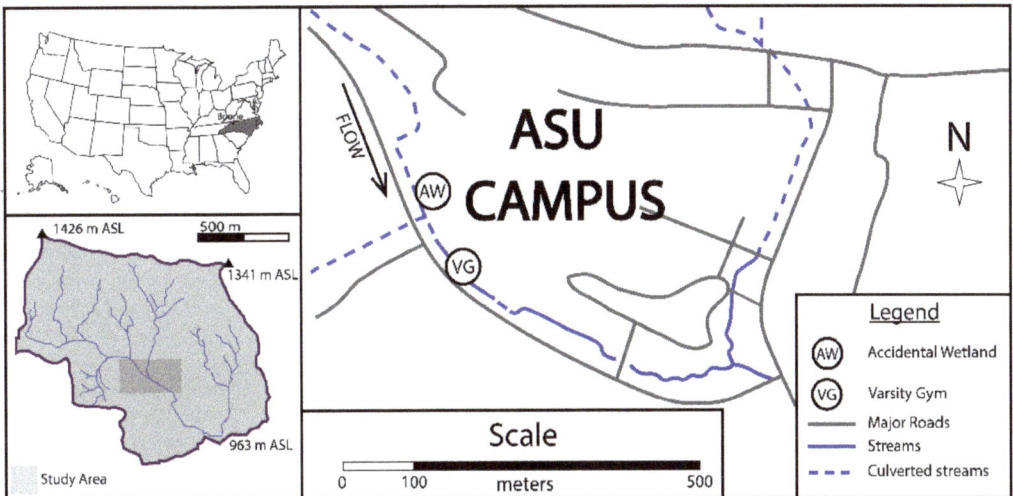

Figure 1. The Accidental Wetland (AW) site is located in Boone, North Carolina, USA (left top insert map) at 36°12′54.6″ N 81°40′57.0″ W. The solid blue line denotes the surface water of Boone Creek and the dashed blue line signifies culverted sections of Boone Creek. The circles indicate data collection sites, Accidental Wetland (AW) and Varsity Gym (VG). The bottom left insert map shows the Boone Creek watershed (heavy purple line) with the black triangles indicating peak elevations within the watershed. The arrow represents the direction of flow. ASL denotes above sea level. Adapted from ref. [20].

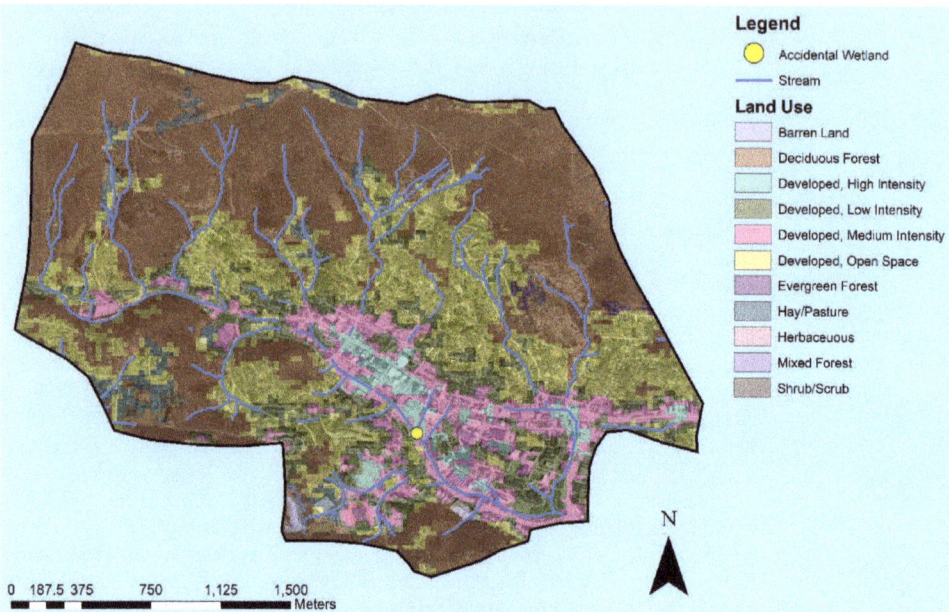

Figure 2. Land use map of the Boone Creek watershed. The Accidental Wetland site is located in the light blue and pink regions, which designate medium and high intensity development, respectively. The stream includes culverted and nonculverted segments. The map was created using ArcGIS® software by Esri, Redlands, CA, USA. Accessed on 10 February 2021. ArcGIS® and ArcMap™ are the intellectual property of Esri and are used herein under license. Copyright © Esri. All rights reserved. For more information about Esri® software, please visit www.esri.com, accessed on 10 February 2021.

Boone Creek (Figure 1) runs through several culverts, with the longest being about 600 m long [20]. The area experiences 89 cm of snow per year, requiring 6×10^5 kg (600 metric tons) of deicing salt applied to the impervious surfaces on average annually [12]. Electrical conductivity data, converted to equivalent chloride, indicate that the stream regularly violates the US Environmental Protection Agency's (EPA) chronic threshold of a four-day average chloride concentration, exceeding 230 mg/L, and also exceeds the acute threshold of a one-hour average chloride concentration, exceeding 860 mg/L on a regular basis in the winter salting season [20,35]. Between 1 July 2014 and 1 July 2015, Boone Creek exceeded the EPA's chronic threshold for 10% of the time, or 36 days, and exceeded the acute threshold for the equivalent of 8.5 days [20].

Our accidental wetland site is unique because it was created in a constrained concrete channel and did not form on an existing substrate. Subsequently, there is no direct connection between the wetland and the natural groundwater reservoir. We are treating the accidental wetland as an isolated "aquifer" that is disconnected from the natural groundwater flow system. Throughout the rest of this paper, our references to groundwater flow and solute transport are solely related to this process in the small accidental wetland that formed in the concrete culvert that ultimately feeds water to Boone Creek through seepage at its downgradient end. The only sources of water to the accidental wetland are from runoff from impervious surfaces or direct precipitation. Figure 3 shows the total area of the drainage basin to the accidental wetland, delineating pervious areas and impervious areas, such as buildings and sidewalks. The 30-year normal precipitation in Boone as measured at the BOONE 1 SE, NC US weather station between 1981 and 2010 is 1338 mm, with roughly equal precipitation during each season (https://www.ncdc.noaa.gov/cdo-web/datatools/normals, accessed on 21 April 2021). As Figure 4a–f show, the 600 m culvert containing Boone Creek is open for five meters to allow stormwater to enter from a concrete conveyance that drains part of campus, as well as part of the Town of Boone.

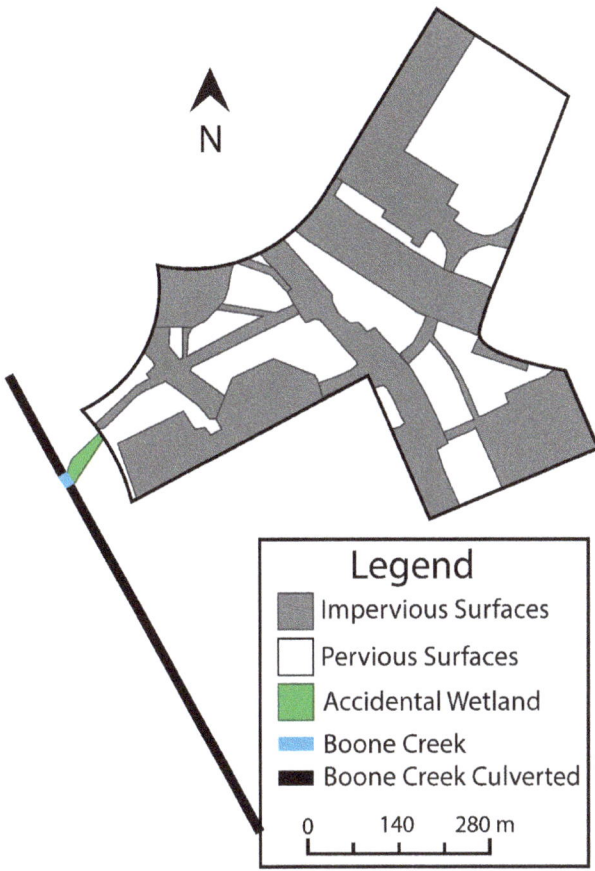

Figure 3. Total area of the drainage basin for the Accidental Wetland (AW) site. Gray areas indicate impervious surfaces such as sidewalks, roads, and rooftops and the white areas represent green spaces.

Figure 4. Evolution of the Accidental Wetland (AW) from 2005–2019. The photos were taken on the bridge over the wetland, facing downstream toward Boone Creek. The wetland was removed in September of 2018.

The focus of this study was not on how accidental wetlands form, because the exact timing for this specific accidental wetland was unclear and did not factor in our findings. It may have originated in 2011–2012 from large storm events, carrying significant sediment into the culvert. By 2015, authors Anderson and Cockerill observed that the channel had become a heavily vegetated wetland (Figure 4b), and we began observing it regularly. Up to that point, we had paid little attention to the site and the few photos available before 2015 show spotty vegetation in 2005 (Figure 4a) and we observed some sediment accumulation along one edge by 2012. Since 2015 the system evolved to include diverse vegetation throughout the entire channel and a meandering "stream" through the vegetation.

A storm event in October 2017 deposited a large amount of sand/soil at the culvert end of the wetland. At that time, authors Maas and Anderson decided to assess what was happening more formally. They installed three wells and an electrical conductivity logger (HOBO Conductivity Data Logger, U24-001) along the length of the wetland in 2017. The monitoring wells showed large fluctuations in water levels throughout the year, with high levels during intense storm events during warm months and during the wet and cold winter. In September 2018 the university removed all of the vegetation and accumulated sediment in the accidental wetland as part of flood control measures on campus (Figure 4e,f).

2. Materials and Methods

2.1. Field Data Collection

We collected handheld probe measurements in the accidental wetland from 7 September 2017 through 6 February 2020. We used a Yellow Springs Instrument (YSI) 556 MPS (YSI, Yellow Springs, OH, USA) to measure salinity, dissolved oxygen, temperature, and other parameters. These data allowed us to gain an understanding of the salinity distribution within the wetland both spatially and temporally, and guided our conceptual model of transport through the sediment. For examples of studies using the YSI 556 MPS, see [36,37]. We collected handheld measurements on a regular basis at four locations along the wetland and at one location in Boone Creek (Figure 5). The YSI probe measured salinity in PSU, and we converted these data to equivalent chloride concentrations by multiplying by 606.6 mg/L Cl$^-$ per PSU [38]. PSU is approximately equivalent to parts per thousand (ppt) of NaCl; therefore, for every 1 g/L of NaCl, 606.6 mg/L of the solution is Cl$^-$ and 393.4 mg/L is Na$^+$ [38]. All subsequent data in this study are presented in terms of equivalent chloride concentrations.

Datalogging at the Varsity Gym (VG) monitoring site has collected electrical conductivity (EC) data and stage-discharge data at 15-min intervals since July 2014, approximately 200 m downstream of the wetland (Figure 1). EC is a proxy for salinity [20]. The stream's salinity is less than that of the wetland due to the mixing of runoff after winter storm and precipitation events and the constant input of baseflow, which most often has a lower salinity than winter runoff. These data are scaled to provide a continuous dataset of boundary conditions for the groundwater flow and transport simulations, which are described later in the manuscript.

We used the Hazen method [39] on five soil samples collected from throughout the wetland to analyze grain size distributions. We used these data to estimate the accidental wetland's hydraulic conductivity using

$$K = C\,(d_{10})^2, \qquad (1)$$

where d_{10} is the diameter of the 10% finer grains (cm) and C is a coefficient based on grain size and sorting. We averaged the samples to estimate a bulk hydraulic conductivity for the wetland, and we further calibrated this value with subsequent numerical modeling experiments.

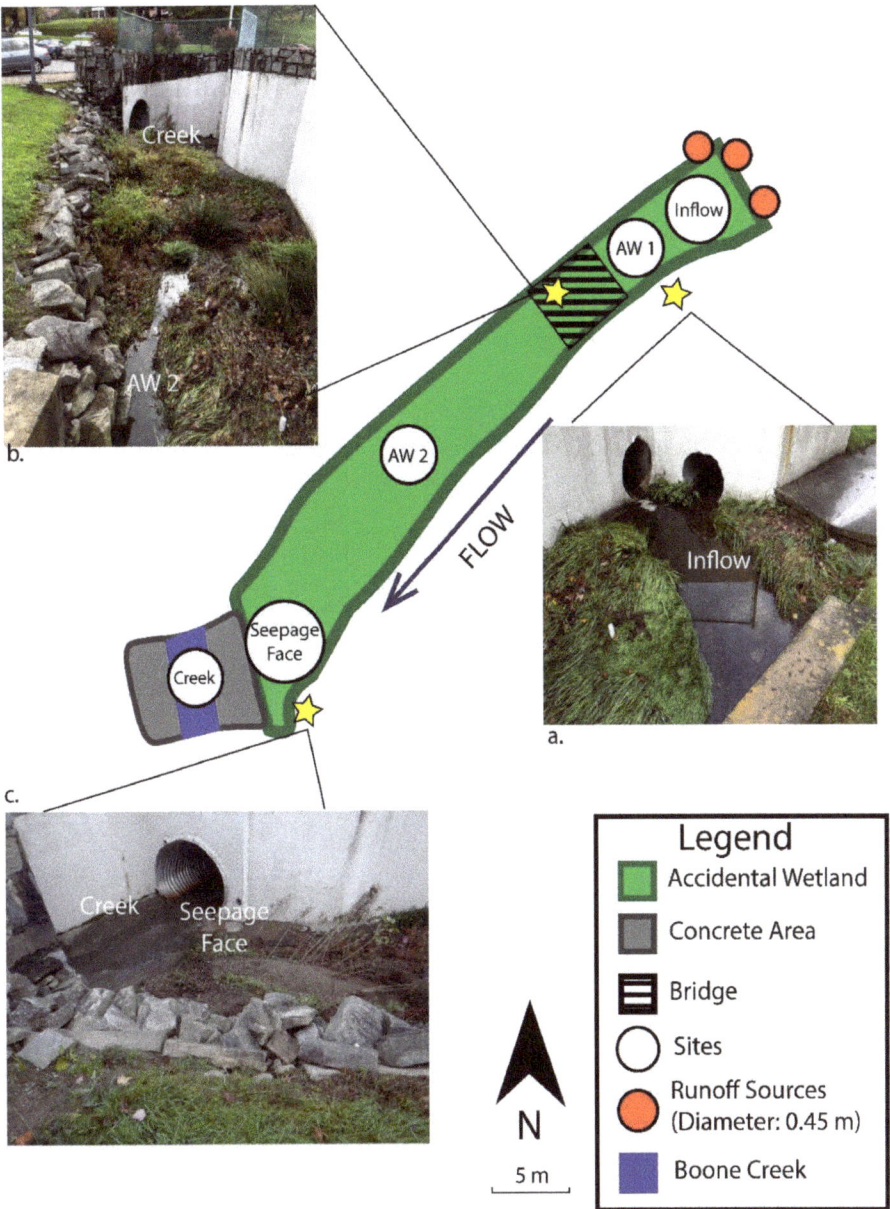

Figure 5. A map of the Accidental Wetland (AW). The stars indicate the locations from which the pictures were taken. Red circles indicate runoff surfaces at the inflow area of the wetland. The sites, denoted by white circles, were named Inflow, Accidental Wetland Site 1 (AW 1), Accidental Wetland Site 2 (AW 2), Seepage face, and Boone Creek. (**a**) Facing upstream towards the stormwater culverts, or the inflow sources. (**b**) Standing on bridge facing downstream towards seepage face and Boone Creek. (**c**) On grass facing Boone Creek stormwater culvert and the seepage face. The diameter of the culvert is 1.5 m.

2.2. Numerical Modeling of Groundwater Flow and Solute Transport

In this study, we used FEFLOW (DHI-WASY, DHI Group, Horsholm, Denmark) to simulate groundwater flow and solute transport through the wetland, utilizing salinity and runoff data from 2018 to formulate boundary conditions. FEFLOW is a three-dimensional finite-element groundwater flow and solute transport model that is well-documented in the literature and has been used to study subsurface solute transport (e.g., [40,41]). It has also been used in previous studies of the Boone Creek watershed [20]. The numerical model is not an attempt to recreate the empirical values collected from the accidental wetland through our sampling efforts; rather, the goal is to demonstrate with the model the groundwater flow and transport processes that are causing the decrease in salinity along the length of the wetland. The model does not account for any uptake of salt by the wetland plants or short-term retention by the wetland sediment, so the model output should be considered a conservative estimate of the effectiveness of the accidental wetland in delaying the arrival of salt-laden runoff.

Figure 6 shows the model domain that was used in groundwater flow and solute transport simulations for this study. The two-dimensional simulations represent a 21.2-meter-long by 0.2 m cross section of the accidental wetland. The base of the model is a no-flow boundary and represents the interface between the wetland and the concrete culvert. The stormwater inflow boundary at the right of the model domain, including two meters of the upper surface, is a time-dependent head and salinity boundary, the values of which are scaled to measurements taken at a long-term monitoring site on Boone Creek (see Figure 1 for the location of the Varsity Gym stream gauge relative to the accidental wetland). The seepage boundary on the left, representing the flow of water out of the wetland, is a time-dependent head boundary, with high water levels occurring only when a high stream stage in Boone Creek extends to the accidental wetland. A fluid flux boundary covers most of the top boundary of the model domain and represents the recharge to the mini-aquifer. Five simulated monitoring wells along the base of the model were used to collect output data for this paper. It was at these locations that we performed a semi-quantitative model calibration of our simulations to the field data.

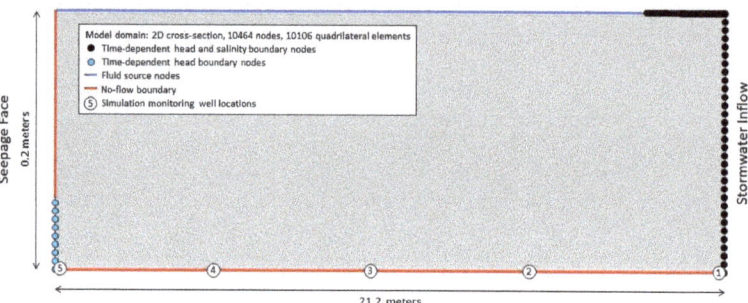

Figure 6. Model domain used in the flow and solute transport models. The flow enters on the right side of the figure, represented by black circles, and moves through the wetland to the seepage face in the left bottom corner, indicated by blue circles. The no-flow boundary (red line) represents the concrete culvert.

3. Results

3.1. Field Data

As shown in Figure 7, chloride values were highest at the upstream end of the wetland (at the inflow and AW1 and AW2 sites) and decreased with increasing distance into the accidental wetland. The highest salinities occurred at the first three sampling locations, and the lowest salinities left as discharge through the seepage face at the end of the accidental wetland (Table 1). As expected, the salinities decreased as the runoff moved downstream through the wetland, indicating that the wetland mitigated salinity by reducing peak values and delaying their arrival to the stream. About 47% of the time, the salinity at the seepage face was lower than the salinity at the inflow. This depended on the timing of the sampling because there may have been times when a fresh rainfall/runoff event occurred as a lagged salt event was still working its way through the wetland. Overall, however, as shown in Figure 7 and Table 1, the average and median chloride concentrations were highest at the upstream end of the wetland and were lowest at the downstream end of the wetland, near the seepage face.

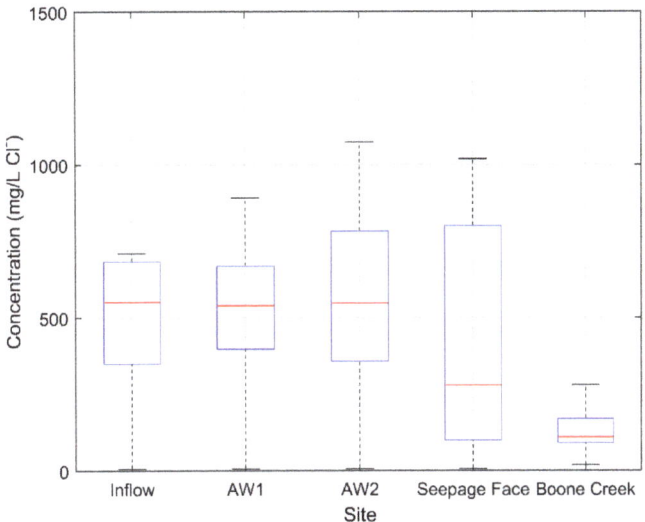

Figure 7. Boxplots of field chloride concentrations at each site throughout the length of the wetland. The middle red line marks the average chloride concentration. Chloride concentrations decrease towards the seepage face.

Table 1. Summary of chloride concentration statistics of the field measurements.

	Inflow (mg/L Cl−)	AW1 (mg/L Cl−)	AW2 (mg/L Cl−)	Seepage Face (mg/L Cl−)	Boone Creek (mg/L Cl−)
Mean	1637.7	1032.5	1060.2	316.3	254.3
Median	552.0	539.9	539.9	48.5	109.2
STD	3781.8	1995.7	1535.2	542.6	408.0

3.2. Synthetic Boundary Conditions

Salinity data, collected at the inflow point to the AW, were collected with a handheld probe. Salinity data were also collected on a continual basis at the VG stream gauging site, just downstream of the AW. A comparison of these data suggests that salinity values at the AW inflow during melt events were 4.5 times greater than those measured at the VG site (Figure 8). Because we did not have continuous data from the AW inflow, we created a synthetic dataset of accidental wetland inflow salinities by taking the VG salinity data and multiplying those values by 4.5 (Figure 9). In doing so, we are acknowledging that we are not studying the exact conditions arriving at the AW, but are instead using a synthetic time series that is consistent with the observed salinities at the inflow point. These scaled salinities were used in the numerical model as the inflow boundary conditions.

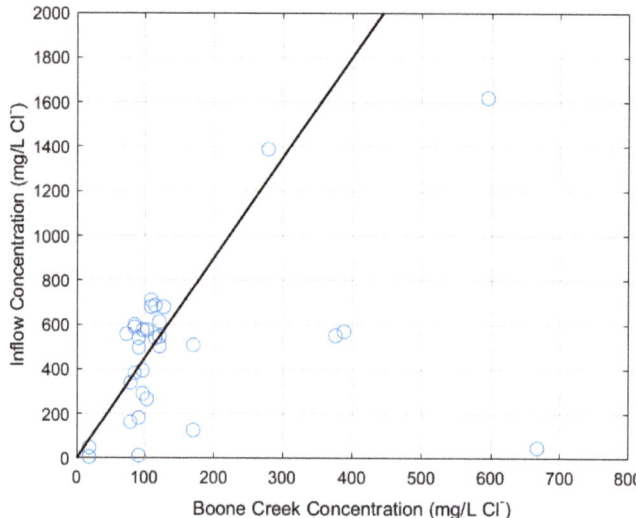

Figure 8. Relationship between the inflow concentrations of the Accidental Wetland and Boone Creek sites from hand samples. The black line represents the factor of 4.5 between the Boone Creek and inflow data. The open blue circles are the Boone Creek chloride concentration compared to inflow concentration collected during sampling.

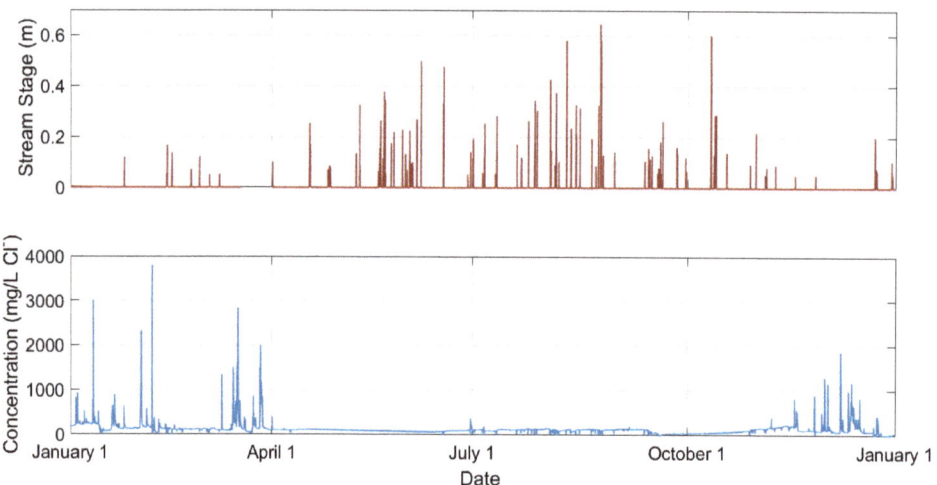

Figure 9. Stream stage–time plot (top panel) and concentration–time plot (bottom panel) from the Varsity Gym (VG) site from 1 January 2018 to 1 January 2019. The salinity data (bottom panel) were scaled by 4.5 to match concentrations observed in the Accidental Wetland and were used in the numerical model as inflow boundary conditions.

3.3. Numerical Modeling of Groundwater Flow and Solute Transport

The model output using the synthetic boundary condition data shows that the accidental wetland at the end of the stormwater culvert reduced peak salinity concentrations and delayed their arrival to Boone Creek. Dispersive transport properties in the wetland reduced peak salinity concentrations as the runoff water flowed as groundwater through the wetland. Here, groundwater refers to the flow of shallow subsurface water in the wetland substrate. There was no connection between the flow in the wetland and the natural water table due to the concrete culvert. The wetland groundwater velocities varied with the influx of water during storm and meltwater events. The gradients became high and in turn, increased the velocities, thus lowering the lag times; the storage of salt was of longer duration during dryer periods.

Initial simulations were carried out to examine the potential for a multi-year build-up of salt in the accidental wetland. To assess this, we ran a five-year simulation repeating our years' worth of scaled boundary data five times. Figure 10 shows the results of this simulation. The upper panel shows the scaled inflow data and the output seepage data. The lower panel shows the other simulated monitoring wells at a different vertical scale. The salinity concentration decreases through the wetland and approaches a concentration of zero during the summer months. The model output demonstrates that there is little, if any, salt accumulation in the wetland from season to season over the five years. This result is not surprising, given the small size of the wetland and the relatively fast circulation of water through the system.

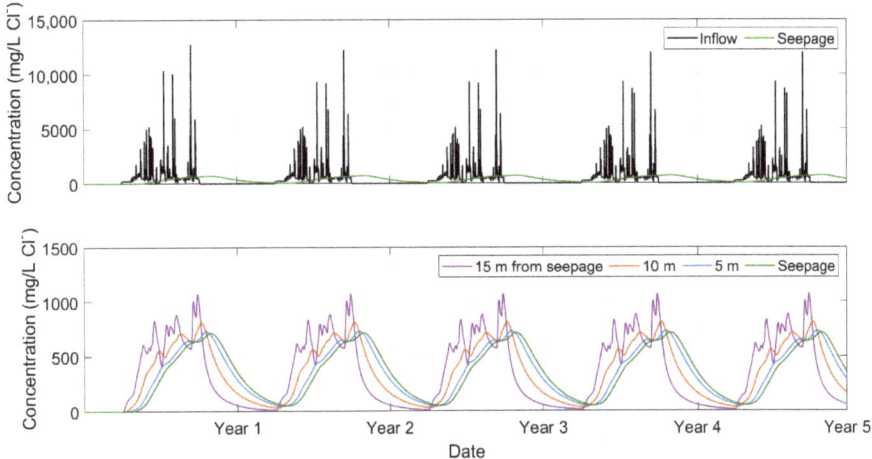

Figure 10. Concentration–time plots of output from the numerical model over five years of simulation time. The top panel shows the differences in salinity between the inflow to the wetland (black line) and the seepage face (dark green). The bottom panel shows the output from the simulated monitoring wells. Salinities decrease with distance travelled through the wetland, with the lowest salinities occurring at the seepage face.

Figure 11 shows simulation output, demonstrating the reduction in peak salinities as dispersion spreads out the salt plume with a transport distance downgradient. Figure 11 focuses on one year of the simulation output from Figure 10. It also demonstrates that there is a notable delay in the arrival of the center of mass of the plume to Boone Creek. The model output in the Figure compares the raw input data, represented by monitoring well 1, to the output data of the seepage face. Peak salinity values at the inflow lag up to approximately 45 days and have a peak salinity reduction of up to 94% by the time the solute reaches the seepage face. These reduced salinities and delayed arrival times, caused by typical solute transport processes in the accidental wetland, prevent the stormwater culvert from being a source of the high-salinity surge that is typical of meltwater events during 'salt season.'

We also performed model sensitivity analyses to understand the effects of hydraulic conductivity, which we had only estimated from Hazen Method calculations and adjusted through model calibration, on modeled salinity concentrations at the seepage face. We varied hydraulic conductivity in the modeled wetland while keeping other parameters, such as boundary conditions and longitudinal and transverse dispersivities, equivalent to the base case scenarios. As expected, higher hydraulic conductivity values increase the groundwater velocity, which in turn results in a quicker pulse of salinity through the wetland system and higher seepage salinities (Figure 12). Solute transport processes suggest that macrodispersion increases with groundwater velocity; however, the small scale of the aquifer limits the time available for dispersion to take place, and the result is an advection-dominant transport process. For example, at a simulated hydraulic conductivity of 19.8 m/d, peak concentrations rise above 730 mg/L Cl$^-$ and have a similar lag time to the base case simulation. Lower hydraulic conductivity values result in lower groundwater velocities, which increase the residence time of the salt in the aquifer. The low-permeability simulation, utilizing a hydraulic conductivity of 5 m/d, lagged the other sensitivity simulations by three months and had a peak concentration of roughly half of the base and high-permeability simulations.

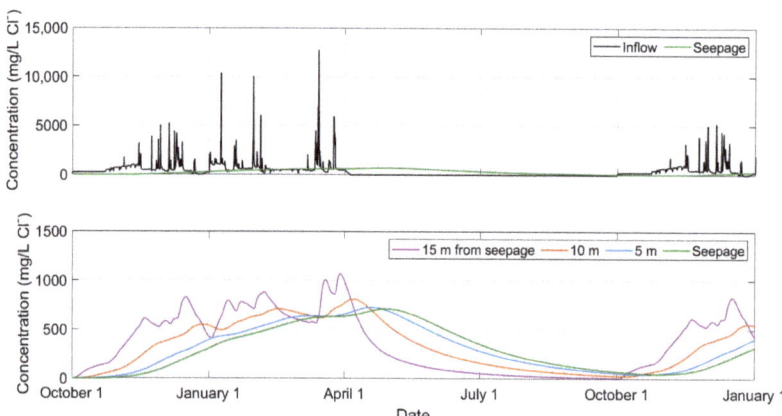

Figure 11. Concentration–time plot from the start of the typical salt season to show the mitigation of salt from the inflow to the seepage face. The top panel plots the inflow (black line) and seepage face (dark green line) salinities. The bottom panel plots the salinities of the different monitoring wells from the numerical model. Closer to the seepage face, lower salinities can be observed.

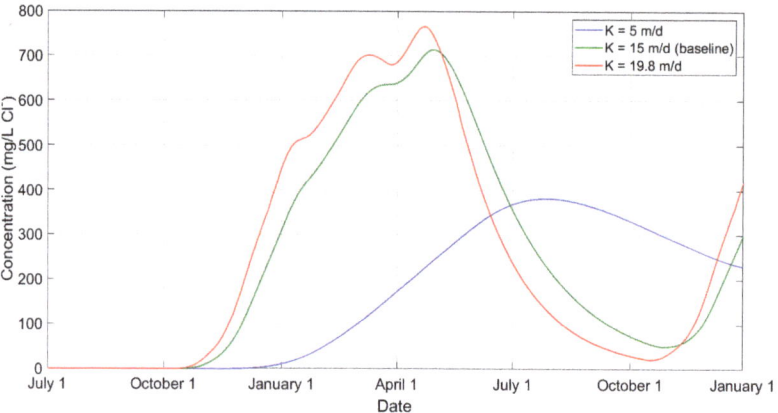

Figure 12. Concentration–time plot for the sensitivity analysis of the numerical model of salinities at the seepage face. K = 15 m/d (dark green) was the hydraulic conductivity used in the above simulations. The model is sensitive to changes in hydraulic conductivity.

4. Discussion

Our study demonstrates that accidental wetlands help remediate salt-laden stormwater runoff by increasing the residence time of the salt and delaying its arrival to a receiving stream, thereby decreasing the peak concentration of the salinity pulse but not the overall quantity of salt that is delivered to the stream.

Our field-sampled data support this conclusion. The median and mean values at the seepage face at the end of the wetland were lower in concentration than those at the upper end of the wetland, where salt-rich runoff entered the wetland during salt events. Our groundwater modeling and solute transport simulations also support this conclusion in a comprehensive way. The accidental wetland slows the delivery of salt-laden runoff, thereby lowering simulated salinities by up to 94 percent and lagging the timing of the peak concentrations by up to 45 days. This type of retention system does not remove the salt, but by slowing down its migration to the adjacent stream, it can lower the chance of chronic chloride violations and greatly reduce the potential for acute salinity spikes. This distinction between acute and chronic is important to stream health in the same way that dosage is relevant to any toxin. As noted previously, in their review of salinity impacts, Hintz and Relyea [17] found significant variability across species and across salinity levels. For example, they found that some species adapt to chronic salinity levels, although acute spikes are lethal. Because road salt will continue to enter streams, finding ways to at least lower the intense spikes of salt should be beneficial. If enough accidental wetlands are encouraged within urban watersheds, rather than removed from them, this could have a positive influence on the overall water quality in streams.

Furthermore, sensitivity analyses with the model showed that hydraulic conductivity values typical of sands are ideal in order for this to take place. This was the primary material in the accidental wetland at our study site and served as the base case for our simulations. It is safe to assume, however, that materials with properties too far outside of this range, such as gravels or clays, would not be effective. The hydraulic conductivity of highly-permeable materials such as gravels would allow rapid migration through the accidental wetland, preventing much of a reduction in peak concentrations or an increase in lag times. Conversely, silts and clays would have so little permeability that surface flow would dominate and would allow the quick delivery of the salt-laden urban runoff to the stream.

We think that there are likely other accidental wetlands similar to the one studied here in other environments; that is, a small scale, concrete substrate wetland forming in an area "designed" for another purpose. The formation of multiple accidental wetlands along a watershed would retain a higher percentage of the road salt, thus decreasing the concentration of the salinity pulses and preventing higher salinities in streams themselves. We can potentially improve the water quality of our urban streams by watching for and then encouraging accidental wetlands in multiple areas throughout a watershed.

4.1. Other Accidental Benefits

In addition to salinity concerns, accidental wetlands also influence other dynamic hydraulic issues such as stream flashiness and temperature surges in urban environments. Stream flashiness is caused by high percentages of impervious surfaces in the watershed, especially in the lower reaches in the Town of Boone and on the ASU campus, and results in rapid stream-stage increases and reversed-gradient events, which drive road salt into riparian aquifers along the length of urban stream reaches. These processes lead to both chronic and acute stream contamination over much longer time frames than what occurs in natural, unurbanized streams [42]. In our study location, urban, flashy runoff provides a significant source of contamination to the riparian aquifer along Boone Creek, a focus of years of urban hydrology research, and these groundwater–surface water interaction processes are especially important in relation to the chronic and acute chloride contamination of Boone Creek.

When stormwater runoff encounters an accidental wetland, the amount of runoff entering the stream decreases and lags according to the travel time that the wetland requires. Temperature surges occur when heated runoff enters a stream and quickly increases the stream temperature [20]. Depending on the formation of the accidental wetland, any detained water has time to cool, thereby reducing the temperature of hot runoff in temperature surges. If enough of these wetlands form, temperature surges will perhaps be greatly reduced or eliminated within a watershed. Long-term average stream temperatures are also improved by accidental wetlands via natural shading [3]; in fact, several small trees, along with other knee-high vegetation, did grow in the accidental wetland we studied. In addition to reducing impacts from salt, our case suggests that accidental wetlands in urban areas can reduce the effects of urban stream syndrome by mitigating flashiness and heat.

4.2. Management Implications

As already noted, campus facilities management removed the accidental wetland in 2018. Removing this wetland aligns with historic management approaches as well as continued perceptions about stormwater, flooding, and vegetated waterways. Traditional stormwater management has focused on shunting water as quickly as possible to the nearest stream to avoid flooding. Although new approaches, including constructing wetlands, bioswales, and retention systems, have been implemented, there remains a lack of understanding and misperceptions about these kinds of management techniques. In his overview of the literature, Everett [43] concluded that while there is some public awareness of alternative management approaches, it is not yet mainstream knowledge. People tend to focus on the amenity value or the aesthetics of entities like wetlands or bioswales rather than on their function. In some cases, this creates a negative feedback loop, in which people do not understand what the infrastructure is designed to do—or in this case what it accidentally does—and they therefore do not manage or maintain it appropriately, which leads to further negative views as aesthetics and/or function decline.

In our Boone Creek case, a misperception that the wetland was reducing the channel's ability to handle runoff and thereby increasing flooding risk was the rationale to remove it. This mistaken perception potentially reduced, rather than improved, both flood management and water quality. Because all of the vegetation and sediment was removed, runoff entering the culvert now goes straight into the stream, carrying any salt with it (Figure 2f). Because accidental wetlands arise without intention and are not part of explicit management plans, they can be ephemeral and can be quickly removed. If paying attention to when and where accidental wetlands form became part of overall urban stormwater management and planning, it could offer reduced flooding, improved water quality, and economic benefits, as the wetland costs nothing to establish.

5. Conclusions

The accidental formation of a wetland in a cold, urban environment inspired us to study the impacts of this system on mitigating road salt. Returning to our research questions for this conceptual study, we conclude that the accidental wetland decreased the peak concentrations of salt and chloride contamination and delayed the arrival of salt to the stream. Simple numerical simulations demonstrated that the saline pulse from urban runoff was delayed by up to 45 days and peak concentrations declined by up to 94%. Although the salt is not removed from the system, its arrival was delayed and the peak concentrations were lowered, ultimately improving water quality and potentially reducing negative consequences for some aquatic species.

We also think that accidental wetlands are likely to serve several functions and potentially offer multiple benefits to stream quality. For example, our model provided a conservative estimate regarding peak salinity reduction because we did not include uptake by plants. Future research assessing the role of vegetation and/or soil in contributing to reducing salinity impacts in accidental wetlands is warranted. Additionally, our results suggest that accidental wetlands may be effective in lowering runoff-induced thermal pollution due to the delay in the runoff entering the stream. Although this delay does not actually remove the salt, it most likely does reduce the temperature, offering a potentially significant improvement in cold-water streams. Again, more detailed analyses of this potential contribution from accidental wetlands would be valuable. Our work highlights the potential for accidental wetlands to improve water quality issues in urban streams. The only way to decrease salinization issues in cold regions is to discontinue the use of road salt during winter storm events. Because of safety concerns, totally eliminating salt is not practicable and therefore, we need to better understand and explore reactive measures to retain salt and delay its delivery into streams. Our study builds on previous work, suggesting that accidental wetlands cost nothing but offer multiple benefits, which are potentially as valuable as some stream restoration or intentional stormwater retention measures.

Author Contributions: Conceptualization, C.M.M. and W.P.A.J.; methodology, C.M.M. and W.P.A.J.; software, C.M.M. and W.P.A.J.; validation, C.M.M., W.P.A.J., and K.C.; formal analysis, C.M.M. and W.P.A.J.; investigation, C.M.M. and W.P.A.J.; resources, C.M.M. and W.P.A.J.; data curation, C.M.M. and W.P.A.J.; writing—original draft preparation, C.M.M. and W.P.A.J.; writing—review and editing, C.M.M., W.P.A.J., and K.C.; visualization, C.M.M. and W.P.A.J.; supervision, W.P.A.J. and K.C.; project administration, W.P.A.J. and K.C.; funding acquisition, C.M.M. and W.P.A.J. All authors have read and agreed to the published version of the manuscript.

Funding: This research received no external funding.

Institutional Review Board Statement: Not applicable.

Informed Consent Statement: Not applicable.

Data Availability Statement: Data are contained within the article and more detailed data may be requested from the authors.

Acknowledgments: The authors thank the Department of Geological and Environmental Sciences, the College of Arts and Science, and the Office of Student Research at Appalachian State University for support of the work. We thank undergraduate researchers Emily Fedders and Andrew Barringer for their help in maintaining the stream monitoring network. We also thank the comments and suggestions of four reviewers of a previous version of this paper.

Conflicts of Interest: The authors declare no conflict of interest.

References

1. Suchy, A. Denitrification in Accidental Urban Wetlands: Exploring the Roles of Water Flows and Plant Patches. Available online: https://core.ac.uk/download/pdf/79587123.pdf (accessed on 11 November 2020).
2. Sueltenfuss, J.P.; Cooper, D.J.; Knight, R.L.; Waskom, R.M. The Creation and Maintenance of Wetland Ecosystems from Irrigation Canal and Reservoir Seepage in a Semi-Arid Landscape. *Wetlands* **2013**, *33*, 799–810. [CrossRef]
3. Palta, M.M.; Grimm, N.B.; Groffman, P.M. "Accidental" Urban Wetlands: Ecosystem Functions in Unexpected Places. *Front. Ecol. Environ.* **2017**, *15*, 248–256. [CrossRef]
4. Brooks, B.W.; Riley, T.M.; Taylor, R.D. Water Quality of Effluent-Dominated Ecosystems: Ecotoxicological, Hydrological, and Management Considerations. *Hydrobiologia* **2006**, *556*, 365–379. [CrossRef]
5. Scheffers, B.R.; Paszkowski, C.A. Amphibian Use of Urban Stormwater Wetlands: The Role of Natural Habitat Features. *Landsc. Urban Plan.* **2013**, *113*, 139–149. [CrossRef]
6. Bateman, H.L.; Stromberg, J.C.; Banville, M.J.; Makings, E.; Scott, B.D.; Suchy, A.; Wolkis, D. Novel Water Sources Restore Plant and Animal Communities along an Urban River. *Ecohydrology* **2015**, *8*, 792–811. [CrossRef]
7. Downing, J.A. Emerging global role of small lakes and ponds: Little things mean a lot. *Limnética* **2010**, *29*, 9–24.
8. Van Meter, K.J.; Basu, N.B. Signatures of Human Impact: Size Distributions and Spatial Organization of Wetlands in the Prairie Pothole Landscape. *Ecol. Appl.* **2015**, *25*, 451–465. [CrossRef]
9. Suchy, A.K.; Palta, M.M.; Stromberg, J.C.; Childers, D.L. High Potential Nitrate Removal by Urban Accidental Wetlands in a Desert City: Limitations and Spatiotemporal Patterns. *Ecosystems* **2020**, *23*, 1227–1242. [CrossRef]
10. Palta, M. Urban "Accidental" Wetlands Mediate Water Quality and Heat Exposure for Homeless Populations in a Desert City. *AGU Fall Meet. Abstr.* **2015**, *21*, H21J-1528. Available online: https://ui.adsabs.harvard.edu/abs/2015AGUFM.H21J1528P/abstract (accessed on 10 November 2020).
11. Atkinson, R.B.; Cairns, J. Possible Use of Wetlands in Ecological Restoration of Surface Mined Lands. *J. Aquat. Ecosyst. Stress Recov.* **1994**, *3*, 139–144. [CrossRef]

12. Gu, C.; Cockerill, K.; Anderson, W.P.; Shepherd, F.; Groothuis, P.A.; Mohr, T.M.; Whitehead, J.C.; Russo, A.A.; Zhang, C. Modeling Effects of Low Impact Development on Road Salt Transport at Watershed Scale. *J. Hydrol.* **2019**, *574*, 1164–1175. [CrossRef]
13. Jin, L.; Whitehead, P.; Siegel, D.I.; Findlay, S. Salting Our Landscape: An Integrated Catchment Model Using Readily Accessible Data to Assess Emerging Road Salt Contamination to Streams. *Environ. Pollut.* **2011**, *159*, 1257–1265. [CrossRef] [PubMed]
14. Kelly, V.R.; Lovett, G.M.; Weathers, K.C.; Findlay, S.E.G.; Strayer, D.L.; Burns, D.J.; Likens, G.E. Long-Term Sodium Chloride Retention in a Rural Watershed: Legacy Effects of Road Salt on Streamwater Concentration. *Environ. Sci. Technol.* **2008**, *42*, 410–415. [CrossRef] [PubMed]
15. Lilek, J. Roadway Deicing in the United States. Available online: https://www.americangeosciences.org/geoscience-currents/roadway-deicing-united-states (accessed on 11 November 2020).
16. Kelly, V.R.; Cunningham, M.A.; Curri, N.; Findlay, S.E.; Carroll, S.M. The Distribution of Road Salt in Private Drinking Water Wells in a Southeastern New York Suburban Township. *J. Environ. Qual.* **2018**, *47*, 445–451. [CrossRef] [PubMed]
17. Hintz, W.D.; Relyea, R.A. A Review of the Species, Community, and Ecosystem Impacts of Road Salt Salinisation in Fresh Waters. *Freshw. Biol.* **2019**, *64*, 1081–1097. [CrossRef]
18. Burgis, C.R.; Hayes, G.M.; Henderson, D.A.; Zhang, W.; Smith, J.A. Green Stormwater Infrastructure Redirects Deicing Salt from Surface Water to Groundwater. *Sci. Total Environ.* **2020**, *729*, 138736. [CrossRef]
19. Oswald, C.J.; Giberson, G.; Nicholls, E.; Wellen, C.; Oni, S. Spatial Distribution and Extent of Urban Land Cover Control Watershed-Scale Chloride Retention. *Sci. Total Environ.* **2019**, *652*, 278–288. [CrossRef]
20. Cockerill, K.; Anderson, W.P.; Harris, F.C.; Straka, K. Hot Salty Water: A Confluence of Issues in Managing Stormwater Runoff for Urban Streams. *JAWRA J. Am. Water Resour. Assoc.* **2017**, *53*, 707–724. [CrossRef]
21. Kratky, H.; Li, Z.; Chen, Y.; Wang, C.; Li, X.; Yu, T. A Critical Literature Review of Bioretention Research for Stormwater Management in Cold Climate and Future Research Recommendations. *Front. Environ. Sci. Eng.* **2017**, *11*, 16. [CrossRef]
22. Barbier, L.; Suaire, R.; Duricković, I.; Laurent, J.; Simonnot, M.-O. Is a Road Stormwater Retention Pond Able to Intercept Deicing Salt? *Water Air Soil Pollut.* **2018**, *229*, 251. [CrossRef]
23. Fanelli, R.M.; Prestegaard, K.L.; Palmer, M.A. Urban Legacies: Aquatic Stressors and Low Aquatic Biodiversity Persist despite Implementation of Regenerative Stormwater Conveyance Systems. *Freshw. Sci.* **2019**, *38*, 818–833. [CrossRef]
24. Cooper, C.A.; Mayer, P.M.; Faulkner, B.R. Effects of Road Salts on Groundwater and Surface Water Dynamics of Sodium and Chloride in an Urban Restored Stream. *Biogeochemistry* **2014**, *121*, 149–166. [CrossRef]
25. Søberg, L.C.; Viklander, M.; Blecken, G.-T. Do Salt and Low Temperature Impair Metal Treatment in Stormwater Bioretention Cells with or without a Submerged Zone? *Sci. Total Environ.* **2017**, *579*, 1588–1599. [CrossRef] [PubMed]
26. Lange, K.; Österlund, H.; Viklander, M.; Blecken, G.-T. Metal Speciation in Stormwater Bioretention: Removal of Particulate, Colloidal and Truly Dissolved Metals. *Sci. Total Environ.* **2020**, *724*, 138121. [CrossRef]
27. Taguchi, V.J.; Weiss, P.T.; Gulliver, J.S.; Klein, M.R.; Hozalski, R.M.; Baker, L.A.; Finlay, J.C.; Keeler, B.L.; Nieber, J.L. It Is Not Easy Being Green: Recognizing Unintended Consequences of Green Stormwater Infrastructure. *Water* **2020**, *12*, 522. [CrossRef]
28. Brydon, J.; Roa, M.C.; Brown, S.J.; Schreier, H. Integrating Wetlands into Watershed Management: Effectiveness of Constructed Wetlands to Reduce Impacts from Urban Stormwater. In Proceedings of the Environmental Role of Wetlands in Headwaters; Krecek, J., Haigh, M., Eds.; Springer: Dordrecht, The Netherlands, 2006; pp. 143–154.
29. Mangangka, I.R.; Egodawatta, P.; Parker, N.; Gardner, T.; Goonetilleke, A. Performance Characterisation of a Constructed Wetland. *Water Sci. Technol.* **2013**, *68*, 2195–2201. [CrossRef]
30. Land, M.; Granéli, W.; Grimvall, A.; Hoffmann, C.C.; Mitsch, W.J.; Tonderski, K.S.; Verhoeven, J.T.A. How Effective Are Created or Restored Freshwater Wetlands for Nitrogen and Phosphorus Removal? A Systematic Review. *Environ. Evid.* **2016**, *5*, 9. [CrossRef]
31. Zu Ermgassen, S.O.S.E.; Baker, J.; Griffiths, R.A.; Strange, N.; Struebig, M.J.; Bull, J.W. The Ecological Outcomes of Biodiversity Offsets under "No Net Loss" Policies: A Global Review. *Conserv. Lett.* **2019**, *12*, e12664. [CrossRef]
32. Copeland, S. On Serendipity in Science: Discovery at the Intersection of Chance and Wisdom. *Synthese* **2019**, *196*, 2385–2406. [CrossRef]
33. Rice, J.S.; Anderson, W.P., Jr.; Thaxton, C.S. Urbanization Influences on Stream Temperature Behavior within Low-Discharge Headwater Streams. *Hydrol. Res. Lett.* **2011**, *5*, 27–31. [CrossRef]
34. NC DEQ: Classifications. Available online: https://deq.nc.gov/about/divisions/water-resources/planning/classification-standards/classifications#DWRPrimaryClassification (accessed on 19 November 2020).
35. Benoit, D. *Ambient Water Quality Criteria for Chloride—1988*; EPA: Washington, DC, USA, 1988; Volume 988, p. 47.
36. Yu, Z.; Qi, Z.; Hu, C.; Liu, W.; Huang, H. Effects of Salinity on Ingestion, Oxygen Consumption and Ammonium Excretion Rates of the Sea Cucumber Holothuria Leucospilota. *Aquac. Res.* **2013**, *44*, 1760–1767. [CrossRef]
37. Stringer, C.E.; Rains, M.C.; Kruse, S.; Whigham, D. Controls on Water Levels and Salinity in a Barrier Island Mangrove, Indian River Lagoon, Florida. *Wetlands* **2010**, *30*, 725–734. [CrossRef]
38. Palmer, J.J. *How to Brew: Everything You Need to Know to Brew Great Beer Every Time*; Brewers Publications: Kent, OH, USA, 2017; ISBN 978-1-938469-35-0.
39. Hazen, A. Discussion of Dams on Sand Foundations by A. C. Koenig. *Trans. Am. Soc. Civ. Eng.* **1911**, *73*, 199–203.
40. Trefry, M.G.; Muffels, C. FEFLOW: A Finite-Element Ground Water Flow and Transport Modeling Tool. *Groundwater* **2007**, *45*, 525–528. [CrossRef]
41. Sivakumar, C.; Elango, L. Application of Solute Transport Modeling to Study Tsunami Induced Aquifer Salinity in India. *J. Sci. Technol. Environ.* **2010**, *15*, 33–41. [CrossRef]

42. Anderson, W.P.; Storniolo, R.E.; Rice, J.S. Bank Thermal Storage as a Sink of Temperature Surges in Urbanized Streams. *J. Hydrol.* **2011**, *409*, 525–537. [CrossRef]
43. Everett, G. Public Perceptions of Sustainable Drainage Devices. In *Sustainable Surface Water Management*; John Wiley & Sons, Ltd.: Hoboken, NJ, USA, 2016; pp. 285–297, ISBN 978-1-118-89769-0.

Article

Modeling Shallow Urban Groundwater at Regional and Local Scales: A Case Study in Detroit, MI

Sadaf Teimoori *, Brendan F. O'Leary and Carol J. Miller

Department of Civil and Environmental Engineering, College of Engineering, Wayne State University, 5050 Anthony Wayne Drive, Detroit, MI 48202, USA; ax9873@wayne.edu (B.F.O.); ab1421@wayne.edu (C.J.M.)
* Correspondence: sadaf.teimoori@wayne.edu

Abstract: Groundwater plays a significant role in the vitality of the Great Lakes Basin, supplying water for various sectors. Due to the interconnection of groundwater and surface water features in this region, the groundwater quality can be affected, leading to potential economic, political, health, and social issues for the region. Groundwater resources have received less emphasis, perhaps due to an "out of sight, out of mind" mentality. The incomplete characterization of groundwater, especially shallow, near-surface waters in urban centers, is an added source of environmental vulnerability for the Great Lakes Basin. This paper provides an improved understanding of urban groundwater to reduce this vulnerability. Towards that end, two approaches for improved characterization of groundwater in southeast Michigan are employed in this project. In the first approach, we construct a regional groundwater model that encompasses four major watersheds to define the large-scale groundwater features. In the second approach, we adopt a local scale and develop a local urban water budget with subsequent groundwater simulation. The results show the groundwater movement in the two different scales, implying the effect of urban settings on the subsurface resources. Both the regional and local scale models can be used to evaluate and mitigate environmental risks in urban centers.

Keywords: groundwater flow; urban groundwater; numerical modeling; water budget

Citation: Teimoori, S.; O'Leary, B.F.; Miller, C.J. Modeling Shallow Urban Groundwater at Regional and Local Scales: A Case Study in Detroit, MI. *Water* **2021**, *13*, 1515. https://doi.org/10.3390/w13111515

Academic Editor: C. Radu Gogu

Received: 24 April 2021
Accepted: 25 May 2021
Published: 28 May 2021

Publisher's Note: MDPI stays neutral with regard to jurisdictional claims in published maps and institutional affiliations.

Copyright: © 2021 by the authors. Licensee MDPI, Basel, Switzerland. This article is an open access article distributed under the terms and conditions of the Creative Commons Attribution (CC BY) license (https://creativecommons.org/licenses/by/4.0/).

1. Introduction

Groundwater serves a vital role in supplying drinking water and providing essential services such as cooling water for power generation, irrigation water for farms and landscapes, and industrial water in the Great Lakes Basin (GLB). Groundwater quality issues can lead to environmental and human health problems and can challenge environmental management. Groundwater exists in hidden natural reservoirs (aquifers) that gradually deliver water to many other surface water resources such as lakes, streams, and wetlands. Due to groundwater's slow transport rates and the lack of a visible interface, there is often a tremendous lag between a contamination incident and the recognition of a problem [1]. To that end, protecting and remediating groundwater is often far more complex and costly than the same efforts for surface water bodies [2–4]. Regulatory efforts on groundwater primarily focus on contamination from hazardous chemicals and typically rely on a site-by-site approach. The high financial cost and piecemeal approach to groundwater remediation has led to a research gap in understanding of urban groundwater movement in cities that rely on surface water in the GLB [5]. This study reviews known and expanding information about groundwater resources in the Detroit metro region, the principal urban area in Michigan, as well as one of the largest in the GLB. In this region, groundwater is recharged almost entirely from rainfall and snowmelt that infiltrates down to the water table. The hydrologic interconnection between the groundwater and surface water systems leads to the mixing of subsurface and surface contaminants. Examples of these subsurface contaminants are volatile organic compounds (VOCs), polychlorinated biphenyls (PCBs), polynuclear aromatic compounds (PNAs), and metals [6]. Groundwater, therefore, can serve as a pollutant transport mechanism and can ultimately expose Detroiters to contaminants such as VOCs via numerous routes, including vapor intrusion and ingestion [7–9] (Figure 1). Ingestion can occur through direct consumption of vegetation "fed" by contaminated vadose zone water or by consuming water impacted by groundwater contamination [10–12].

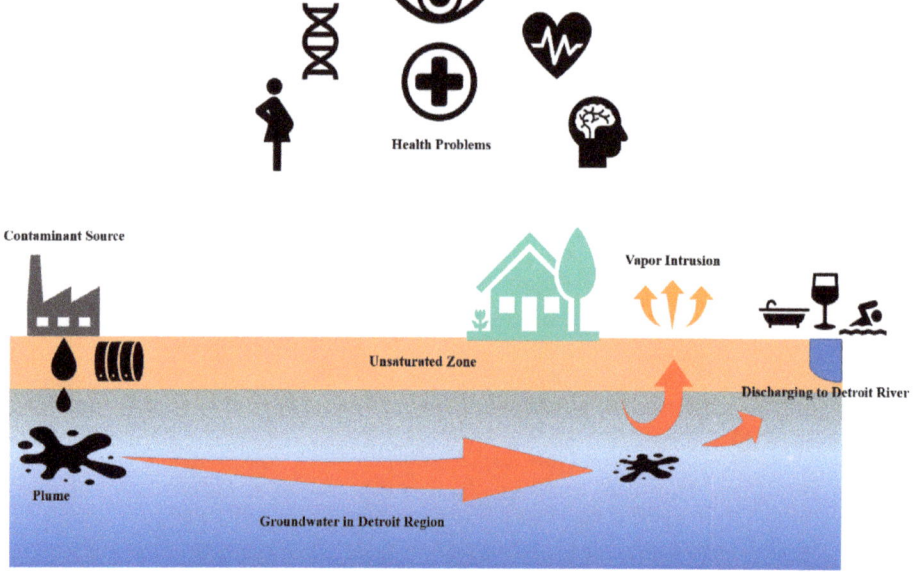

Figure 1. Hypothetical scheme of hydrogeological cycle in the Detroit region.

The quality of urban groundwater is critical as it can influence human health issues. Groundwater in areas such as Detroit with shallow urban groundwater systems is physically near human activities and can contribute to short exposure routes due to the short vertical distances. Urban disturbances influence subsurface hydrologic pathways, which make understanding subsurface exposure routes difficult [13,14]. While there has been some research performed on the groundwater quality in southeast Michigan [15–20], further studies are required to evaluate migration pathways and contaminants in shallow urban groundwater. For example, the VOCs group is one of these chemical pollutants, of which the adverse impacts of their pathways on human health are well-documented [21–25]. Preterm birth is a crucial health issue within the Detroit region [26]. Detroit has the nation's highest preterm birth rate among America's largest cities in 2018 [27]. A recent study links VOC exposure to higher preterm birth rates [28]. Therefore, a comprehensive investigation of groundwater to evaluate groundwater flow direction, transmissivity, and the depth of the water table is imperative to explain urban water pathways.

This paper aims to describe the urban shallow groundwater of the metro Detroit region by utilizing public data from multiple sources to generate a series of models at differing resolutions. This paper includes a review of the geologic setting and the development of two groundwater models at regional and local scales aimed to provide an initial assessment of urban water movement in coastal Great Lakes cities. Through developing an understanding of groundwater, potential pollutant exposure pathway risks posed to urban environments and human health are better understood.

2. Materials and Methods

Regional and local modeling provide a baseline to assess urban groundwater in southeast Michigan. Regional datasets for southeast Michigan were assessed to provide a general understanding of groundwater flow direction, discharge, and depth to groundwater. A narrower, neighborhood-scale evaluation was completed for RecoveryPark, providing a case study to evaluate localized urban water budgets and local-scale groundwater flow. Both datasets provide useful but different contributions to the understanding of urban groundwater movement.

The paper highlights the use of readily available data from multiple sources to understand groundwater movement at the regional and local scales. While there is no comprehensive database for urban shallow subsurface information, there are several studies and data centers that provide information regarding urban groundwater movement. This paper brings data together from historical well logs, current ongoing field studies, and simulations to provide both a regional and local understanding of shallow subsurface groundwater movement in Metro Detroit. MODFLOW [29] was used to simulate the steady-state groundwater head distributions for both the regional- and local-scale applications. ESRI ArcMap 10.7.1 was used for data interpolation and mapping.

2.1. Southeast Michigan Regional and Local Field Setting

Within the Detroit regional watershed, the subsurface sediment layers are dominated by a large clay diamicton [6,30,31], which supports the presence of shallow groundwater. The amount of water transferring through this shallow system is not well-quantified since most of the historic water resource studies have focused on deep groundwater systems occurring at depths of more than six meters. Although the shallow groundwater does not supply a significant source of drinking water for the residents in Detroit, the position of the groundwater table can significantly affect the design and requirements of sewer systems, drinking water networks, and surface infrastructures. More importantly, the depth of the groundwater table directly controls the thickness of the vadose zone and the separation distance between surface features and groundwater features.

2.1.1. Regional Scale: General Depositional Environment and Drainage Characteristics

The Detroit metropolitan area predominantly overlies glacial lacustrine deposits producing scattered small to moderate quantities of water in some locations [20]. Detroit's till dates from the Wisconsin age and varies in thickness and composition [32]. As shown in Figure 2, the thickness of glacial sediments gradually decreases to the southeast and the area along south to the east is mainly clay mixed with isolated beaches, terraces, and lenses of gravel and sand [30,31]. The glacial drift of the region consists of irregular beds of gravel, sand, silt, and clay [6] and may be cemented by iron oxide or carbonate. In southeast Michigan, the bedrock consists of Devonian age Antrim Shale, Traverse Group, which is a mix of shale and limestone, Dundee Limestone, and Detroit River Dolomite. Given the thickness of the clay lacustrine deposits, clay composition hydrologically generates a disconnected media in the deposits underground. The bedrock underlying the Detroit metropolitan area is not a reliable source of groundwater due to the low permeability of the soil and poor water quality.

Historical hydrogeological studies relied on water well logs and surficial and bedrock geologic maps. These studies were primarily conducted for groundwater exploitation and did not typically address near-surface aquifers because they were not generally considered potential potable water sources. Leverett [33] conducted the first detailed study of the groundwater resources of southeastern Michigan. Leverett's study was prompted by a potential shortage of available groundwater resources in the central and eastern portions of Wayne County. Hydrogeologic studies of southeastern Michigan rely primarily on water well logs. The reliability of these water well logs is limited by the lack of uniformity among drillers, absence of detailed subsurface lithologic descriptions, and insufficient geographic distribution of wells. The stratigraphic correlation of near-surface sediments in southeastern Michigan is complicated because of the complex nature of sedimentation by both ice and water [34]. Groundwater flows toward the surface water within the sand units, and if present, is generally found at a depth ranging from ~one to three meters from the land surface [6,31]. However, in Metro Detroit, the perched or discontinuous groundwater is usually encountered in the upper one meter due to the predominant underlying clay unit. Mozola [30] suggested that the source of groundwater recharge for sand, gravel, and coble layer may be the Defiance Moraine; however, this has not been confirmed. Howard [32] presented a surficial geologic map of the Detroit quadrangle, which can serve as a framework for assessing and redeveloping future urban sites. The final maps, including stratigraphic sections and soil layers across Metro Detroit, are useful for expanded studies on groundwater and the vadose zone beneath this region.

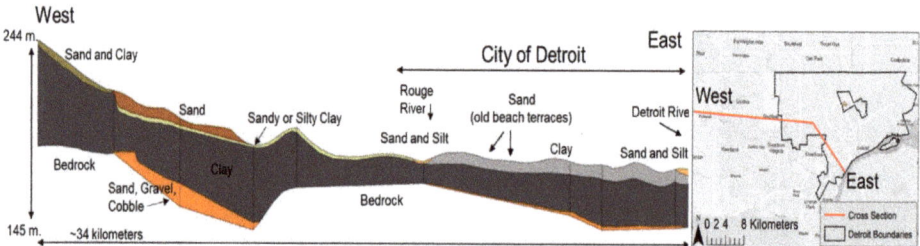

Figure 2. Interpolated surficial deposits cross-section of metropolitan Detroit (based on a cross-section from [6]).

2.1.2. Local Scale: RecoveryPark Field Site

RecoveryPark is a nonprofit urban farm in the Poletown neighborhood of Detroit that provides local agricultural employment for residents [35] (Figure 3). While RecoveryPark is located in an urban neighborhood, a majority of the lots are cleared of debris and vacant. Apart from the existing or historic utility infrastructure and building basements, no other major subsurface features are present.

The United States Geological Survey (USGS) and the United States Environmental Protection Agency (USEPA) collaborated to study the effectiveness of green stormwater infrastructure (GSI) at this site since 2014. A water cycle monitoring approach was used to assess the role of GSI in a larger hydrologic context. The property includes a weather station and the only publicly available groundwater wells in the city of Detroit. The weather station was initially located north of the study (first location: 2014 to 2017) and then was moved to its current onsite location during fall 2017 (second location: 2017 to present). Soil boring data and in-pipe flow meter data from the site are available from the Detroit Water and Sewage Department (DWSD). Overall, there are 23 groundwater wells and seven in-pipe sewer meters spread over a 1.686×10^{-1} km^2 area (Figure 4). The onsite sewer lines appearing in Figure 4 transported only local sewage (no off-site areas contributed), and the sewer output all flowed through meters E and F before exiting the study boundaries. Soil moisture sensors were added to the field equipment in 2017 and are located near the weather station. The location and head value of observation wells are also presented in Table 1.

Figure 3. Location map for RecoveryPark. City of Detroit boundaries are shown by the dashed lines, and RecoveryPark location is shown by the red dot.

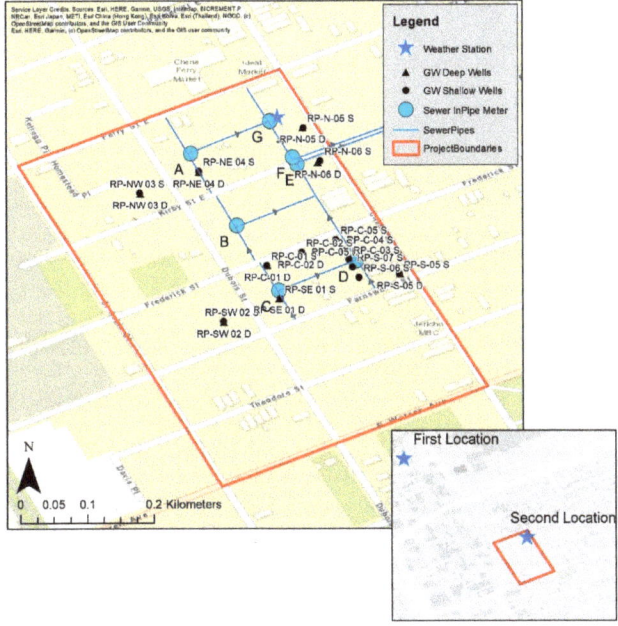

Figure 4. Site map at RecoveryPark with shallow wells, deeper wells, sewer in-pipe meter locations, and weather station. Wells start with RP, and in-pipe sewer meters are alphabetically labeled.

RecoveryPark is a unique urban study area in that it includes one of the only urban well fields in the Great Lakes Basin [36]. The insights gleaned from this location offer a detailed understanding of one localized neighborhood but still does not remedy the larger regional knowledge gap. The lack of publicly available urban well data presents issues for groundwater mapping both at the regional and local scales. This problem is not unique to the metropolitan Detroit region. Most major cities in the United States lack open-source groundwater field data [5].

Table 1. Location and head value of observation wells at RecoveryPark (USGS [37]).

Type	Name	Alias	X [1]	Y [1]	Top Elevation [2,3] (m)	Depth to Water Table (m)	Observed Value (m)
Deeper Observation Wells	RP-C-01 D	C1D	−83.0437139	42.3672917	192.27698	1.8212	190.4558
	RP-C-02 D	C2D	−83.0432556	42.3674583	192.10325	1.0790	191.0243
	RP-C-05 D	C5D	−83.0427917	42.3676222	191.9417	1.3122	190.6295
	RP-N-05 D	N5D	−83.0432250	42.3691139	192.32948	2.4201	189.9094
	RP-N-06 D	N6D	−83.0430194	42.3686500	191.44782	1.4356	190.0122
	RP-S-05 D	S5D	−83.0419250	42.3671778	191.65519	1.9111	189.7441
	RP-SE-01 D	SE1D	−83.0435528	42.3668583	192.49644	1.3457	191.1507
	RP-SW-02 D	SW2D	−83.0443111	42.3665417	193.4044	2.2906	191.1138
	RP-NE-04 D	NE4D	−83.0446444	42.3685250	192.13373	1.9126	190.2211
	RP-NW-03 D	NW3D	−83.0430194	42.3686500	192.90198	No data recorded	No data recorded
Shallow Observation Wells	RP-C-01 S	C1S	−83.0437194	42.3672889	192.27394	0.6614	191.6125
	RP-C-02 S	C2S	−83.0432444	42.3674611	191.28435	0.1798	191.1045
	RP-C-03 S	C3S	−83.0426111	42.3673639	192.60767	1.7450	190.8627
	RP-C-04 S	C4S	−83.0427000	42.3674972	191.92951	1.0973	190.8322
	RP-N-05 S	N5S	−83.0432194	42.3691056	192.40397	2.0604	190.3435
	RP-N-06 S	N6S	−83.0429889	42.3686722	191.44349	0.9876	190.4559
	RP-NE-04 S	NE4S	−83.0446444	42.3685333	192.48642	1.4722	191.0142
	RP-NW-03 S	NW3S	−83.0454500	42.3682472	192.35318	1.1841	191.1690
	RP-S-05 S	S5S	−83.0419333	42.3671750	191.63081	0.8915	190.7393
	RP-S-06 S	S6S	−83.0424778	42.3671306	192.50219	1.7054	190.7968
	RP-SW-02 S	SW2S	−83.0443139	42.3665472	193.04669	1.4402	191.6065
	RP-SE-01 S	SE1S	−83.0435444	42.3668500	192.50558	0.7254	191.7802
	RP-S-07 S	S7S	−83.0425611	42.3672667	192.8624	1.9888	190.8736
	RP-C-05 S	C5S	−83.0427833	42.3676250	191.9824	1.0287	190.9537

[1] NAD83. [2] above NAVD88. [3] Land surface elevation has been adjusted based on the surface elevation raster data applied to the model.

2.2. Regional Scale Groundwater Model: Metro Detroit Watersheds

The focus of this investigation is metropolitan Detroit, with an area of 3850.32 km^2 and encompassing four major watersheds, all discharging to the Detroit River. The four watersheds (Clinton, Lake St. Clair, Rouge, and Ecorse Creek) are shown in Figure 5. In this region, sink and source elements, including precipitation, evapotranspiration, pumping wells, rivers, and lakes, affect the groundwater flow, quantity, and quality. The values of hydraulic conductivity are obtained from the borehole logs dataset [6,37] and presented in Table 2 for all five units simulated in the model. The ratio of anisotropy (K_H/K_V) is assumed to be equal to 1 for all five soil units. Precipitation and evapotranspiration values are available from the USGS [37] data sources; average annual rates are used in these simulations. In order to simulate the groundwater recharge coming from surface waters, the head stages for the two main rivers of the region, Clinton River and Rouge River, are obtained from the USGS [37] and applied to the model. Furthermore, lake water surface and bottom elevation datasets are collected from the Michigan Department of Natural Resources [38] and applied to all 47 lakes, which have an area greater than 0.3 km^2. Pumping well data were obtained from groundwater datasets for Michigan [39] and categorized into five groups, including industrial, irrigation, household, commercial/institutional, and public-supply wells. Within the study area, there are 12,866 active pumping wells, mostly located in the northwestern portion of the study area.

Table 2. Hydraulic conductivity of stratigraphic units assumed in the model.

Stratigraphic Unit	Hydraulic Conductivity (m/day)
Moraine Unit	8.64×10^{-1}
Sand Clay Unit	8.64×10^{-2}
Sand Unit	8.64
Sandy and Silty Clay Unit	8.64×10^{-2}
Clay Unit	8.64×10^{-3}

The developed conceptual model of regional groundwater contains the shape, discharge and recharge sources, and boundary conditions, including general head boundary (representing water elevation at Lake St. Clair and Detroit River) and no-flow boundary. The model grid is set up in one layer with 10,584 active cells of various sizes and shapes (Ugrid–Voronoi [40]). The surface elevation ranges from ~173 m in the southeast to ~325 m in the northwest (NAVD88), while the bottom elevation ranges from ~144 m to ~282 m. Due to the lack of transient data of pumping wells, the simulation is conducted in a steady state.

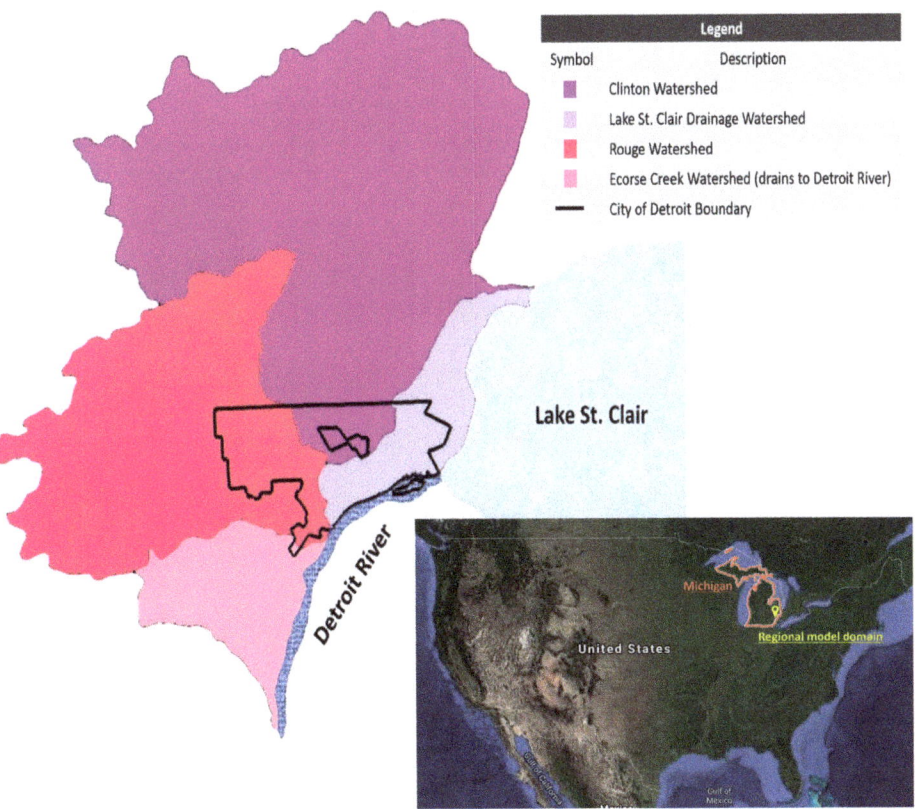

Figure 5. The boundary of Detroit and watersheds in the study area (regional model domain) [41].

The regional groundwater model has been developed to evaluate the groundwater head distribution based on the control volume finite difference (CVFD) formulation using MODFLOW-USG [42]. Groundwater Modeling System 10.3.2 (GMS) was used as the pre- and post-processing software tool to construct and run the MODFLOW model. Datasets needed for this model include meteorological (precipitation, evaporation, etc.), hydraulic and geochemical attributes, and topographic elevation.

2.3. Local Scale Model: RecoveryPark

RecoveryPark is outfitted with a real-time weather station that includes multiple instruments. The weather station is comprised of a Campbell Scientific evapotranspiration station (ET-107), solar radiation sensor (CS305-ET), air temperature and relative humidity probe (HMP60-ETS), tipping bucket rain gage (TE525-ET), and wind set (034B-ETM), combining a three-cup anemometer and vane into a single integrated package to measure wind speed and direction.

The well groundwater level was recorded using a Schlumberger Diver pressure transducer at 23 individual wells nested with deeper (~six meters below ground surface) and shallow (~three meters below ground surface) wells (Figure 6). The DWSD performed the in-pipe flow measurements at six different locations with meters E and F measuring the discharge out of the RecoveryPark field site. DWSD used V-notch weirs with velocity and stage sensors to generate velocity and discharge values [36]. Eight Campbell Science 655 TDR probes were situated at three different intervals close to the weather station. These were positioned near the surface, at 0.3 m below the surface level and 0.5 m below the surface level. Figure 6 shows the vertical distribution of the monitored subsurface features.

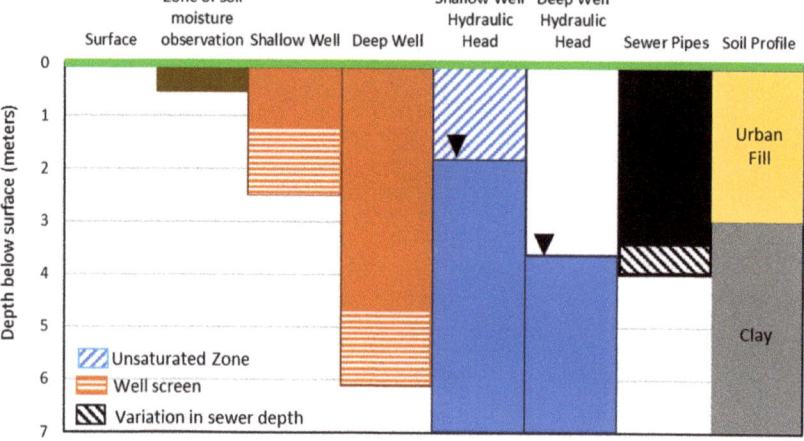

Figure 6. Subsurface vertical profile of observed features at the RecoveryPark study site. Hydraulic head is an example hydraulic head based on the average October 2015 well data.

2.3.1. Urban Water Budget

Characterizing urban water budgets is a challenge because of the overlap between groundwater movement and subsurface anthropogenic structures such as sewer lines. In order to better understand these changes, an urban water budget model was developed for the RecoveryPark field site. Periods of rainfall and seasonal variability impact groundwater. In order to account for seasonality and consistency from precipitation events, water budgets were constructed for each quarter around a rainfall event. Four separate rain events were picked to evaluate changes in RecoveryPark's groundwater storage.

This study applied a mass balance approach to quantify the water budget at the RecoveryPark study area. The area has been extensively reworked to channel surface water flow into sewer drains and onsite GSI. Runoff over the study area is assumed to flow into the sewer drains or is infiltrated into groundwater. Given the locations of the GSI within the study region, interactions between groundwater infiltration from the surface and fluxes in and out of the sewer piping is assumed to be contained within well data and sewer flow data. The shallow wells, located in urban fill, and the deeper wells, located in mostly clay-rich sediments, showed similar fluctuations in hydraulic head levels, indicating that the well sets are hydrologically connected. Referring to Howard and Olszewska [43], the urban fill materials are compacted loamy texture probably originated from a combination of sandy surficial deposits and clayey diamicton family. This urban fill was later mixed with various construction and demolition artifacts due to the urban redevelopment in Detroit city. These artifacts and waste building materials were weathered over the years and left the soil with some significant contaminants. It is worth mentioning that, at RecoveryPark, groundwater has been observed through two sets of observation wells with two different lengths. This observation reflects the field data, where nested shallow and deeper wells show similar changes in hydraulic head. This is also as the shallow well data was used for the water budget analysis. Changes in soil moisture in the vadose zone impact groundwater movement, especially in shallow systems such as Detroit. The changes to hydraulic head in the shallow wells did not account for soil moisture. However, soil moisture influence is reviewed by post-processing. Parameters in the mass balance include the change in groundwater storage (ΔS) in m^3, precipitation (P) in m^3, evapotranspiration (ET) in m^3, and volume of water exiting the site via the sewer network (O) in m^3. The following equation was used to generate a change in groundwater storage estimate scaled to the study area.

$$\pm \Delta S = P - ET - O \quad (1)$$

An urban water budget was calculated at RecoveryPark for four different precipitation events, each occurring during the period between October 2015 and September 2016. The specific precipitation events selected for this analysis were chosen were based on the availability of field data and the accuracy of sewer data. Start times of 12:00 a.m. and 12:00 p.m. were chosen based on the start of the precipitation event to ensure baseline data was collected before a precipitation event and contained five days of continuous data from this start time. Table 3 shows a breakdown of available field data for each period studied. Precipitation data (Figure 7), evapotranspiration data (Figure 8), sewer data (Figure 9), and groundwater well data (Figure 10) were included in the example of water budgets.

Inverse distance weighting (IDW) models were used to represent the spatial variation of hydraulic head distribution for both the pre-and post-rain time periods. The pre-rain time period corresponds to head measurements taken immediately before the rainfall, while the post-rain time period corresponds to head measurements taken five days after the beginning of the study period. At least one well was operational during the four precipitation events. For the period of 27 to 31 October 2015, the single shallow well is assumed to represent the study area, and an IDW was not generated for this event. The well values were inputted into the ESRI ArcMap, and the ArcMap Spatial Analysis Extension tool was used to create the inverse distance weighting maps for this time period (Figure 11). The layers were then converted to a raster file, and difference maps were generated from the pre-precipitation raster and post-precipitation raster to show net gains and net losses. An effective porosity (n) of 0.3 was used as an average porosity for the RecoveryPark study area based on typical values associated with soils present at RecoveryPark [36]. Change in groundwater storage (ΔS) is the porosity (n) times the net volume gain or loss (m^3) over the study period (Equation (1)).

Table 3. Overview of field data used over the one-year period.

Dates	Start Time	Precipitation	Evapotranspiration	Sewer Data	Shallow Wells Online	Deeper Wells Online
27–31 October 2015	12:00 p.m.	Continuous 5-min data	Continuous 60-min data	Meter E, Meter F	1	4
13–17 March 2016	12:00 a.m.			Meter E, Meter F	14	8
10–14 May 2016	12:00 p.m.			Meter E, Meter F	14	9
11–15 August 2016	12:00 a.m.			Meter E, Meter F	4	8

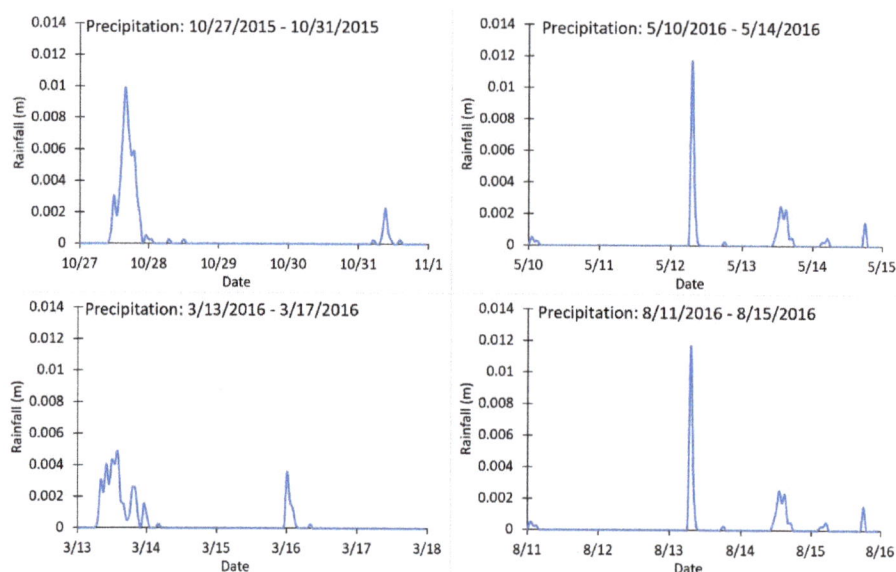

Figure 7. Daily totals of precipitation at RecoveryPark.

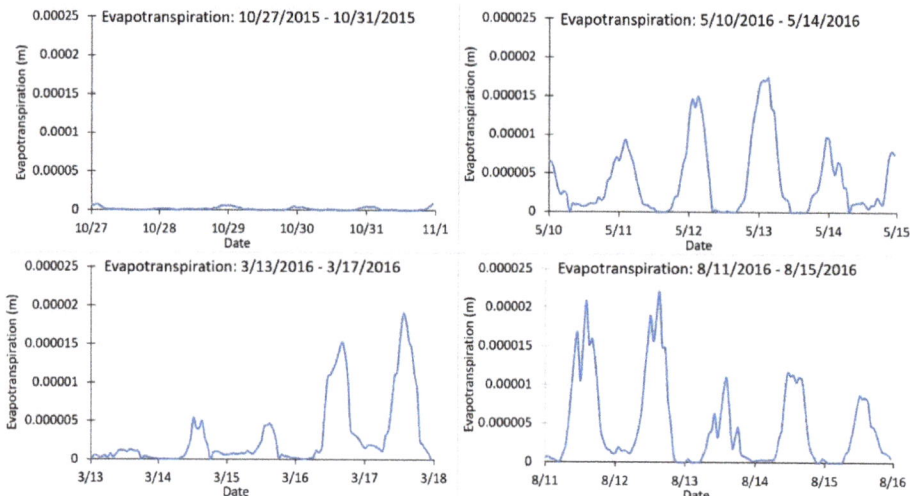

Figure 8. Evapotranspiration values for RecoveryPark over the four study periods.

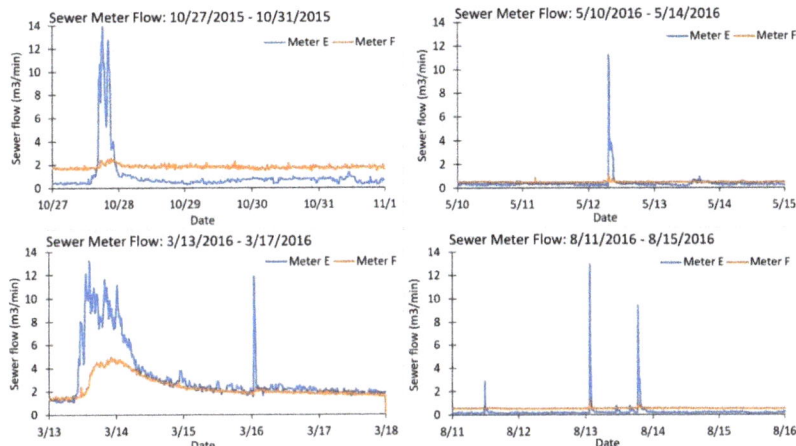

Figure 9. In-pipe sewer flow meter data for RecoveryPark over the four-study period.

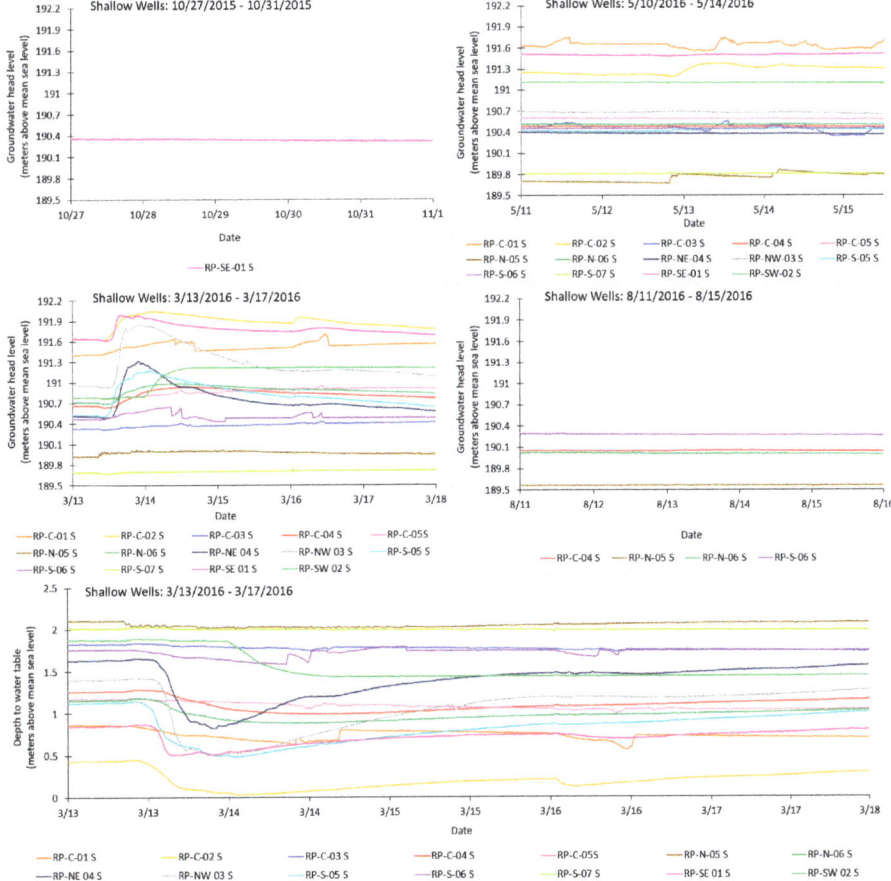

Figure 10. Groundwater head levels at shallow wells at RecoveryPark over the four study periods.

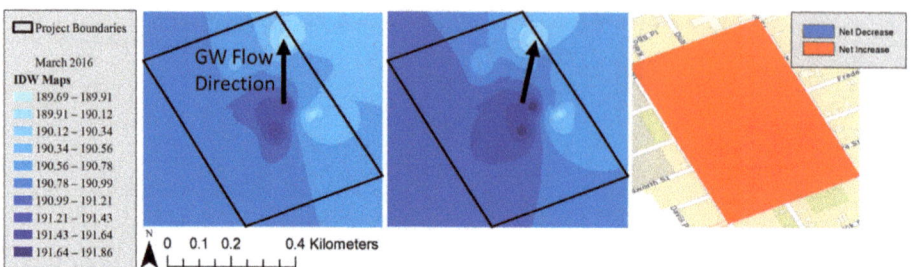

Figure 11. Inverse distance weighted model of groundwater above mean sea level (meters) for shallow wells on 13 March 2016 (left), and 17 March 2016 (center). The legend for these models is on the left side. The change in the hydraulic head is measured by the net decrease and net increase map on the right with the legend located in the top right of the map. In this example, there was no net decrease observed.

Soil moisture was not initially reviewed because the sensors were not in place during the initial study period during 2015 and 2016. Soil moisture plays a critical role in water retention and, ultimately, in the water budget. It was determined that soil moisture is critical to understanding an urban water budget, and four model rainfall events in 2017 and 2018 were chosen to give an estimated idea of the scale of water retained during rainfall events. While the data was taken for different rainfall events, it still provides valuable information regarding the volume of water retained in the top 0.5 m at RecoveryPark. Given the close proximity of the soil moisture sensors, the averages of sensors placed near the surface, at 0.3 m below the surface, and 0.5 m below the surface were used to estimate soil moisture at RecoveryPark (Figure 12).

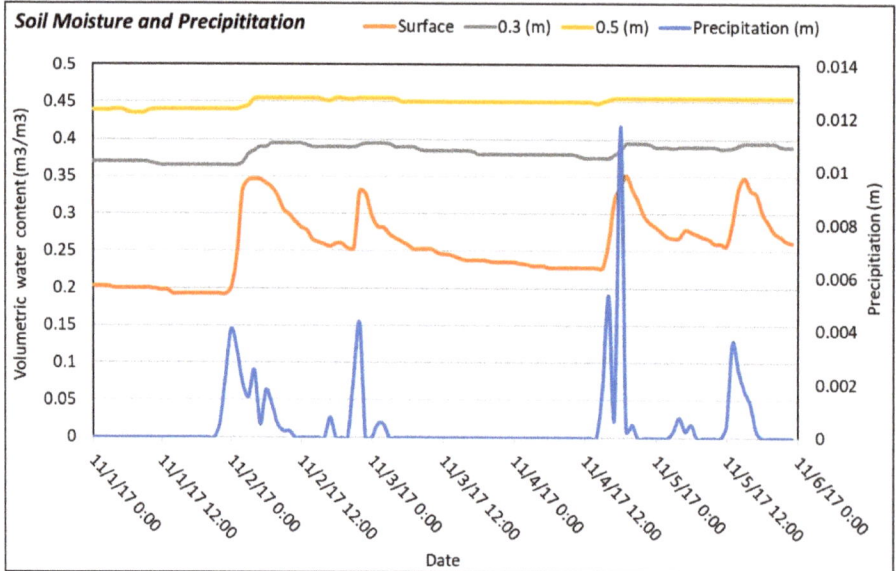

Figure 12. Example figure showing changes in volumetric soil moisture over a five-day period with a rain event. Volumetric soil moisture is on the left axis, while precipitation is on the right axis. The orange line is soil moisture near the surface, the grey line is at ~0.3 m below the surface, and the yellow line is ~0.5 m below the surface.

2.3.2. Groundwater Model

Similar to the regional simulation, GMS 10.3.2 (MODFLOW 2000) is used to identify groundwater head in the steady-state condition for the local scale model. Since there are two sets of groundwater observed data in RecoveryPark, i.e., shallow and deeper observation wells, two conceptual models for each observed dataset are created to investigate groundwater through shallow and deeper observation wells. In other words, at RecoveryPark, groundwater is investigated through two separate models with two different lengths of the observation wells. Both conceptual models are one-layer models with 41,184 active structured cells. The surface elevation ranges from ~195 m in the north to ~189 m in the south (NAVD88), while the bottom elevation is assumed to be ~150 m for the entire area. The local model is a part of the regional one as it is located in the southeastern part of the regional model. The boundary conditions applied in the local model are adapted to the regional model outputs. The local boundary conditions were chosen based on the regional model results. A constant head of ~188 to 193 m is also assumed for RecoveryPark boundaries.

Sewer drains and evapotranspiration are the discharging sources for groundwater, while precipitation is the only source for recharging the groundwater. In the RecoveryPark 3-D model, we obtained sewer line data from [36,44] and divided the area into six zones, as shown in Figure 13. Referring to the samples collected by Rogers [6], the clay unit appears to be uniformly deposited within Detroit city with a hydraulic conductivity range of 8.64×10^{-4} to 1.1 (m/d). Furthermore, based on the soil-boring data in RecoveryPark [37,39], six different hydraulic conductivities of clay are recognized for RecoveryPark soil. So, we assumed six types (zone) of hydraulic conductivity in the local model and subdivided the model into six zones to make sure that the model correctly represents the hydrogeological features of RecoveryPark. The hydraulic conductance and bottom elevation of pipes are applied to the sewer lines in the local model based on available data of sewerage at RecoveryPark. The sewer network at RecoveryPark is a combined sewer system [44]. Based on the available data, the depth of the sewer pipes junctions is assumed to be ~2 to 3 m in the model. The sewer pipes have been placed deeper in the south to southwest of the local model, while they can be found at a smaller depth in the center-north to the northeast of the model. A hydraulic conductance of 0.0025 $((m^2/d)/m)$ is also assumed for all sewer pipes through the Drain package in the MODFLOW model.

An essential part of any groundwater modeling exercise is the model calibration process. In this project, horizontal and vertical hydraulic conductivity were selected as the calibration variables since they are the least well-known data in this region. The horizontal and vertical hydraulic conductivity values were fine-tuned using the PEST calibration process (Table 4) and the 16 March 2016 observations of groundwater head for the shallow and deeper wells at RecoveryPark. During the PEST calibration process, the groundwater system inputs are estimated from the system results by comparing the model outputs and observed head values. We applied the zone-based approach in the model calibration and selected each hydraulic conductivity zone as a parameter. In RecoveryPark modeling, PEST tracks how the model responds to the hydraulic conductivity changes and calculates the residual for computed and observed head values. In each round of parameter estimation, PEST determines new hydraulic conductivity values for each zone, interpolates new values to the MODFLOW model, and updates the input files for the next MODFLOW run. Each time, the latest output of the model is compared to the observed head values. PEST repeats this process and adjusts the hydraulic conductivity values until the minimum value of residual is achieved.

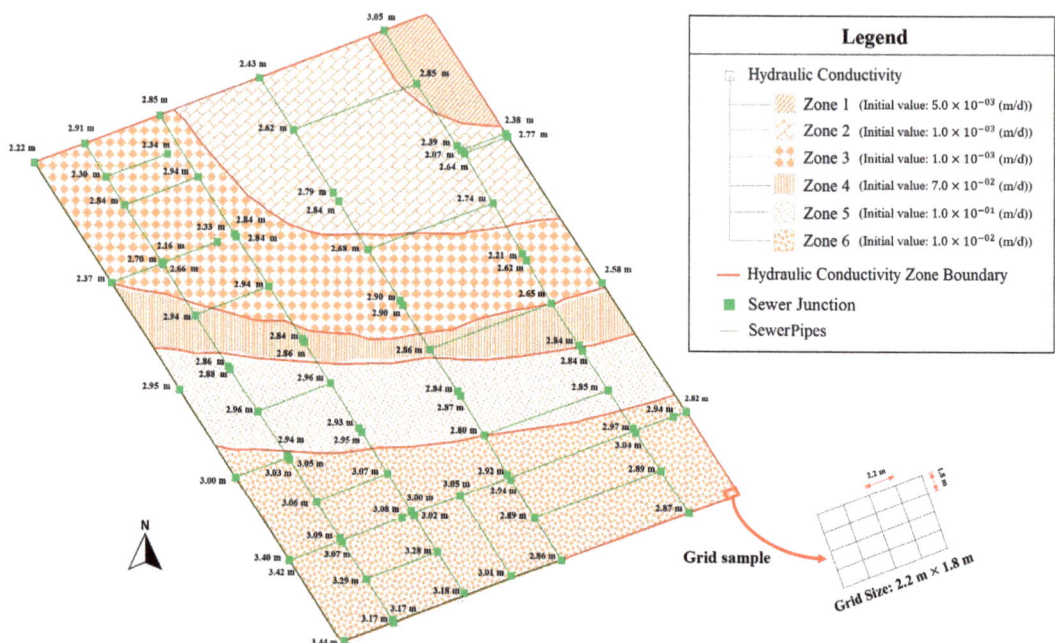

Figure 13. Sewer pipes' location and hydraulic conductivity zones in the RecoveryPark model. The numbers are the depth at which the sewer junctions are located.

Table 4. Details of sink/sources inputs for calibrated RecoveryPark model.

Parameter	Values
Precipitation (m) [1]	1.524×10^{-3}
Horizontal hydraulic conductivity (m/d) (obtained after calibration process)	Zone 1: 1.818×10^{-3} Zone 2: 8.738×10^{-3} Zone 3: 3.141×10^{-3} Zone 4: 3.830×10^{-2} Zone 5: 1.0 Zone 6: 9.298×10^{-3}
Vertical hydraulic conductivity (m/d) in all zones (obtained after calibration process)	0.5
Evapotranspiration rate (m/d) [1]	4×10^{-6}

[1] USGS [37] data.

3. Results

3.1. Regional Scale Groundwater Model: Metro Detroit Watersheds

The study area's initial groundwater model (Figure 14) shows that in the northwest of the region, the groundwater head level reaches a maximum of ~320 m in the northwest, where the maximum elevation of the ground surface is ~325 m. The minimum groundwater head level of ~170 m is also found at the southeast parts of the region, where the ground surface has a minimum elevation of ~173 m. The difference between the ground surface elevations and groundwater head levels demonstrates that the depth to the groundwater gradually decreases toward the southeast, and groundwater is found at a deeper depth in the northwest of the region. As reviewed in the literature, the groundwater head level within the Detroit urban area, along the cross-section of Figure 2, is at ~one to three meters below the ground surface.

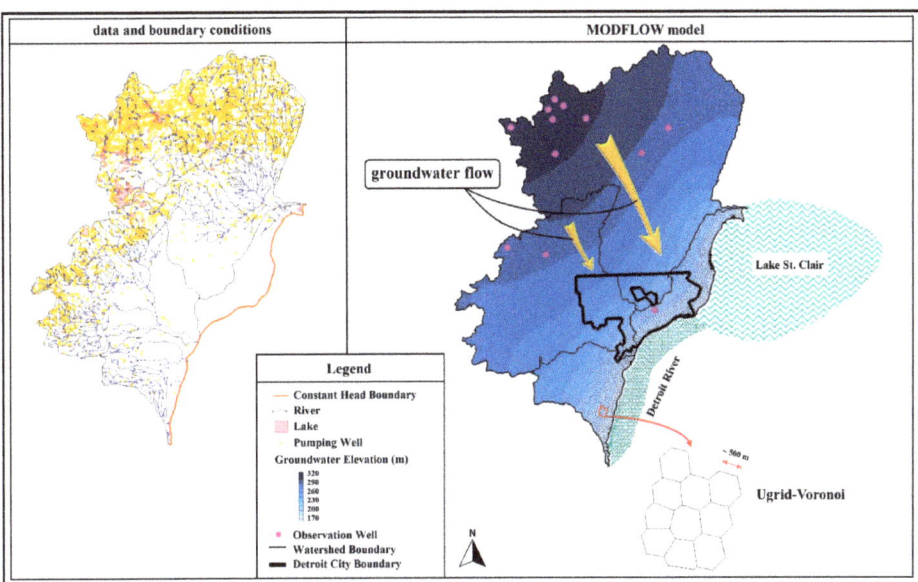

Figure 14. The initial model of groundwater flow in the region of Detroit watersheds.

Furthermore, the water table is found at a higher level in the northwest parts of the model than the southeast of the region. Therefore, groundwater flows in a southeasterly direction and discharges to the Detroit River and Lake St. Clair.

3.2. Local Scale Model: RecoveryPark

3.2.1. Urban Water Budget

Water budget variables were evaluated for four precipitation events: 27 to 31 October 2015; 13 to 17 March 2016; 10 to 14 May 2016; and 11 to 15 August 2016. These included precipitation (Figure 7), evapotranspiration (Figure 8), in-pipe flow meter (Figure 9), and changes in groundwater storage (Figure 10). The totals are showed in Table 5 and compared in Figure 15. Equation (1) calculated the total change in storage with a measured change in storage based on the hydraulic head. The unaccounted water value was determined to be the difference between the computed storage and the measured change in storage, which is defined as excess in Table 5.

The sewage flows below the study area were monitored during each evaluation period at meters E and F (Figure 9). Meter E showed consistent surges of water during each precipitation event, while meter F showed slight increases. It is most likely due to the fact that meter E's outlet conveyed sewer water for a larger portion of the study area. The sewer flow decreased rapidly after the initial storm surge.

Table 5. Totaled values for each precipitation event.

Date	27–31 October 2015	13–17 March 2016	10–14 May 2016	11–15 August 2016
Units	m^3	m^3	m^3	m^3
Precipitation	9509.10	7859.12	4863.10	8032.80
Evapotranspiration	−39.97	−62.74	−85.15	−97.65
Sewer Flow	−1370.63	−2715.42	−1561.84	−2130.72
Change in Storage	−2188.24	8550.26	119.67	−719.84
Excess	5910.269	13,631.22	3335.781	5084.59

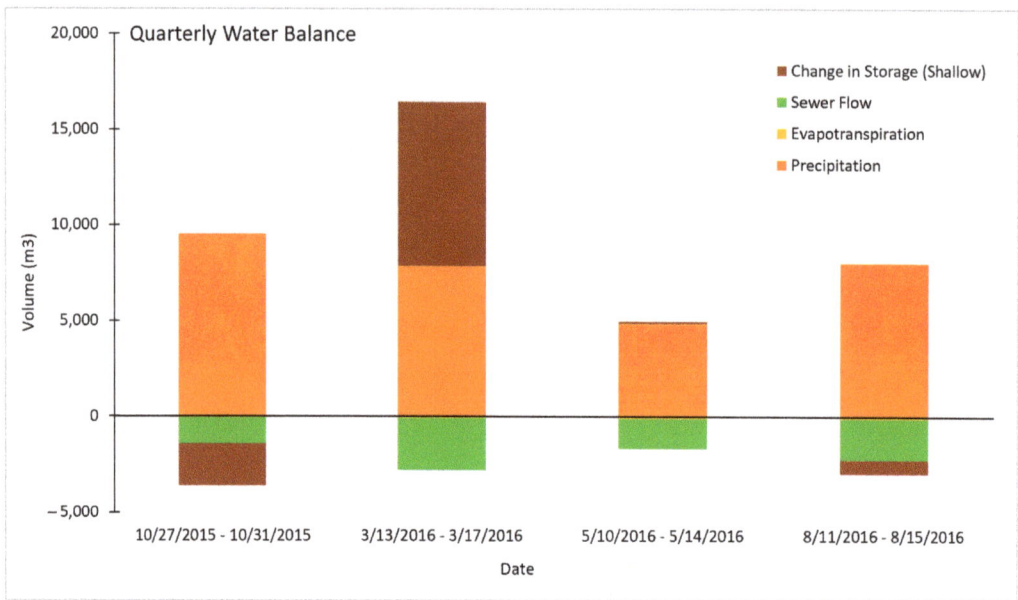

Figure 15. Quarterly calculated water balances at RecoveryPark. Values are shown next to the bar graphs. Evapotranspiration is <100 m³ and not shown in text form.

The measured hydraulic head in the shallow wells exhibited similar trends during each observed time period (Figure 10). The October and August time periods showed a consistent drop in groundwater. The March and May time periods showed responses to the rainfall events during the observed time periods. The difference maps generated from the IDW models produced a change in storage values for the shallow well interval for the four study periods (Table 5).

Quarterly water budgets were assessed for each time period (Figure 15). October 2015, May 2016, and August 2016 showed excess unaccounted water onsite, while March 2016 showed an increase in groundwater storage beyond precipitation (Table 5). This indicates that the measurements taken at RecoveryPark during 2015 and 2016 did not fully capture the water budget during episodic rain events. Soil moisture was evaluated during four different seasonal rain events at RecoveryPark in 2017 to give an indication of the changes in soil water above the water table. Figure 12 represents an example measurement reviewed from 1 November 2017 to 5 November 2017. The volume of the total change in soil moisture calculated in Figure 12 shows a positive change of 2286.41 m³ when scaled to the RecoveryPark study area. Other precipitation events in March 2018 resulted in an increasing volume of 1581.26 m³, June 2018 showed 910.29 m³, and August 2018 showed 2200.94 m³.

3.2.2. Groundwater Model

Figure 16 presents the final results of the shallow and deeper groundwater simulation for RecoveryPark. The detail of assumptions and modeling process have been presented in Section 2.3.2. The results show that the groundwater gradient is to the north of RecoveryPark. Additionally, the shallow and deeper observation wells reflected similar groundwater fluctuations. The groundwater observed through observation wells with two different lengths (shallow and deeper) behaves equally and moves from the south to the north of the RecoveryPark models.

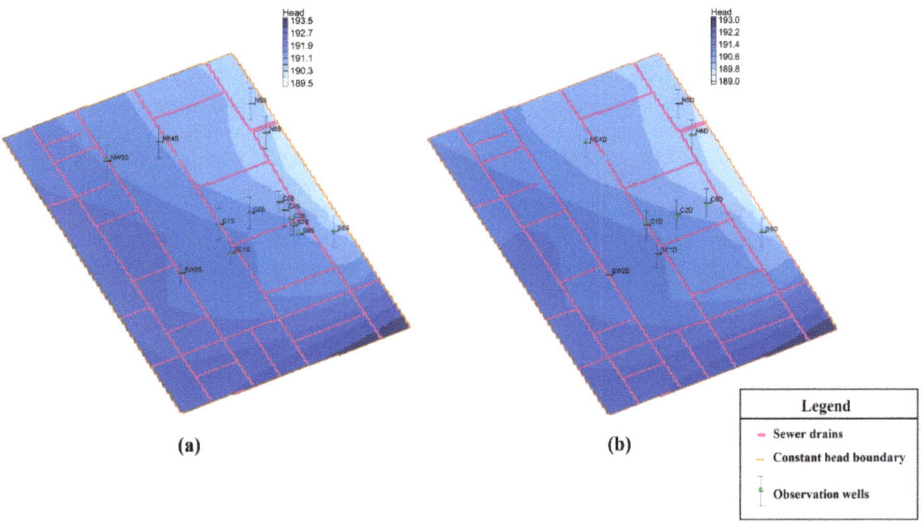

Figure 16. Groundwater head model at RecoveryPark calibrated by (**a**) shallow (**b**) deeper observation wells.

The RecoveryPark groundwater simulation model was calibrated by varying hydraulic conductivity parameters (Table 4) to optimize hydraulic head values. Through using differing hydraulic conductivity parameters, the residual between the observed values and calculated values showed a consistent reduction in mean error, which resulted in a reduction of uncertainty in the model. The final output from the groundwater simulation at RecoveryPark provides a conceptual model representing the behavior of neighborhood-scale groundwater. Figure 17 provides a comparison between the computed head levels and the observed ones. The Nash–Sutcliffe coefficient for the shallow and deeper groundwater model of RecoveryPark are NSE > 0.8 and NSE > 0.69, respectively.

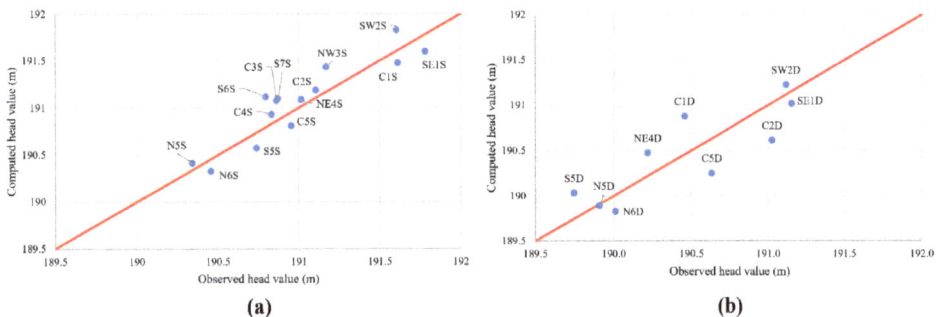

Figure 17. Computed head values vs. observed head values at (**a**) shallow (**b**) deeper groundwater model of RecoveryPark.

4. Discussion

This study provides an initial start to defining urban groundwater movement in coastal urban cities in the Great Lakes Basin. The modeling of field data from the RecoveryPark field site and the broader Detroit metro-area highlights the lack of publicly available datasets while demonstrating pragmatic approaches to evaluate urban groundwater movement. These two approaches of groundwater movement modeling in four-integrated watersheds and localized scale in Detroit show how differences in scale yield diverse insights into assessing urban influences into groundwater flow.

4.1. Regional Scale Groundwater Model: Metro Detroit Watersheds

At the regional scale, both groundwater flows and groundwater depth generally agreed with results reported in past literature [6,31]. The regional modeling indicates that the groundwater in the region of Detroit watersheds is primarily a shallow aquifer placed on impermeable clay units and flows in sand layer units. Additionally, the model shows that groundwater can be found at a depth ranging from ~one to three meters below the ground surface. The probability of encountering groundwater decreases toward the east as it flows toward the southeast of the region [6]. Therefore, the groundwater gradient tends to the southeast direction, discharging into the Detroit River. There are many legacy contaminated sites located throughout the study area that include superfund sites and smaller-scale brownfields [45,46]. The general direction of groundwater flow is towards the Detroit River, creating potential pathways for the pollutants coming from these contaminated sites to reach drinking water sources for Detroiters. Moreover, since the water table is found at only a few meters below the land surface of Detroit city, volatile contaminants existing in the groundwater can easily intrude the upper zones and produce human and environmental health issues for this urbanized area.

4.2. Local Scale Model: RecoveryPark

This paper demonstrates techniques to evaluate the urban groundwater flow and assess an urban water budget. Previous studies at RecoveryPark focused on implementing green stormwater infrastructure for sustainable development [44,47], monitoring urban water cycles for understanding the effectiveness of green stormwater infrastructure [48] and interactions with urban water infrastructure [36]. The hydraulic conductance and bottom elevation of sewer lines applied to the local model manifest the groundwater interaction with the sewer line system as the urban infrastructure available at RecoveryPark. As the local model is built in one layer, the sewer lines are assumed to remain in that one layer to have their full interaction with local groundwater in steady-state modeling. By the full interaction term, we mean that the entire sewer lines are included in the model layer. The urban water budget portion of this paper focuses on evaluating the groundwater, while considering subsurface sewer networks and weather parameters.

4.2.1. Urban Water Budget

The RecoveryPark case study is unique as it accesses one of the few urban datasets housed on an open data platform through the USGS National Water Information System. The RecoveryPark field site provides a broad scope of site-specific data sets. While it does not apply to the larger Detroit area, it is a useful start to understand urban water fluxes and localized flow directions. The urban water budget evaluated in this paper represents an insight into episodic rainfall events, which differs from Hoard, Haefner, Shuster, Pieschek and Beeler [36], where urban water budgets were evaluated on a monthly and seasonal basis.

A review of the datasets at RecoveryPark confirms that the silty clay loam deposits located 3 m below the ground surface are hydraulically connected to the shallow fill zone. These results also show that the wells in the clay layer are horizontally connected, most likely as a result of sand lenses [32]. Evaluating the depth to water on a yearly basis shows a seasonal variation with a low depth to water in the summer. Periods of no data suggest the water table is not present or below the pressure transducers set in the groundwater wells.

The hydraulic head observed in the shallow wells did not always show a response to infiltration from rain events. Potential reasons for this include overland flow being directed into the sewer and not infiltrating and groundwater flow increases. Another consideration is that a large portion of the surface cover is utilized by GSI, which could direct water to areas not covered by the urban well field. March 2016 noted large increases in the hydraulic head at both the shallow and deeper wells, which may be an indication of rainwater melting and elevating the water table. The low changes in groundwater during May 2016 may reflect the drop in the water table during the warmer months of the year. This continued during the drop-in groundwater levels during the August 2016 study period.

The sewer systems did not have as large of an influence on the water outputs off the RecoveryPark site as the groundwater. RecoveryPark was originally designed as an experimental field site for green stormwater infrastructure. Groundwater is the controlling feature for moving precipitation on and off the study location because of this land use. Evapotranspiration also had a small impact on the overall water balance. Factors for evapotranspiration include the small scale of the study area and that the study periods specifically targeted periods of precipitation. Evapotranspiration showed consistent, rapid decreases during rain events at RecoveryPark, which is most likely due to increasing cloud cover and cooler temperatures.

The large excess water noted in Table 5 indicates that Equation (1) did not initially fully encompass all inputs and outputs of the urban water budget mass balance. Notably, soil moisture in the vadose zone above the shallow wells was absent. An initial review of the 2017 and 2018 soil water data demonstrates the importance of water retention in the vadose zone. The observed soil moisture data evaluated in this paper only represents the top 0.5 m at RecoveryPark. The calculated volume can account for a large amount of excess water at RecoveryPark, which would increase if extrapolated down to two to three meters below the surface to the water table.

GSI are important features of urban hydrologic interactions but these systems are contained within the study region. The sewer and groundwater elevation data were assumed to capture the subsurface GSI interactions. Therefore, the inputs and outputs measured in the urban water budget are assumed to be representative of the green stormwater infrastructure. Additionally, the soil cores taken at RecoveryPark show a confining clay layer located directly beneath the GSI. Our model assumes that this clay layer along with the shallow water table enable horizonal water movement that is captured by the wells.

4.2.2. Groundwater Model

The RecoveryPark groundwater simulation provides a city-block-scale estimate of groundwater flow over the study location. The local neighborhood scale modeling revealed a flow that is the opposite of the regional flow direction. Modeling results and field site data showed the higher hydraulic head in the southeast and the lower hydraulic head in the northwest [36]. This contrasting flow direction confirms that spatial location in urban areas impacts groundwater flow. The reasons for local changes in groundwater flow direction include heterogeneity of subsurface, and influences on the hydraulic head such as dewatering pumps may dictate neighborhood-scale flow direction. It demonstrates that local geology, disturbance patterns of the soil, anthropogenic activities, and urban setting, in general, have a large influence on localized groundwater flow patterns.

Furthermore, RecoveryPark models calibrated based on shallow and deeper observation wells show similar hydraulic head fluctuation values for depth to groundwater. This observation reflects the field data, where nested shallow and deeper wells show similar changes in hydraulic head. Since the models of shallow and deeper wells are under the same conditions of recharge, discharge, and boundary conditions, these models suggest that there is no confining condition up to the screened zone of deeper wells, which may cause changes in the groundwater fluctuation pattern at RecoveryPark.

4.3. Larger Impact on Southeast Michigan

High population density and industrial centers increase the susceptibility of shallow groundwater to contamination within the Great Lakes Basin. Since very few data are available in urban areas of the Great Lakes Region, developing groundwater models provide invaluable information for urban groundwater resource management [49]. This paper contributes to addressing the priority science of urban groundwater outlined by the last status report on the Great Lakes Water Quality Agreement provided by Grannemann and Van Stempvoort [49]. The paper represents a pragmatic way to evaluate shallow subsurface groundwater movement in southeast Michigan with the eventual goal of understanding the subsurface fate and transport in urban settings. Remediation sites are not limited to property boundaries, and through understanding regional and neighborhood-scale transport, decision-makers can better advise on public health concerns such as VOC exposure. Since the shallow water table occurs only a few meters below the ground surface, VOCs are able to transmit faster from groundwater to the upper unsaturated zone and ground surface. These hydrogeological characteristics of Detroit's urban groundwater can increase the chances of soil vapor movement.

4.4. Limitations

This paper relies on publicly available data and is limited in its ability to expand these datasets to other areas of the metro-Detroit and Great Lakes Basin. The regional model covers a large segment of southeast Michigan and is not adequately covered by wells. The predictions of groundwater model results may be affected by many factors, for example, the sampling method of datasets, the structure of the groundwater conceptual model, and the deviation resulted from the mathematical solution of the groundwater model. Additionally, the scarcity of datasets, including pumping rate and hydraulic conductivity, creates uncertainty in both regional and neighborhood-scale models. The neighborhood-scale model for RecoveryPark has been calibrated using horizontal and vertical hydraulic conductivity as calibration parameters. However, a limited number of observation wells at the regional scale restrict the full calibration process of the regional model. This irregular distribution of wells limits the ability to support the datasets for further understanding and studies on the groundwater movement as well as transferring of contaminants in the groundwater. However, this model can provide useful results for initial assessments of the groundwater flow within the study area. These results are also beneficial in guiding the activities of data collection and combining large amounts of data for hydrological investigations.

At the RecoveryPark field site, several external factors limited the presented modeling work. The sewer data showed inconsistent readings, and the data were quality controlled only through 2016. Therefore, only data that underwent quality control was used in the municipal water budget. Unlike the precipitation, evapotranspiration, and sewer data, the shallow and deeper well values were derived from time-specific data. Soil moisture field data is limited to after August 2017. External factors, such as interstate highways close to the field site, may have a disproportionate influence on the groundwater flow of the region. The interstate freeways in this region are below ground and operate dewatering pumps.

5. Conclusions

Urban groundwater plays a critical role in the vitality of the Great Lakes Basin and is often difficult to characterize, given anthropogenic changes of subsurface hydrology. While data availability presents issues, this study demonstrates pragmatic methods to utilize existing datasets to start developing a multiscale understanding of near-surface groundwater movement. This study highlights the interconnected nature of natural and urban systems from the regional water drainage characteristics to neighborhood-scale hydrologic water flow. This project is leading research that provides sorely needed insights into general urban groundwater flow and transport in the Great Lakes region.

In this project, we developed a conceptual model of groundwater in the Detroit region located in four major watersheds, analyzed hydraulic head changes, and provide a baseline for understanding groundwater's role in subsurface urban contaminant movement. In this model, we assumed a steady-state condition for the finite volume solution of the groundwater equations and considered some critical hydrogeological data such as precipitation and evaporation rates, rivers head and depth, soil layers with their hydraulic conductivities, and discharging wells. The regional model shows that there is shallow groundwater underlying the Detroit region and flowing from the northwest to the southeast within the study area. At RecoveryPark, the water balance demonstrated the difficulty in generating episodic urban water balances but provides insight into the controlling variables for water flow in urban settings. The neighborhood-scale simulation evaluated the small-scale heterogeneity of urban soils and subsurface infrastructure to provide a modeled flow direction on a city block basis. In the neighborhood-scale simulation, the local movement of groundwater is in the north direction, which is opposite to the regional groundwater flow. The different flow directions in local and regional scales imply the effect of urban settings on the behavior of groundwater resources. Understanding this difference is important in the accurate prediction of contaminant transport pathways and the effective application of remediation practices.

Author Contributions: Conceptualization, C.J.M.; data curation, S.T. and B.F.O.; formal analysis, S.T. and B.F.O.; funding acquisition, C.J.M.; investigation, S.T. and B.F.O.; methodology, S.T. and B.F.O.; project administration, S.T.; resources, S.T. and B.F.O.; software, S.T. and B.F.O.; supervision, C.J.M.; validation, S.T. and C.J.M.; visualization, S.T. and B.F.O.; writing—original draft, S.T., B.F.O. and C.J.M.; writing—review and editing, S.T., B.F.O. and C.J.M. All authors have read and agreed to the published version of the manuscript.

Funding: This study received financial support from Wayne State University through the Office of the Vice President for Research and from Healthy Urban Waters through the Fred A. and Barbara M. Erb Family Foundation and by the National Science Foundation (NSF) under Grant No. 1735038.

Institutional Review Board Statement: Not applicable.

Informed Consent Statement: Not applicable.

Data Availability Statement: Data are contained within the article. More detailed data may be requested from authors. The raw data sets can be accessed at the National Water Information System (USGS Water Data for the Nation available at https://nwis.waterdata.usgs.gov/mi/nwis, (accessed on 5 May 2020)), Michigan Department of Natural Resources Open Data (https://gis-midnr.opendata.arcgis.com/search?collection=Dataset, (accessed on 5 May 2020)), and Department of Environmental Quality (Wellogic System, available at https://secure1.state.mi.us/wellogic/Login.aspx?ReturnUrl=%2fwellogic%2fdefault.aspx, (accessed on 2 May 2020)).

Acknowledgments: The authors gratefully acknowledge the contributions of the United States Geological Survey and the United States Environmental Protection Agency work at RecoveryPark. Specifically, Ralph Haefner and Chris Hoard from USGS and William Shuster from Wayne State University for their work in providing RecoveryPark field data and background information. The authors appreciate RecoveryPark for providing a study location in the city of Detroit, MI.

Conflicts of Interest: The authors declare no conflict of interest. The funders had no role in the design of the study; in the collection, analyses, or interpretation of data; in the writing of the manuscript, or in the decision to publish the results.

References

1. Bachmat, Y. Groundwater and aquifers. In *Encyclopedia of Soils in the Environment*, 1st ed.; Hillel, D., Ed.; Elsevier: Amsterdam, The Netherlands, 2005; Volume 2, pp. 153–168.
2. Barcelona, M.J. Development and applications of groundwater remediation technologies in the USA. *Hydrogeol. J.* **2005**, *13*, 288–294. [CrossRef]
3. Kavanaugh, M.C. *Alternatives for Ground Water Cleanup*, 2nd ed.; National Research Council, National Academies Press: Washington, DC, USA, 1994.
4. Luo, J.; Lu, W.; Yang, Q.; Ji, Y.; Xin, X. An adaptive dynamic surrogate model using a constrained trust region algorithm: Application to DNAPL-contaminated-groundwater-remediation design. *Hydrogeol. J.* **2020**, *28*, 1285–1298. [CrossRef]
5. Howard, K.; Gerber, R. Impacts of urban areas and urban growth on groundwater in the Great Lakes Basin of North America. *J. Great Lakes Res.* **2018**, *44*, 1–13. [CrossRef]
6. Rogers, D.T. *Environmental Geology of Metropolitan Detroit*; Clayton Environmental Consultants Inc.: Novi, MI, USA, 1996.
7. Guo, Y.; Holton, C.; Luo, H.; Dahlen, P.; Johnson, P.C. Influence of Fluctuating Groundwater Table on Volatile Organic Chemical Emission Flux at a Dissolved Chlorinated-Solvent Plume Site. *Ground Water Monit. Remediat.* **2019**, *39*, 43–52. [CrossRef]
8. Qi, S.; Luo, J.; O'Connor, D.; Cao, X.; Hou, D. Influence of groundwater table fluctuation on the non-equilibrium transport of volatile organic contaminants in the vadose zone. *J. Hydrol.* **2020**, *580*, 124353. [CrossRef]

9. Yu, L.; Rozemeijer, J.C.; Van Der Velde, Y.; van Breukelen, B.; Ouboter, M.; Broers, H.P. Urban hydrogeology: Transport routes and mixing of water and solutes in a groundwater influenced urban lowland catchment. *Sci. Total Environ.* **2019**, *678*, 288–300. [CrossRef] [PubMed]
10. Bailey, R.T. Review: Selenium contamination, fate, and reactive transport in groundwater in relation to human health. *Hydrogeol. J.* **2016**, *25*, 1191–1217. [CrossRef]
11. Chakraborti, D.; Rahman, M.M.; Das, B.; Chatterjee, A.; Das, D.; Nayak, B.; Pal, A.; Chowdhury, U.K.; Ahmed, S.; Biswas, B.K.; et al. Groundwater arsenic contamination and its health effects in India. *Hydrogeol. J.* **2017**, *25*, 1165–1181. [CrossRef]
12. Hill, M.K. *Understanding Environmental Pollution*, 3rd ed.; Cambridge University Press: Cambridge, UK, 2010.
13. Beckley, L.; McHugh, T. A conceptual model for vapor intrusion from groundwater through sewer lines. *Sci. Total Environ.* **2020**, *698*, 134283. [CrossRef]
14. Bonneau, J.; Fletcher, T.D.; Costelloe, J.F.; Burns, M.J. Stormwater infiltration and the 'urban karst'—A review. *J. Hydrol.* **2017**, *552*, 141–150. [CrossRef]
15. Çiçek, A.; Bakiş, R.; Uğurluoğlu, A.; Köse, E.; Tokatli, C. The Effects of Large Borate Deposits on Groundwater Quality. *Polish J. Environ. Stud.* **2013**, *22*, 1031–1037.
16. Dumouchelle, D.; Cummings, T.; Klepper, G.R. *Michigan Ground-Water Quality*; US Geological Survey Open-File Report 87-0732; USGS: Denver, CO, USA, 1987.
17. Goovaerts, P.; AvRuskin, G.; Meliker, J.; Slotnick, M.; Jacquez, G.; Nriagu, J. Geostatistical modeling of the spatial variability of arsenic in groundwater of southeast Michigan. *Water Resour. Res.* **2005**, *41*. [CrossRef]
18. Katz, B.G.; Eberts, S.M.; Kauffman, L.J. Using Cl/Br ratios and other indicators to assess potential impacts on groundwater quality from septic systems: A review and examples from principal aquifers in the United States. *J. Hydrol.* **2011**, *397*, 151–166. [CrossRef]
19. Kim, M.-J.; Nriagu, J.; Haack, S. Arsenic species and chemistry in groundwater of southeast Michigan. *Environ. Pollut.* **2002**, *120*, 379–390. [CrossRef]
20. Thomas, M.A. The effect of residential development on ground-water quality near detroit, Michigan. *JAWRA J. Am. Water Resour. Assoc.* **2000**, *36*, 1023–1038. [CrossRef]
21. Miller, C.J.; Runge-Morris, M.; Cassidy-Bushrow, A.E.; Straughen, J.K.; Dittrich, T.M.; Baker, T.R.; Petriello, M.C.; Mor, G.; Ruden, D.M.; O'Leary, B.F.; et al. A Review of Volatile Organic Compound Contamination in Post-Industrial Urban Centers: Reproductive Health Implications Using a Detroit Lens. *Int. J. Environ. Res. Public Health* **2020**, *17*, 8755. [CrossRef]
22. Cassidy-Bushrow, A.E.; Peters, R.M.; Johnson, D.A.; Templin, T.N. Association of depressive symptoms with inflammatory biomarkers among pregnant African-American women. *J. Reprod. Immunol.* **2012**, *94*, 202–209. [CrossRef]
23. Forand, S.P.; Lewis-Michl, E.L.; Gomez, M.I. Adverse Birth Outcomes and Maternal Exposure to Trichloroethylene and Tetra-chloroethylene through Soil Vapor Intrusion in New York State. *Environ. Health Perspect.* **2012**, *120*, 616–621. [CrossRef]
24. Herdt-Losavio, M.L.; Lin, S.; Druschel, C.M.; Hwang, S.-A.; Mauer, M.P.; Carlson, G.A. The Risk of Having a Low Birth Weight or Preterm Infant among Cosmetologists in New York State. *Matern. Child Health J.* **2008**, *13*, 90–97. [CrossRef]
25. Montero-Montoya, R.; López-Vargas, R.; Arellano-Aguilar, O. Volatile Organic Compounds in Air: Sources, Distribution, Exposure and Associated Illnesses in Children. *Ann. Glob. Health* **2018**, *84*, 225–238. [CrossRef]
26. MDHHS. Rate of Live Births by Age of Mother and Prematurity Classification Michigan 2018. Available online: http://www.mdch.state.mi.us/osr/chi/births14/frameBxChar.html (accessed on 25 July 2020).
27. Stafford, K.; Tanner, K.; Guillen, J. As Mayor Mike Duggan Touts Make Your Date's Success, Detroit's Preterm Birth Rate Spikes. Available online: https://www.freep.com/story/news/investigations/2019/11/07/detroit-preterm-birth-rate-make-your-date/4156777002/ (accessed on 21 August 2020).
28. Cassidy-Bushrow, A.E.; Burmeister, C.; Lamerato, L.; Lemke, L.D.; Mathieu, M.; O'Leary, B.F.; Sperone, F.G.; Straughen, J.K.; Reiners, J.J. Prenatal airshed pollutants and preterm birth in an observational birth cohort study in Detroit, Michigan, USA. *Environ. Res.* **2020**, *189*, 109845. [CrossRef] [PubMed]
29. McDonald, M.G.; Harbaugh, A.W. *A Modular Three-Dimensional Finite-Difference Ground-Water Flow Model*; USGS (United States Government Printing Office): Denver, CO, USA, 1988.
30. Mozola, A.J. *Geology for Land and Groundwater Development in Wayne County*; State of Michigan Department of Natural Resources, Geology Survey: Lansing, MI, USA, 1969.
31. Wisler, C.O.; Stramel, G.J.; Laird, L.B. *Water Resources of the Detroit Area, Michigan*; U.S. Department of the Interior: Washington, DC, USA, 1952; Volume 183.
32. Howard, J.L. *Quaternary Geology of the Detroit, Michigan Quadrangle and Surrounding Areas*; Department of Geology, Wayne State University: Detroit, MI, USA, 2013.
33. Leverett, F. *Flowing Wells and Municipal Water Supplies on the Southern Portion of the Southern Peninsula of Michigan*; USGS (United States Government Printing Office): Washington, DC, USA, 1906.
34. Mozola, A.J. *A Survey of Groundwater Resources in Oakland County, Michigan*; Doctoral Dissertation, Syracuse University: Syracuse, NY, USA, December 1953.
35. Henriette, C.V. Anna Kohn of RecoveryPark Talks Jobs, Opportunity and Impact. Available online: https://detroitisit.com/detroit-works-anna-kohn-of-recovery-park-jobs-impact-oppertunity/ (accessed on 25 July 2020).

36. Hoard, C.J.; Haefner, R.J.; Shuster, W.D.; Pieschek, R.L.; Beeler, S. Full Water-Cycle Monitoring in an Urban Catchment Reveals Unexpected Water Transfers (Detroit MI, USA). *JAWRA J. Am. Water Resour. Assoc.* **2019**, *56*, 82–99. [CrossRef] [PubMed]
37. USGS. National Water Information System Data Available on the World Wide Web (USGS Water Data for the Nation), United States Geological Survey. Available online: https://waterdata.usgs.gov/mi/nwis (accessed on 5 May 2020).
38. Michigan Department of Natural Resources. DNR Open Data. Available online: https://gis-midnr.opendata.arcgis.com/search?collection=Dataset (accessed on 5 May 2020).
39. Wellogic System. Department of Environmental Quality (DEQ), State of Michigan's Statewide Groundwater Database. Available online: https://secure1.state.mi.us/wellogic/Login.aspx?ReturnUrl=%2fwellogic%2fdefault.aspx (accessed on 2 May 2020).
40. Aquaveo. GMS: UGrid Module. Available online: https://www.xmswiki.com/wiki/GMS:UGrid_Module (accessed on 15 December 2019).
41. EGLE. Michigan's Major Watersheds. Available online: https://www.michigan.gov/egle/0,9429,7-135-3313_3684_3724---,00.html (accessed on 3 March 2020).
42. Panday, S.; Langevin, C.D.; Niswonger, R.G.; Ibaraki, M.; Hughes, J.D. *MODFLOW–USG Version 1: An Unstructured Grid Version of MODFLOW for Simulating Groundwater Flow and Tightly Coupled Processes Using a Control Volume Finite-Difference Formulation*; 2328-7055; USGS: Reston, VA, USA, 2013.
43. Howard, J.L.; Olszewska, D. Pedogenesis, geochemical forms of heavy metals, and artifact weathering in an urban soil chronosequence, Detroit, Michigan. *Environ. Pollut.* **2011**, *159*, 754–761. [CrossRef]
44. Pieschek, R.; Carpenter, D. Modeling green infrastructure in the support of the re-development of Detroit's neighborhoods. In Proceedings of the World Environmental and Water Resources Congress 2016, West Palm Beach, FL, USA, 22–26 May 2016; pp. 247–254.
45. Aichele, S.S. *Ground-Water Quality Atlas of Oakland County, Michigan*; Water-Resources Investigation 00-4120; USGS: Lansing, MI, USA, 2000.
46. EGLE. Environmetal Mapper. Available online: https://www.mcgi.state.mi.us/environmentalmapper/# (accessed on 2 May 2020).
47. Zhang, K.; Chui, T.F.M. A review on implementing infiltration-based green infrastructure in shallow groundwater environments: Challenges, approaches, and progress. *J. Hydrol.* **2019**, *579*. [CrossRef]
48. Carpenter, D.D.; Pieschek, R.L.; Drummond, C.D. Documenting the Urban Water Cycle and Implications for Determining the Effectiveness of Transforming the Landscape with Green Stormwater Infrastructure. In Proceedings of the 10th Novatech Conferenace, Lyon, France, 1–5 July 2019.
49. Grannemann, G.; Van Stempvoort, D. Groundwater Science Relevant to the Great Lakes Water Quality Agreement: A Status Report. Environment and Climate Change Canada and U.S. Environmental Protection Agency, May 2016. Available online: https://binational.net/2016/06/13/groundwater-science-f/ (accessed on 5 November 2020).

Article

Integration of Numerical Models and InSAR Techniques to Assess Land Subsidence Due to Excessive Groundwater Abstraction in the Coastal and Lowland Regions of Semarang City

Weicheng Lo [1], Sanidhya Nika Purnomo [1,2,*], Bondan Galih Dewanto [3,4], Dwi Sarah [5] and Sumiyanto [2,6]

[1] Department of Hydraulic and Ocean Engineering, National Cheng Kung University, No. 1 University Road, Tainan 701, Taiwan; lowc@mail.ncku.edu.tw
[2] Civil Engineering Department, Universitas Jenderal Soedirman, Jl. HR Bunyamin, Purwokerto 53122, Indonesia; sumiyanto@unsoed.ac.id
[3] Department of Geodetic Engineering, Universitas Gadjah Mada, Jl. Grafika No. 2 Bulaksumur, Yogyakarta 55281, Indonesia; bondan.g.d@ugm.ac.id
[4] Center for Disaster Studies, Universitas Gadjah Mada, Yogyakarta 55281, Indonesia
[5] Research Center for Geotechnology, Research Organization for Earth Sciences, National Research and Innovation Agency (BRIN), Jl. Sangkuriang, Kompleks LIPI, Kota Bandung 40135, Indonesia; sarahpr28@gmail.com
[6] Department of Civil Engineering, Universitas Diponegoro, Jl. Prof. Soedarto, Tembalang, Semarang 50275, Indonesia
* Correspondence: sanidhyanika.purnomo@unsoed.ac.id

Abstract: This study was carried out to assess land subsidence due to excessive groundwater abstraction in the northern region of Semarang City by integrating the application of both numerical models and geodetic measurements, particularly those based on the synthetic aperture radar interferometry (InSAR) technique. Since 1695, alluvial deposits caused by sedimentations have accumulated in the northern part of Semarang City, in turn resulting in changes in the coastline and land use up to the present. Commencing in 1900, excessive groundwater withdrawal from deep wells in the northern section of Semarang City has exacerbated natural compaction and aggravated the problem of land subsidence. In the current study, a groundwater model equivalent to the hydrogeological system in this area was developed using MODFLOW to simulate the hydromechanical coupling of groundwater flow and land subsidence. The numerical computation was performed starting with the steady-state flow model from the period of 1970 to 1990, followed by the model of transient flow and land subsidence from the period of 1990 to 2010. Our models were calibrated with deformation data from field measurements collected from various sources (e.g., leveling, GPS, and InSAR) for simulation of land subsidence, as well as with the hydraulic heads from observation wells for simulation of groundwater flow. Comparison of the results of our numerical calculations with recorded observations led to low RMSEs, yet high R^2 values, mathematically indicating that the simulation outcomes are in good agreement with monitoring data. The findings in the present study also revealed that land subsidence arising from groundwater pumping poses a serious threat to the northern part of Semarang City. Two groundwater management measures are proposed and the future development of land subsidence is accordingly projected until 2050. Our study shows quantitatively that the greatest land subsidence occurs in Genuk District, with a magnitude of 36.8 mm/year. However, if the suggested groundwater management can be implemented, the rate and affected area of land subsidence can be reduced by up to 59% and 76%, respectively.

Keywords: regional land subsidence; groundwater abstraction; numerical simulation; InSAR; Semarang City

Citation: Lo, W.; Purnomo, S.N.; Dewanto, B.G.; Sarah, D.; Sumiyanto Integration of Numerical Models and InSAR Techniques to Assess Land Subsidence Due to Excessive Groundwater Abstraction in the Coastal and Lowland Regions of Semarang City. *Water* 2022, 14, 201. https://doi.org/10.3390/w14020201

Academic Editor: C. Radu Gogu

Received: 6 December 2021
Accepted: 7 January 2022
Published: 11 January 2022

Publisher's Note: MDPI stays neutral with regard to jurisdictional claims in published maps and institutional affiliations.

Copyright: © 2022 by the authors. Licensee MDPI, Basel, Switzerland. This article is an open access article distributed under the terms and conditions of the Creative Commons Attribution (CC BY) license (https://creativecommons.org/licenses/by/4.0/).

1. Introduction

Land subsidence refers to a gradual or sudden vertical deformation of the ground surface, which is usually economically and socially detrimental since it often inevitably leads to structural damages to buildings and public infrastructure, as well as expands the flood inundation area [1–3]. Land subsidence may occur due to natural and anthropogenic (manmade) processes. The former (natural factors) includes fault compaction and tectonic movement, etc., whereas the latter (anthropogenic factors) includes excessive extraction of subsurface fluids (e.g., groundwater, oil, and gas), building loads on the ground surface, and so on [4–9].

Numerous studies have been extensively conducted to explore the physical mechanisms behind land subsidence arising from natural compaction. Gambolati and Teatini [10] employed a 1-D nonlinear finite element model to investigate soil compaction induced by groundwater flow through an isothermal sedimentary basin subjected to a continuous vertical sedimentation process to mimic the evolution of the accreting Quaternary column. Zoccarato et al. [8] used an adaptive finite-element mesh to analyze the development and evolution of the Mekong Delta in Vietnam and then described accretion and natural consolidation to characterize delta dynamics. Aside from natural compaction, another crucial natural factor that brings about land subsidence is tectonic movement. Tectonic subsidence is most common in subduction zones. This can be seen in the Pingtung Plain, Taiwan [11], which has a very active subduction zone and is prone to land subsidence. In addition, groundwater extraction frequently exacerbates land subsidence in this area. The impact of soil texture, cyclic loading, and gravitational body force on one-dimensional consolidation of unsaturated and saturated soils was quantitatively examined in a series of papers by Lo et al. [12–14]. Employing the theory of poroelasticity, Lo et al. [15,16] have recently formulated a boundary-value problem based on a set of coupled partial differential equations to numerically model the spatial and temporal distribution of excess pore water and air pressures in a two-layer soil system with an upper unsaturated zone and a lower saturated zone caused by external loads.

Land subsidence can be induced not only by natural sources, but also by human intervention. Since 1935, Beijing, China's capital city, has been experiencing ground subsidence, with at least five significant sites. Yang and Ke [17] concluded that the rapid urban-area development in Beijing increased the density of multi-story buildings, which accelerated the occurrence of land subsidence, as evidenced by a time series of surface displacement data recorded from the Persistent Scatterer Interferometry Synthetic Aperture Radar (PSInSAR). Excessive groundwater pumping, on the other hand, is perhaps the most significant human intervention that gives rise to land subsidence. In fact, land subsidence resulting from groundwater removal has been a central problem in many regions over the world, including San Francisco Bay and the Florida Everglades in the United States [18], Romagna in Italy [5], central Taiwan [19], Esna City in Egypt [20], Bangkok in Thailand [21], central Saudi Arabia [22], Tehran in Iran [23], and Shanghai in China [24].

Land subsidence is an active study issue that involves various disciplines since it typically gives rise to large economic losses and disperses around the world. Several studies on land subsidence have been conducted using geodetic measurement techniques such as leveling, GPS, and satellite image (e.g., InSAR) to monitor surface deformations. Geodetic measurements are very useful in quantifying and interpreting the magnitude of land subsidence [25–29]. In addition to monitoring ground surface deformation, the technique of GPS and InSAR can also be used to assess the hydraulic head and aquifer system parameters for a regional area suffering from land subsidence [30–34].

Field experiments are another technique to gain physical insight into the behaviors of land sinking [35,36], but, they are often time-consuming and expensive. Therefore, numerical modeling that usually takes little time and is more cost-effective has become a preferred method for understanding and estimating the vertical deformation of the ground surface in local regions [37–39]. The integration of numerical simulation and field measurement is shown to be powerful because the implementation of both approaches can support the assessment process of land subsidence in a comprehensive framework that can systematically link its causes and effects.

The present study focuses on the northern part of Semarang City, which is located on the northern coast of Indonesia's Java Island. Land subsidence is a common phenomenon there and has been monitored using different geodetic measurements, including leveling, GPS, and satellite images [40–44]. Natural compaction and excessive groundwater pumping are the main causes of land subsidence in Semarang City [43,45], which has been taking place for over a century [46]. Land subsidence due to natural compaction in the study area has been investigated by Sarah, et al. [47]. It is revealed that natural compaction is prevalent in the Demak region (east of the study area), but that this is not been the case in Semarang City. The rapid piezometric drop in Semarang city has drained the excess pore pressure developed during deposition so that the current land subsidence is entirely due to groundwater exploitation. Measurement of active tectonic role to land subsidence using GNSS data sets [48] revealed that during 2011–2017, the rate of tectonic subsidence in Semarang City was 3–5 mm/year. Considering that the subsidence rate exceeds the natural displacement rate by an order of magnitude, the calibration process has not been affected by natural displacements.

Several sub-districts in Semarang City have undergone over-exploitation of groundwater resources, which has evidently given rise to extreme declines in groundwater elevations and has thus been regarded as the main cause of land subsidence there [49]. Field observations have indicated that several housing estates in the northern part of Semarang City are found to be experiencing land subsidence. People in the area have no alternative but to move and leave their homes as a result of land subsidence. Figure 1 depicts several houses in Genuk District that have been affected by land subsidence. Figure 1a shows that in the areas where land subsidence is occurring, the ground elevation of the houses is lower than the road elevation, resulting in heavy inundation during the rainy season. The picture clearly points out that newer houses must raise their base elevation to be at or above the road level. Figure 1b illustrates the current condition of an older house whose window elevation is the same as the ground elevation due to land subsidence.

Figure 1. (**a**) An old house in a housing area that has been impacted by land subsidence, thus resulting in the ground elevation of the houses is lower than the road elevation, (**b**) the current condition of an older house whose window elevation is the same as the ground elevation due to land subsidence.

In Indonesia, the availability of groundwater observation data and land subsidence monitoring data is still very limited. The use of satellite imaging, leveling, and GPS in land subsidence monitoring and assessment is also still separate. Furthermore, most groundwater and land subsidence studies in Indonesia, particularly in Semarang City, have been limited to individual hydrogeological, geomatics, and soil compression models, with no study combining different models to test their relationship and predict changes in groundwater that cause variations in land subsidence. Furthermore, a well-structured regulatory mechanism that can effectively and significantly reduce the negative impacts of land subsidence is required. The current study was thus undertaken to start from establishing a physically-consistent framework for numerically modeling the phenomenon of land subsidence caused by groundwater abstraction that occurs in the northern part of Semarang City. Our model was verified by monitoring data of crustal deformation obtained from leveling, GPS, and Sentinel-1 InSAR in Semarang City. The InSAR analysis generates a time series of deformations and corresponding speeds in the study area, which are also used for calibration. Next, the prediction of future land subsidence with and without groundwater management measures was performed based on our well-validated numerical model. Lastly, the effectiveness of the proposed regulatory policies was quantitatively evaluated in terms of reductions in the affected area of land subsidence in spatial and temporal scales.

2. Study Area
2.1. Focus of Study Area

Semarang City is the capital of Central Java Province and also the fifth largest city in Indonesia. Semarang City is divided into 16 sub-districts: Mijen, Gunungpati, Banyumanik, Gajah Mungkur, Semarang Selatan, Candisari, Tembalang, Pedurungan, Genuk, Gayamsari, Semarang Timur, Semarang Utara, Semarang Tengah, Semarang Barat, Tugu, and Ngaliyan. Among these sub-districts, Tugu, Semarang Barat, Semarang Utara, Semarang Tengah, Semarang Timur, Gayamsari, Genuk, and the northern part of Pedurungan are located in the study area (i.e., the northern part of Semarang City). According to field observations, Genuk, the northern part of Semarang Utara, Gayamsari, and Semarang Timur are experiencing land subsidence. The study area, as shown in Figure 2, spans 23.5 km × 10.5 km and is situated between 6°50′–7°10′ S and 109°35′–110°50′ E, near to the Java Sea.

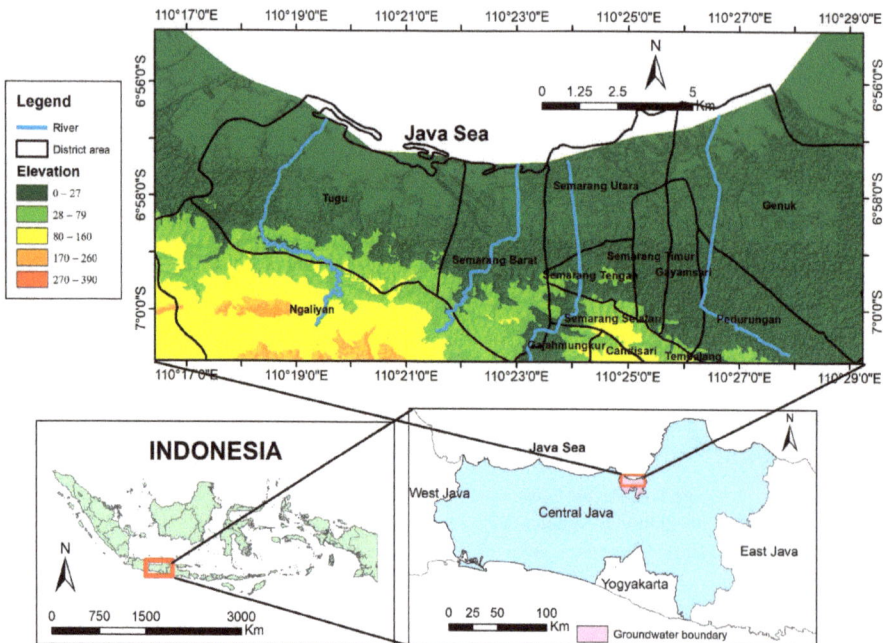

Figure 2. The study area (the northern part of Semarang City).

2.2. Geological Setting

Geologically, the northern part of Semarang city is situated in an alluvial plain that is bounded to the north by the Java Sea and to the south by the Semarang highland. The north plain in this area is composed of a thick sequence of Holocene alluvial deposits (Qa) and the south highland comprises Quartenary volcanic rocks (QTd) and Tertiary sedimentary formation (Tmpk) [50]. The geological formations in the northern area of Semarang City are strongly affected by the sedimentation process and shoreline changes. The heterogeneity of subsurface is influenced by the sea-level fluctuation from the last glacial maximum until the Holocene transgression, as explained in Sarah et al. [47]. The rapid advancement of the shoreline indicates the rapid deposition of the Semarang alluvial deposit, forming soft clayey sediment on the upper part of the subsurface. In the eighth century, Semarang City originated from a cluster of small islands off the coast of Pragota (now Bergota) [51]. The coastline changes result from the sedimentation process. Old topographical maps revealed that the Semarang shoreline advanced 8–12 m/year from 1695–1940. Figure 3a shows that an 884 m advancement took place over 1847–1991 [52]. It appears that accretion of 884 m occurred over 144 years, implying that sedimentation in the northern part of Semarang City was extensive during this time period.

Figure 3b provides an evolution of the coastal accretion and retreat processes for changes in shoreline from 1984 to 2016 based on satellite images. The coastal shoreline in 1984 was accreting towards the sea. This phenomenon was influenced by natural sedimentation and intensive reclamation for land development that began in the 1980s [53]. The shoreline position in 1994 was similar to that in 1984. Slight advancement was found near the west canal (Kanal Barat), possibly due to an increase of sedimentation discharge from the headwater. An examination of the shoreline in 2004 showed that land subsidence may have affected the coastal morphology. Shoreline retreat was observed at the western and eastern sections. The combined influence of sea-level rise and land subsidence caused the seawater to transgress the land; thus, the west and east coastal lands are susceptible to inundation. This was not seen in the center part, because some flood protection systems were built for the city center. The dyke, canals, and polder pumping system help prevent the city from the multiple disasters of subsidence, coastal flood (rob), and seasonal flood during the rainy season. In 2016, the shoreline retreated further in the western part, implying a more aggravating impact of land subsidence and sea-level rise. The 2016 shoreline position in the city center remained similar to 2014 and only some slight advancements were observed near the west canal, Semarang Port, and east Kaligawe industrial complex. The small shoreline advances were due to sediment discharge from the west and east canals, as well as reclamation work in the port, trade zone, housing, and industrial complex.

Figure 4 presents a geological map of Semarang City, which is one part of the Semarang-Demak groundwater basin [50]. The lithological point and axes are elaborated in Figure 5. The northern region of Semarang City is mainly composed of alluvial deposits (Qa) due to the existence of a sedimentation process that also causes changes in the coastline. The thickness of the alluvial deposit ranges from 20 to 100 m, becoming thicker to the north and east (Figure 5). The bulk of the sediments in the basin consists essentially of grey, very soft clay, and silt, with abundant calcareous and shell fragments. Tons of sands and gravels are found at various depths from 5 to 80 m (Figure 5a,b). At the southern highland, thin clay and silt overlay alluvial fans of sand and gravel (Figure 5b). Apart from Qa, Semarang City has Pleistocene sediments (in the form of the Damar Formation, Qtd), which were generated by sedimentations due to variations in sea level in at least the past 500 years or approximately in the middle of the 14th century [54], caused by transgression and regression processes that form deltas and tidal deposits [45].

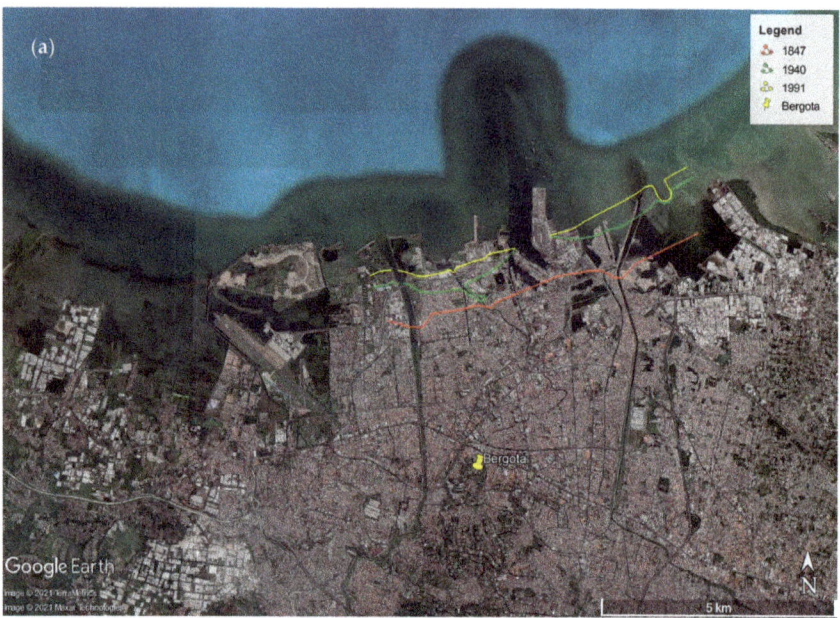

Figure 3. (**a**) Coast line changes of an 884 m advancement that took place over 1847–1991, (**b**) evolution of the coastal accretion and retreat processes for changes in shoreline from 1984 to 2016.

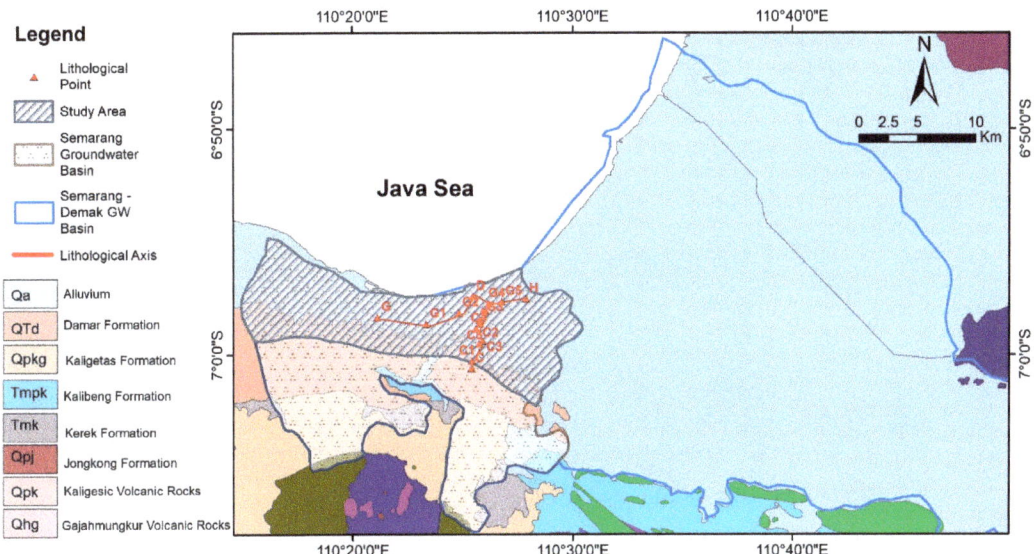

Figure 4. Geological map of Semarang City.

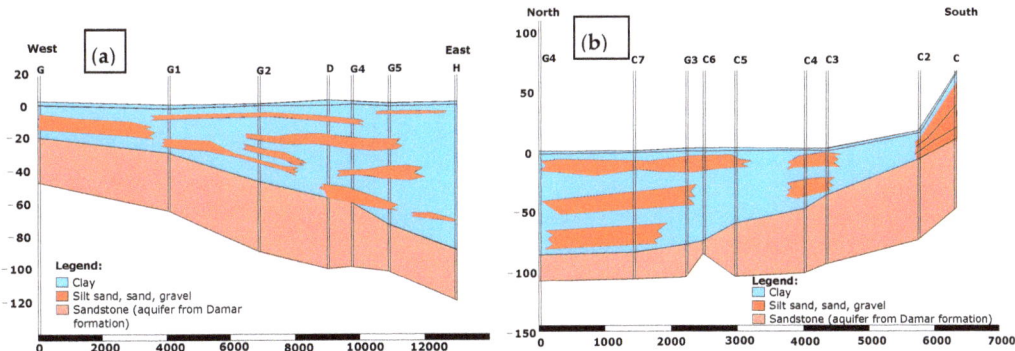

Figure 5. (a) The lithology of the northern region of Semarang City from west to east, (b) The lithology of the northern region of Semarang City from north to south.

The stratigraphy in Figure 5 is divided into three units, clay to silty clay (Unit 1), sand lenses (Unit 2), and volcanic sandstone (Unit 3). The clay unit has a soft to medium consistency with an N-SPT value of 1–9. The sand lenses are medium dense (n-spt value 17–22), and the volcanic sandstone is dense to very dense (n-spt value >30). The geotechnical properties are derived from borehole investigations in Terboyo and North Semarang, and laboratory analysis [45]. Compressibility of the aquitard units was derived from 1-D oedometer tests [55] and hydraulic conductivity was obtained from the falling head permeability test [56]. Properties of the coarse-grained sediments are characterized based on their grain-size composition [57]. The geotechnical properties for the subsurface units are presented in Table 1.

Table 1. The geotechnical properties.

Stratigraphic Unit	Natural Unit Weight (γ_n) (kN/m^3)	Initial Void Ratio (e_0)	Compressibility Index (c_c)	Coefficient of Recompression (c_r)	Modulus of Elasticity (E) (kPa)	Hydraulic Conductivity (k) (m/s)
Clay to silty clay (Unit-1)	15–17	1.2–1.7	0.35–0.74	0.10–0.17	1468–2000	1.68×10^{-10}–6.54×10^{-9}
Sand lenses (Unit-2)	17–19	1.5–2.0	0.41–0.77	0.11–0.18	4000–5000	1.59×10^{-6}–5.03×10^{-5}
Volcanic sandstone (Unit-3)	21–25	1.56–2.16	0.45–0.8	0.13–0.2	6900–7000	1.20×10^{-6}–2.20×10^{-5}

2.3. Hydrogeological Setting

The groundwater basin of the northern region of Semarang City, a part of the Semarang-Demak groundwater basin, is composed of unconfined and confined aquifers. The former (unconfined aquifer) is near the ground surface, with the groundwater table in direct contact with the atmosphere. The latter (confined aquifer) is separated from the unconfined aquifer by an impermeable barrier and is in compressed or semi-stressed status, consisting of lenses of sand and gravel that are covered by a layer of clay or sandy loam [53].

A hydrogeological evaluation to quantify interstitial fluid movements in aquifer layers needs a numerical representation of the groundwater flow regime. The hydrogeological dataset being inputted into our numerical models was collected from a geoelectric field study conducted in February 2019, as well as borehole works from a variety of sources, including the Balai Besar Wilayah Sungai (BBWS) Pemali-Juana, and Marsudi [53]. The lithology of the Semarang groundwater basin is illustrated in Figure 5.

The confined aquifer in the northern part of Semarang City is made up of two formations, i.e., the sandstone and conglomerates of the Damar formation along with the alluvial fan, as well as lenses of sand and gravel of the Garang deltaic deposit. The Garang delta was developed from the former river channels during the deposition of the deltaic deposit. In the Damar delta, groundwater flows from the volcanic rocks in the southern hills to the sedimentary basin in the north Semarang. The Damar formation has long served as a reliable source of fresh groundwater, mostly exploited by industries using deep wells at depths of 60–180 m. The Garang delta is also exploited by wells, but to a lesser degree due to its limited lateral extent. The numbers of registered wells and their output capacity are presented in Figure 6.

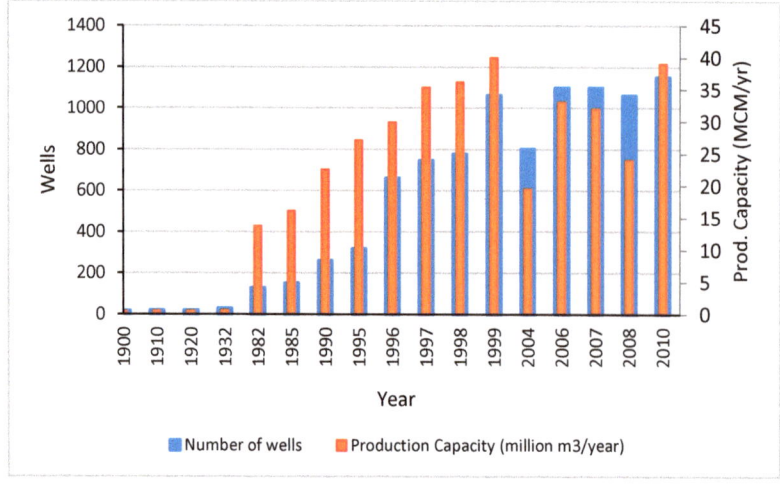

Figure 6. The number of registered deep wells in Semarang City and their output capacity.

In Semarang City, field investigation reveals that regional groundwater flows from the south-southwest to the north-northeast through the deposits of gravel lens and sand lens. As a result, the groundwater level tends to fall to the north-northeast, with the conical decline's center pointing towards LIK Kaligawe. Pump wells have been considerably used to extract groundwater in Semarang City and have two types, i.e., dug wells (shallow wells) and boreholes (deep wells). Boreholes, with depths ranging from 60 to 180 m, are positioned in a confined aquifer and often utilized for industrial purposes, whereas dug wells are typically installed by locals for their daily needs. In 1900, the number of pump wells in Semarang City was first reported, but these data are only available until 2010. Figure 6 shows the production capacity and the total number of groundwater wells [53,58–60], indicating that both the number and capacity of wells are increasing every year. Previous groundwater monitoring showed that the groundwater condition in Semarang city is already stressed. Over-exploitation of groundwater resources caused the formation of a cone of depressions that was first observed in 1984, but widened southward in 2010. Groundwater over-withdrawal also leads to the lowering of the piezometric pressure, thus increasing the effective stress of the aquifer system. However, due to the low permeability of the aquitard layer, the dewatering process is delayed, thus giving rise to gradual land subsidence. When the pore pressure decrease is smaller than the preconsolidation stress of the aquitard, the phenomenon of small, elastic, reversible subsidence occurs [61]. On the contrary, when the pore pressure drop is larger than the preconsolidation stress, it induces large, inelastic, irreversible subsidence.

3. Ground Deformation

3.1. Leveling and GPS Survey for Land Subsidence Monitoring

Land subsidence measurements were taken in the northern part of Semarang City by several institutes. Land subsidence was measured at 137 locations intermittently from 1991 to 2019 [62]. However, because the measurements (leveling and GPS) were carried out by different agencies and not synergized with each other, the measurement data at each point were not recorded continuously, but instead only for very short periods of time. Accordingly, in the current study, the data from the monitoring points of land subsidence that are close to each other, which are measured at different years, are used for calibration of the numerical model. Figure 7 depicts the time scale and period of leveling and the GPS survey points for land subsidence monitoring.

3.2. Interferometric Synthetic Aperture Radar (InSAR) Data and Processing

The most widely applied technologies for monitoring ground subsidence may be leveling and GPS, although their usage is still restricted due to a lack of geographic samples and expensive cost [63], which accordingly makes it difficult to undertake long-term monitoring of land subsidence. InSAR technology has been developed in recent years to tackle this issue [25]. Although this satellite-derived radar imaging has the potential to become a formidable tool for providing low-cost, continuous ground movement data over a vast area [64], the method of assessing ground deformation using InSAR image suffers from low accuracy because of the spatial-temporal decorrelation of the distance between objects and satellite orbits, high land cover complexity, and signal interference caused by air conditions [63–66].

DInSAR (Differential InSAR) is a technique for generating a large-scale map of line-of-sight (LOS) components using highly precise displacements [67]. A multitemporal deformation map, as well as many differential interferograms of the same zone from separate tempo acquisitions, must be considered. The short baseline subset (SBAS) and the persistent scatterer interferometric synthetic aperture radar (PSInSAR) are two sophisticated DInSAR algorithms that can be used to monitor the deformation of the earth's surface over time. SBAS is carried out by using a mixture of differential interferograms produced by data pairs with a modest orbital separation (baseline) [68]. PSInSAR is based on phase characteristics and detects low-amplitude pixels with phase stability that are not found by conventional amplitude-based methods. It also uses spatially-correlated phases rather than historical phases so that the variables can be examined across time [69]. Consequently, the SBAS and PSInSAR algorithms are applied to address the problem of inaccuracies that may occur while using InSAR to detect land subsidence.

Figure 7. Leveling and GPS measuring points.

The InSAR data used in this study was taken from Sentinel-1 between 2015 and 2020 (https://comet.nerc.ac.uk/comet-lics-portal/, accessed date: 25 June 2020), with frame ID 076D_09725_121107 covering Central Java Province. The boundary of the area covers 6°50′0″–7°10′0″ S and 110°12′0″–110°37′0″ E, and more than 100 InSAR images were used in the analysis. An open-source program called LiCSBAS was applied in the present study to execute the InSAR time series analysis using LiCSAR products [70]. Generic Atmospheric Correction Online Service (GACOS) was used to correct atmospheric inaccuracies appearing in SBAS-InSAR [71]. Due to the correctness and coverage of unwrapped data, as well as loop closure control, inaccurate unwrapped data must be recognized and removed in the time series analysis. GACOS atmospheric products employ the interpolation technique of iterative tropospheric decomposition (ITO) to remove elevation-related and turbulent signals from the zenith total delay (ZTO) and then provide high-resolution tropospheric delay maps for InSAR and other data [71].

SBAS-InSAR is an analytical method for the multi-image InSAR time series to generate an estimation of the deformation of the earth's surface by combining interferometric pairs from small time-space baselines. Land subsidence detection and monitoring using SBAS-InSAR has now become predominant. The analysis starts with $N + 1$ SAR pictures taken at ordered times $(t_0, \ldots, t_N)$ and is based on the same area. Assuming that each acquisition can interfere with other images, each SB subset thus must have at least two acquisitions, giving rise to the following inequality for the number of potential differential interferograms M [68]:

$$\frac{N+1}{2} \leq M \leq N\left(\frac{N+1}{2}\right) \tag{1}$$

Using the estimated generic j-interferogram of the SAR acquisition at times t_A and t_B, the topographic phase component removal in the azimuth pixels and coordinate range (x, r) can be described as follows:

$$\Phi_j(x,r) = \Phi(t_B, x, r) - \Phi(t_A, x, r) \approx \frac{4\pi}{\lambda}[d(t_B, x, r) - d(t_A, x, r)] \tag{2}$$

where $d(t_B, x, r)$ and (t_A, x, r) are the cumulative line-of-sight (LOS) deformations at time t_A and t_B with respect to the reference instant t_0 and λ is the transmitted signal's center wavelength. Consequently, one can obtain $d(t_0, x, r) \equiv 0$, and it is fair to identify $d(t_i, x, r)$ as the required deformation time series, with $i = 1, \ldots, N$. Assume $\Phi(t_i, x, r)$ to be the corresponding phase component; therefore, we have $\Phi(t_i, x, r) \approx \frac{4\pi d(t_B, x, r)}{\lambda}$.

Based on a series of displacement results, an SB inversion was performed on the interferogram network to estimate the velocity of a surface pixel over time. It is assumed that an M-unwrapped interferogram stack $d = [d_1, \ldots, d_M]^T$ was generated from N images acquired at $(t_0, \ldots, t_{N-1})$ incremental displacement vector $m = [m_1, \ldots, m_{N-1}]^T$ (i.e., m_i is the incremental displacement between time t_{i-1} and t_i), and can be extracted by solving Equation (3):

$$d = Gm \tag{3}$$

where G is an $M \times (N-1)$ zero architecture matrix representing the interferogram network relationship with incremental displacements, and the unwrapped interferogram (difference between two acquisitions) is the sum of the corresponding incremental displacements [72]. Cumulative displacements (i.e., displacement time series) are calculated by adding the incremental displacements for each acquisition. The mean displacement velocity is then computed based on at least the quadrature of the cumulative displacements.

The NSBAS approach [73] was used, which imposes a temporary limitation to obtain a more practical time series of the displacement even with a disconnected network.

$$\begin{bmatrix} d \\ 0 \end{bmatrix} = \begin{bmatrix} \begin{matrix} & G & & 0 & 0 \\ 1 & 0 & \cdots & 0 & -t_1 & -1 \\ \vdots & \ddots & \ddots & \vdots & -t_2 & \vdots \\ 1 & \cdots & 1 & \ddots & \vdots & \vdots \\ \vdots & & \vdots & \ddots & 0 & \vdots \\ 1 & \cdots & 1 & \cdots & 1 & -t_{N-1} & -1 \end{matrix} \end{bmatrix} \begin{bmatrix} m \\ v \\ c \end{bmatrix} \tag{4}$$

A linear displacement ($d = vt + c$) is assumed if "γ" is the scaling (weighting) element in the temporal constraints. The low time limit has little effect on the solution within the network's linked components (e.g., 0.0001). As a result, the time restriction component only affects the connection via network gaps. Equation (4) can be used for pixels with fully connected networks as well as pixels with gaps.

Due to the limitation of interferometric phases, phase unwrapping was applied to quantify the absolute value of the phase with respect to a reference point inside the interferogram [74]. In the loop closure and mask time series phase, the mask was created utilizing multiple noise indices acquired at earlier stages for the time series and speed of displacement. If the value of any noise indicator for that pixel was greater than or less than the threshold set, that pixel was masked. The generated time series also includes several noise-related conditions, such as residual tropospheric noises, ionospheric noises, and orbital errors. A space-time filter can be utilized to isolate these components from the displacement time series (i.e., high-pass time and low-pass space) [75].

3.3. Land Deformation Mapping

High-precision subsidence mapping can be obtained using a satellite-based technique based on SAR using the DInSAR method. The DInSAR method is based on analyzing SAR images to identify surface changes to sub-centimeters along the sensor's line of sight to the target, or Line of Sight (LOS), in order to calculate the LOS displacement value. LOS is the surface displacement between the satellite and the ground pixel in the azimuth direction along the flight path. The method is commonly used to estimate vertical displacement is to divide the LOS displacement obtained from DInSAR by the cosine of the incidence angle, assuming no horizontal movement occurs [76]. Furthermore, by using the least-squares algorithm, which is a matrix approach that aims to estimate the unknown parameters, during the inversion process, the displacement time series (in millimeters) can be obtained. After obtaining the displacement time series, a regression is performed to determine the annual average rate of land subsidence.

Before mapping land deformations, InSAR vertical displacement must be validated with GPS station measurements. There are several GPS points in the InSAR image frame, but many of them do not have enough temporal overlap with the InSAR 2015–2020 time series. Only the GPS points of K371, KOP 8, SMK 3, N259, BM11, 259, SFCP, and 1114 have temporal overlap with InSAR time series. Figure 8 depicts an InSAR validation for 2015–2020 with displacements monitored using GPS for 2015–2019. When compared to GPS monitoring results, it appears that in Figure 8, the InSAR displacement is in good agreement with an RMSE of 0.810 cm and R^2 of 0.949.

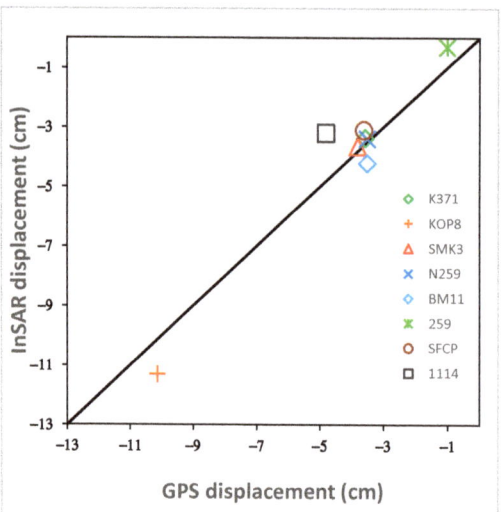

Figure 8. InSAR vertical displacement vs. GPS displacement.

Figure 9 gives a map of the deformation velocity generated by the InSAR analysis in the northern section of Semarang City from 2015 to 2020. Inspecting Figure 9, it can be noted that the most significant deformation occurs in the north-east part of Semarang City, with a rate above 80 mm/year, whereas the west and southern parts of Semarang City are dominated by an uplift rate up to 20 mm/year. Figure 10 illustrates a time series of the deformation with and without a spatial time filter at the black dot in Figure 9 in more detail. The time series of the black dot in Figure 9 reveals that the magnitude of land subsidence was roughly 300 mm from 2015 to 2020.

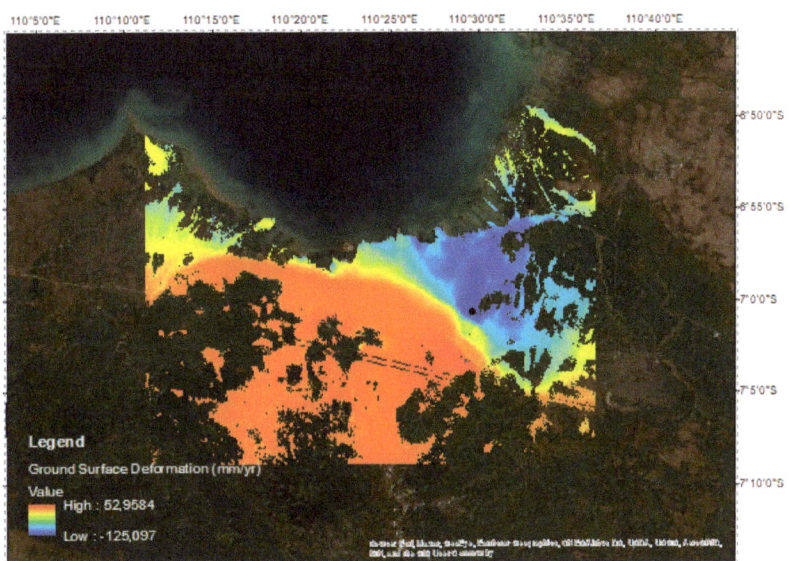

Figure 9. The deformation velocity map.

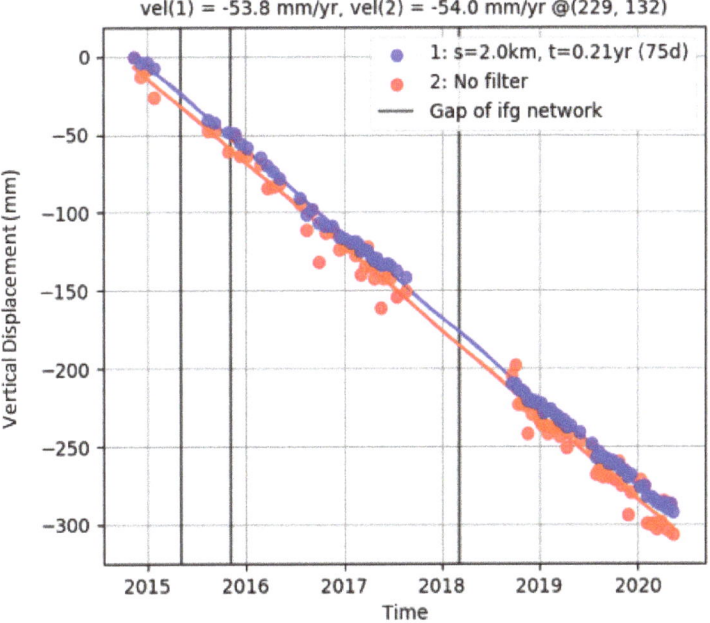

Figure 10. A time series of the deformation.

A plot was constructed from the base map to determine the position of land deformation and uplift in northern Semarang City, creating a deformation map as shown in Figure 11. By inspecting Figure 11, one can observe that the highest deformation occurs in Genuk District, followed by Semarang Utara District and parts of Semarang Timur and Gayamsari Districts. Meanwhile, Semarang Timur and Gayamsari, as well as practically all of Tugu, Semarang Barat, Semarang Tengah, Semarang Selatan, Pedurungan, Gajahmungkur, Candisari, Tembalang, and Ngaliyan's sub-districts experience the uplift status. The vertical uplift in the southwest part of Semarang City may be related to thrust faults in Semarang City, which is the part of the Baribis–Kendeng active fault [77].

Figure 11. The deformation map in the northern part of Semarang City based on InSAR.

4. Groundwater and Land Subsidence Numerical Model

4.1. Groundwater and Geotechnical Subsidence Equation

Groundwater and land subsidence behaviors have been evaluated extensively and quantitatively using numerical modeling. In the current study, the groundwater flow analysis was performed with the flow package in MODFLOW, while the land subsidence analysis was carried out with the SUB (Subsidence and Aquifer-System Compaction) package. The SUB package can be applied to simulate both elastic (recoverable) and inelastic (non-recoverable) interbed compactions with either no or delayed drainage.

The partial differential equation that combines Darcy's Law with mass balance for describing a three-dimensional groundwater movement in MODFLOW takes the form:

$$\frac{\partial}{\partial x}\left(K_{xx}\frac{\partial h}{\partial x}\right) + \frac{\partial}{\partial y}\left(K_{yy}\frac{\partial h}{\partial x}\right) + \frac{\partial}{\partial z}\left(K_{zz}\frac{\partial h}{\partial z}\right) - W = S_s \frac{\partial h}{\partial t} \qquad (5)$$

where x, y, and z express Cartesian coordinates; K_{xx}, K_{yy}, and K_{zz} designate the tensor components of hydraulic conductivity in the x, y, and z axes; W represents the volumetric flux of water sources and (or) sinks per unit volume; S_s denotes specific storage; and h signifies hydraulic head.

The mechanism of vertical soil deformation induced by groundwater pumping is similar to that of soil consolidation acted upon by vertical load compression, in which the cause of consolidation is an increase in vertical effective stress and is then followed by a reduction in pore volume. However, in the former, excessive groundwater extraction leads to a decrease in pore water pressure and in turn raises effective stress, thus bringing about a decrease in pore volume [33,78].

According to Terzaghi's effective stress concept, this can be mathematically written as:

$$\sigma'_{ij} = \sigma_{ij} - \delta_{ij}u \qquad (6)$$

where σ'_{ij} is the effective stress tensor component; σ_{ij} is the total geostatic stress tensor component; δ_{ij} is the Kronecker delta function where $\delta_{ij} = 1 \, if \, i = j$ or $\delta_{ij} = 0 \, if \, i \neq j$; u is the fluid pore pressure; i and j represent the Cartesian coordinates x, y, and z. For 1D vertical stress problems, the equation reduces to:

$$\sigma'_{zz} = \sigma_{zz} - u \tag{7}$$

The general equation that evaluates the thickness of compaction or expansion, Δb, between intervals t_{n-1} and t_n can be written as

$$\Delta b = \frac{0.434 b_0}{(1+e_0)\sigma'} [C_n(\sigma'_n - \sigma'_{c,n-1}) + C_r(\sigma'_{c,n-1} - \sigma'_{n-1})]$$
$$C_n = \begin{cases} C_c, \sigma'_n > \sigma'_{c,n-1} \\ C_r, \sigma'_n < \sigma'_{c,n-1} \end{cases} \tag{8}$$

where e_0 is an initial void ratio, b_0 is initial thickness, σ' is effective stress, σ'_{n-1} and σ'_n are the effective stresses at t_{n-1} and t_n, respectively; $\sigma'_{c,n-1}$ is the preconsolidation stress at t_{n-1}. The relationship of σ'_n to $\sigma'_{c,n-1}$ is used to decide whether the value of C_n is C_c or C_r. The equation provides a quantitative estimate of settlements in over-consolidated sediments, those in normally-consolidated sediments, and those in sediments under transition from the over-consolidated state to the normally-consolidated state.

4.2. Groundwater and Geotechnical Subsidence Equation

The land subsidence process involves the hydro-mechanical coupling between the flow of groundwater and the deformation of the solid matrix in the aquifer system. Our models are rigorously established in a physically-consistent manner with the land subsidence monitoring results from GPS and InSAR. A predictive model for future development of land subsidence is then constructed based on groundwater usage regulations adopted by Lo et al. [49]. In contrast to Lo et al. [49], who used a grid measuring 250 m × 250 m to perform a numerical simulation of the groundwater model and its management, the simulation in this study used a grid measuring 100 m × 100 m. Furthermore, more aquifer parameter data were collected and input for modeling land subsidence. Since the northern part of Semarang City is part of the Semarang groundwater basin, a conceptual model for the groundwater and land subsidence model was made for the entire Semarang groundwater basin. Figure 12 is a schematic of the computational cell of a groundwater model for the Semarang City groundwater basin. The height of the cell is equivalent to the vertical depth of the soil layer at each site. Following the geological investigation presented in Figure 5, the model's z-axis complies with the stratigraphic condition that is made up of three layers, consisting of (1) upper aquitard intercalated with (2) sand and gravel lenses, and (3) the Damar Formation at the base.

The boundary conditions for the numerical simulation of the Semarang groundwater basin are prescribed as (1) a no-flow boundary in some southern parts that reflect the existence of impermeable rocks and reverse faulting, (2) a groundwater divide boundary in the southwest part (extending from the south to the north), and (3) a constant head boundary in the north, west, and east parts due to sea and rivers.

The topographic data used in the numerical simulation are cited from DEM produced by DEMNAS Indonesia with a resolution of 0.27 arc-sec ≈ 8.325 m, while the data for the coordinates of the well are taken from the Mining and Energy Agency of Central Java Province, and the tidal data is the average tide of 1.07 m from the Meteorology and Geophysics Agency.

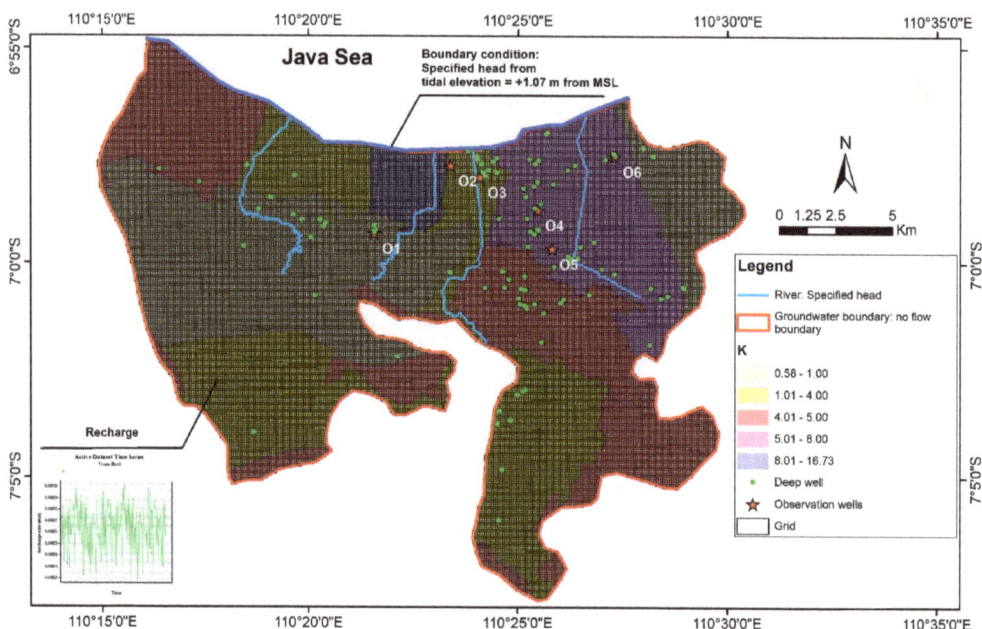

Figure 12. Schematic of the computational cell of the Semarang groundwater basin.

Semarang, similar to other Indonesian cities, has two seasons, i.e., dry and wet (rainy). The dry season lasts from April to September, while the wet season begins in October and ends in March. The average annual temperature in Semarang City is 28.08 °C, with a humidity of 76.61%. In Semarang City, annual rainfall ranges from 1500 to 3000 mm. Rainfall during the rainy season recharges groundwater and also causes river overflows and floods [79]. However, we did not include the effects of river overflow and flooding as specified head conditions due to insufficient data. Rainfall data recorded by the Indonesian Water Resource Agency, Public Work Department, are used for recharge input data from 1970 to 2010. Mangkang Barat/West Mangkang River, Mangkang Timur/East Mangkang River, Garang River, and Canal Timur/East Canal River are all included in the model and are prescribed as the boundary conditions (IBOUND). The piezometric level of the Damar aquifer in Semarang City drops rapidly and has surpassed the aquitard past maximum effective stress, and the demand for groundwater from the Damar aquifer was assumed constant during the dry and wet season.

The aquifer of Semarang City is separated into 20 hydrogeological polygons to input the hydrogeological parameters into the numerical model, as shown in Figure 12. The hydrogeological polygons separating sections were initially obtained by borehole and laboratory data in Table 1. The values of the elastic skeletal storage coefficient (*Sfe*) and inelastic skeletal storage coefficient (*Sfv*) are used for the no-delay interbed layer, whereas vertical *K* (hydraulic conductivity), elastic particular storage, and inelastic specific storage are integrated into the delay interbed layer. For modeling purposes, the initial properties were adjusted to satisfy the model calibration, as summarized in Table 2. In this modeling, a delayed bed was assigned for Unit-1 as justified by its behavior as an aquitard. Unit-2 and 3 were assigned non-delay beds and *Sfe* and *Sfv* were set equal to establish elastic behavior.

Table 2. Hydrogeological model parameter.

Stratigraphic Unit	Hydraulic Conductivity (k) (m/s)	Sfe	Sfv
Clay to silty clay (Unit-1)	1.68×10^{-10}–6.54×10^{-9}	5.56×10^{-3}	1.31×10^{-2}
Sand lenses (Unit-2)	1.59×10^{-6}–5.03×10^{-5}	2.45×10^{-3}	2.45×10^{-3}
Volcanic sandstone (Unit-3)	1.20×10^{-6}–2.20×10^{-5}	1.42×10^{-3}	1.42×10^{-3}

4.3. Groundwater Numerical Model for the Steady-State Flow Model

The steady-state flow model was utilized to ensure that the hydrogeological input data are rational before they are next applied as the initial condition for the transient flow model. In the steady-state flow model, the values of appropriate input parameters for all grids, such as recharge and hydraulic conductivity, are determined to guarantee that the computed head in the model is able to match the observed head accurately. The steady-state flow model was performed based on the period of 1970 to 1990. Since the steady-state conditions do not take into account the temporal changes, the calibration of the steady-state process used the average observed data from 1970–1990 obtained from observation wells to assess whether the developed model is in agreement with field measurements. There are 54 observation wells in Semarang City. Unfortunately, much recorded data at observation wells have been lost. To this end, the model in this work was calibrated and verified using data from six observation wells, i.e., Prawiro Jaya Baru (O1), PRPP (O2), SMKN 10 (O3), Wot Gandul (O4), Santika Hotel (O5), and LIK. Kaligawe (O6).

To reduce the difference between the calculated and measured heads, a number of trial-and-error attempts were made to calibrate the hydraulic conductivity and recharge within the range of their theoretical values. The calibration results between the computed and observed values are depicted in Figure 13. It can be seen that values are close to the 45° line on the graph, mathematically pointing out a perfect correspondence between the computed head and the observed head. The colored bars represent the magnitude of the error at each observation well, with green color indicating that the error value is less than 1 m, yellow color indicating that the error value is between 1 m and 2 m, and red color indicating that the error value is greater than 2 m.

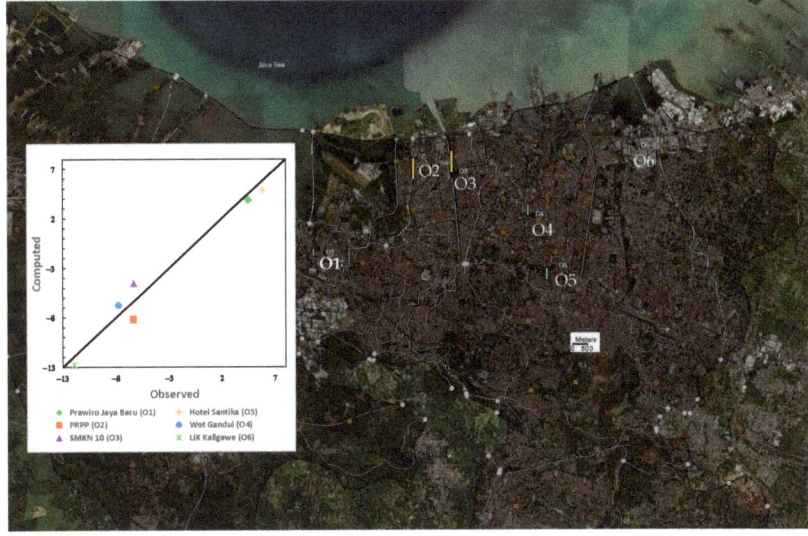

Figure 13. Calibration results of the steady-state flow model.

Figure 13 shows that the correlation value is very close to the 45° line at all observation wells. The value of Root Mean Square Error (RMSE) computed for all observation wells is 1.270 m. The individual discrepancy in water head for Prawiro Jaya Baru (O1), PRPP (O2), SMKN 10 (O3), Wot Gandul (O4), Hotel Santika (O5), and LIK Kaligawe (O6) are 0.618 m, 1.805 m, 1.812 m, 0.995 m, 0.980 m, and 0.597 m, respectively. Among them, the values at Wells O2 and O3 are relatively larger. Considering that the distance of the O2 and O3 observation wells to the boundary conditions is 0.29 km and 1.56 km, respectively, it is possible that the specified head boundary has a strong influence on the observation wells.

4.4. Groundwater and Land Subsidence in the Past

After the steady-state flow model was precisely calibrated, simulation and calibration were then undertaken for the model of transient flow and land subsidence to analyze the effect of groundwater flow due to pumping on land subsidence. The recharge rate utilized in the transient flow model was based on precipitation data from 1990 to 2010. Calibration of the transient flow model was conducted based on the period from 1990 to 2010, whereas the period from 2010 to 2020 was subsequently applied for validation. As achieved in the steady-state flow model, the computed groundwater level was thus compared with the observed level in the historical record of six observation wells from 1990 to 2010. The calibration of land subsidence magnitude was performed with the recorded data at the GPS measurement stations around the four observation wells (O2, O3, O4, and O6) since the other two observation wells are far from the GPS measurement points. Land subsidence data from 1990 to 2014 were utilized for land subsidence calibration. In addition to the GPS measurement, InSAR data were also used for calibration and validation. The use of InSAR can bridge the gaps between the leveling and GPS methods, and can provide the latest deformation because the satellite image is available until 2020. Figure 14 shows the subsidence contour map at the end of 2010 based on the simulation result, as well as the comparison of the computed and observed groundwater heads from 1990 to 2010. Figure 15 presents a graphic representation of the simulation result of land subsidence at O2, O3, O4, and O6, as well as the calibration with various geodetic measurements.

Figure 14 reveals that the most significant drop in land subsidence and groundwater level in 2010 occurred in Semarang Utara and Genuk sub-districts, whereas there was a vertical uplift in the southwest part of Semarang City. The values of RMSE for the calibration of groundwater level calculated in the transient flow model at O1 to O6 are 0.42 m, 1.91 m, 1.82 m, 1.76 m, 0.86 m, and 1.1 m, respectively, while the coefficient of determination (R^2) takes the value of 0.6, 0.84, 0.91, 0.8, 0.75, and 0.98 at O1 to O6, correspondingly. One can also find from Figure 15 that the highest land subsidence in 2020 still took place in the northeast region of Semarang City. The RMSE values for calibrating the simulation of land subsidence at O2, O3, O4, and O6 are 5.88 cm, 7.34 cm, 7.32 cm, and 2.17 cm, respectively, while the R^2 values at these four observation wells are 0.91, 0.97, 0.99, and 0.99, correspondingly. This implies in the mathematical sense that the land subsidence model is in good agreement with field measurements. InSAR was also utilized to calibrate the land subsidence model when no land subsidence monitoring data from leveling or GPS was available, as done in Figure 15.

Figure 15 also reveals the fact that O2 (PRPP) at the northwest region of Semarang City and O6 (Kaligawe) at the northeast suffered from the greatest land subsidence. It can also be observed that the linear portion of the subsidence curves in Figure 15 represents the fast subsidence rate during 1990–2000. The subsidence patterns follow the groundwater withdrawal trend as observed in Wotgandul and Kaligawe. Different patterns are shown in the northwest area (PRPP and SMKN); the fast subsidence rate occurred from 1990 to 2000, but subsidence appears to decelerate afterward while groundwater levels were still falling. These differences are possibly due to the heterogeneity of the subsurface stratigraphy and hydraulic conductivity. In addition, a physical explanation behind this phenomenon can be explored by revisiting Figure 5, which shows that the sand lenses within the alluvial deposit are more intensive in the northwestern part. These sand layers act as drainage paths that accelerate pore pressure dewatering of the clayey soil, causing fast subsidence during the early period of a piezometric drop. Further decrease in piezometric levels thus creates smaller subsidence due to the dissipation depletion of excess pore water pressure. Therefore, land subsidence alleviated after 2010, partially because further lowering of piezometric levels did not surpass the soil preconsolidation pressure, in turn leading to a smaller subsidence rate. Preconsolidation pressures of the subsurface clay were measured by oedometer tests from undisturbed samples retrieved from borehole SMG-01 (northeast Semarang) and SMG-02 (northwest Semarang). The clay layers in SMG-01 are underconsolidated to normally consolidated, while the clay layers in SMG-02 are overconsolidated up to 40 m depth and normally consolidated at the depth of 50–60 m. The critical groundwater level required by the clay layer to undergo inelastic compaction was calculated from its preconsolidation pressure. It was shown that the piezometric groundwater level in the northeast has surpassed the critical groundwater level; therefore, inelastic compaction occurs. Meanwhile, in the northwest, the piezometric groundwater level is still above the critical groundwater level; hence, the subsidence is small.

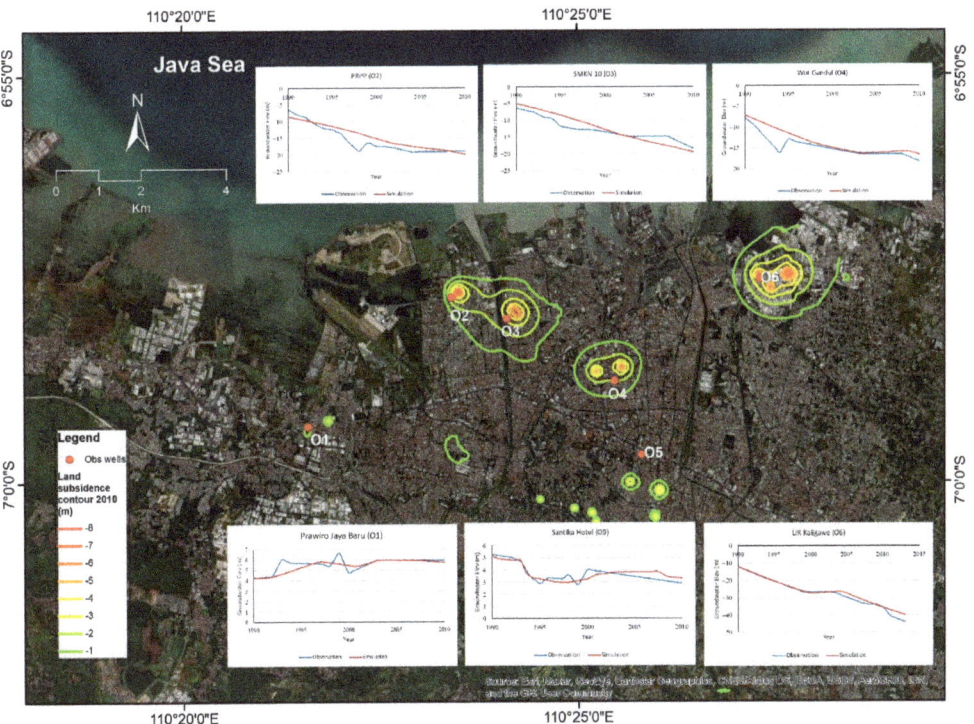

Figure 14. Land subsidence contour map in 2010 and comparison of computed groundwater levels with measurements.

Figure 15 also depicts that in Wot Gandul and Kaligawe, land subsidence was found to accelerate after 2010. This can be physically illustrated by the fact that groundwater piezometric levels have dropped beyond the recent preconsolidation pressure, thus causing fast, irreversible subsidence. A detailed examination of Figure 14 indicates that every 1 m drop of a piezometric level can result in an approximately −0.1 m vertical settlement.

To validate the simulation results, simulated subsidence was compared to subsidence based on InSAR analysis at points other than observation points. Figure 16 depicts sample points for detecting land subsidence using InSAR and simulated subsidence.

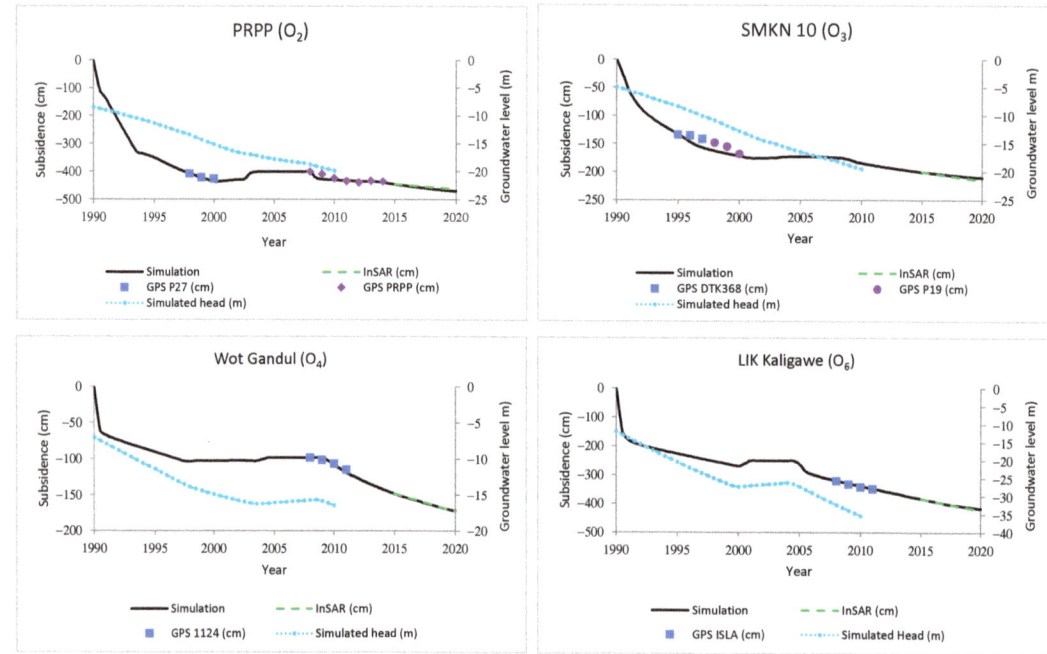

Figure 15. Calibration for the simulation of land subsidence.

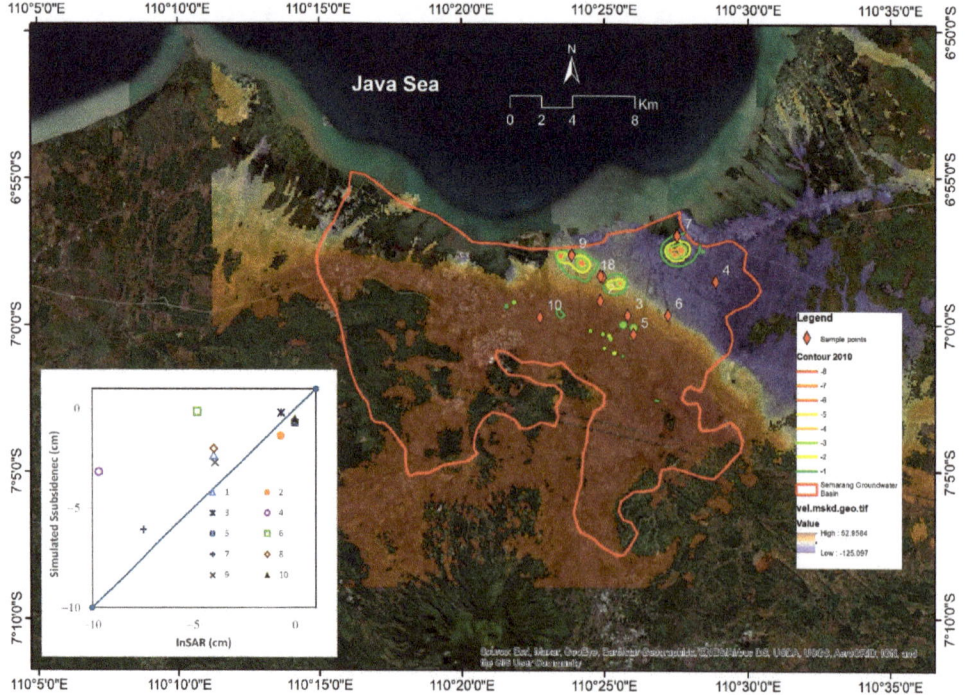

Figure 16. Simulated subsidence vs. InSAR.

According to Figure 16, the simulated subsidence versus InSAR in the northern part of Semarang is close to the 45° line and has a small RMSE value (below 2 cm), whereas the RMSE values in the east-south area (Points 4 and 6) are quite large (above 2 cm) and quite far from the 45° line. The east-south compaction rate has a high RMSE value, which could be attributed to the onset of the subsidence process and the difference in time spans used in the two types of data sets in the area. The land subsidence simulation was carried out by taking a period of 30 years starting in 1990 and ending in 2020, whereas InSAR only examined the period 2015–2020. Based on the land subsidence simulation, the piezometric pressure in the east-south part of the simulation was drastically reduced at the start of the simulation. This is consistent with the piezometric contours of the Damar Formation aquifer [80], as well as the very large decrease in piezometric pressure of the Damar Formation aquifer until 1997, followed by a slower decrease [45]. However, when compared to the GPS measurement results for subsidence, the simulation results in the area have similar values.

4.5. Future Projection of Land Subsidence

This section concerns whether the problem of land subsidence will worsen if groundwater over-extraction continues in the northern part of Semarang City. A further assessment was thus conducted to predict the future development of land subsidence, after which the countermeasures for managing the problem of drops in regional groundwater level and resulting land subsidence are recommended. Lo et al. [49] have recently conducted a numerical study to provide future predictions of groundwater levels under different scenarios based on the implementation of various management measures. Their results show that among three proposed management measures (i.e., TS2, TS3, and TS4), a regulation strategy to reduce by 10% both the number and production capacity of deep wells from 2035 to 2050 has the most significant effect on groundwater restoration.

The same two scenarios (TS3, and TS4) for controlling groundwater pumping used by Lo et al. [49] were adopted and then performed in the land subsidence simulation in this study. The TS3 and TS4 scenarios were compared with the TS1 scenario without any management measure to quantify the effectiveness of these two management strategies on land subsidence. The TS1 scenario is a direct projection of land subsidence from the trend line (calculated in Lo et al. [49]) of the output capacity of deep wells from 2010 to 2050. The TS3 and TS4 scenarios are the representations as a consequence of a reduction in the number of deep wells and their production capacity by 5% and 10%, respectively, annually from 2025 to 2034, and then maintaining their number and capacity from 2035 to 2050. Figure 16 depicts the simulation results of land subsidence from 2010 to 2050 at O2, O3, O4, and O6. The contour maps of land subsidence in 2050 are presented in Figure 16 for TS1, TS3, and TS4, respectively.

A comparison of Figures 15 and 17 reveals that the subsidence trend at PRPP (O2) and SMKN 10 (O3) subject to the TS1 scenario until 2050 takes a similar path to Figure 15, with rates of 10.8 mm/year and 11.98 mm/year, respectively. Higher rates of land subsidence were found at Wot Gandul (O4) (25.83 mm/year) and Kaligawe (O6) (36.8 mm/year). When the control measures (i.e., TS3 and TS4) are put in place, the subsidence degree at PRPP (O2) shows a rebound of around 10 to 27 cm at the end of 2050 as compared to 2020. A slowing down of subsidence rates from 2020 to 2050 can be also observed at SMKN 10 (O3) (26.03% and 51.42% under the TS3 and TS4 scenarios, respectively), Wot Gandul (O4) (27.27% and 53.03%, respectively), and LIK Kaligawe (O6) (27.87% and 59.08%, respectively). However, at PRPP (O2), we found that variations in effective stress due to changes in groundwater level do not surpass its preconsolidation pressure; therefore, the reduction in subsidence rates when the groundwater level is raised under the TS3 and TS4 scenarios would not be significant, while in other places, the rebound of the groundwater level is capable of reducing the subsidence rate.

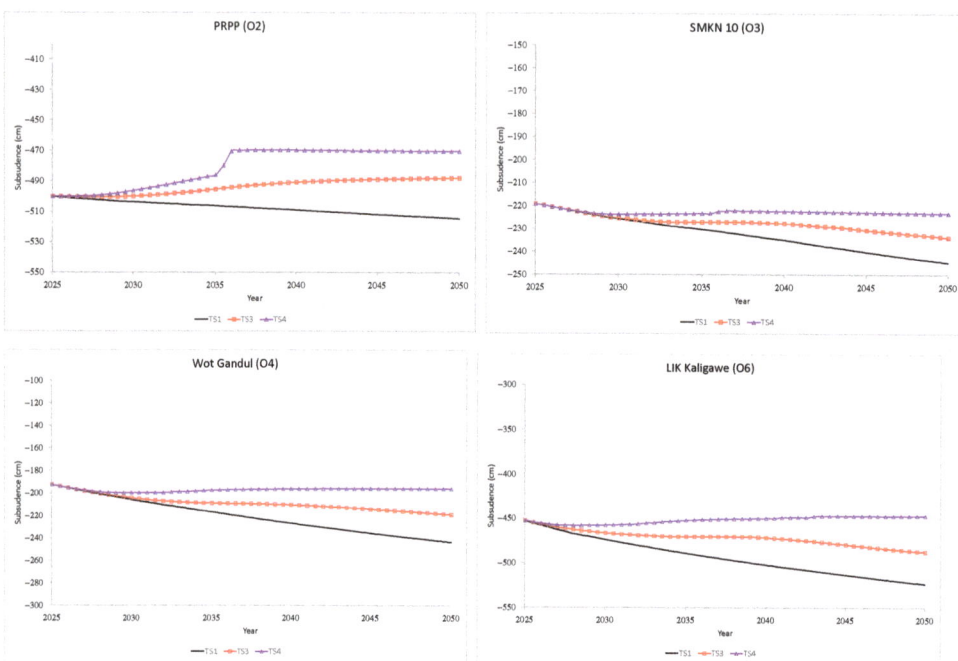

Figure 17. The simulation results of land subsidence from 2010 to 2050 at O2, O3, O4, and O6.

Particular attention should be directed again to Figures 17 and 18, where two groundwater management strategies are demonstrated to indeed mitigate the further development of land subsidence. A quantitative analysis based on these figures is presented in Table 3, summarizing the results of the specified analysis for each scenario in Figures 15 and 16. The affected area of land subsidence is divided into four regions: Region 1 is around Prawiro Jaya Baru (O1), Region 2 is around PRPP (O2), SMKN 10 (O3), and Wot Gandul (O4), Region 3 is around Hotel Santika (O5), and Region 4 is around LIK Kaligawe (O6). Among these regions, Region 2 has the greatest subsidence area. The percentage of decrease in the affected area (an area that has subsidence of more than 1 cm) of land subsidence is 2–75% if the TS3 scenario is taken, while the TS4 scenario is more effective with the percentage being 11–76%. Irrespective of the TS3 and TS4 scenarios, the greatest decrease in the affected area of land subsidence among all regions occurs in Region 1, with values up to 74.85% and 75.60% under the TS3 and TS4 scenarios, respectively. However, reducing land subsidence in Region 2 is critical and can yield the greatest economic benefits, since many commercial centers and government agencies, such as national-scale, city-scale, and neighborhood-scale trade zones, industry, and government offices, as well as high-density housing, are located in this region.

Table 3. Reduction in the affected area of land subsidence.

Region	The Affected Area of Land Subsidence (km^2)			Reduction in the Affected Area TS3–TS1 (%)	Reduction in the Affected Area TS4–TS1 (%)	Land Use
	TS1	TS3	TS4			
1	0.47	0.12	0.12	−74.85%	−75.60%	Government offices, city-scale trade zone, and low-density housing
2	10.91	10.55	9.53	−3.30%	−12.64%	National-scale trade zone, city-scale trade zone, neighborhood-scale trade zone, industry, government offices, high-density housing
3	1.06	0.43	0.29	−59.55%	−72.38%	Sub-city-scale trade zone, neighborhood-scale trade zone
4	6.32	6.16	5.60	−2.47%	−11.28%	Industry, higher education, medium-density housing

Figure 18. *Cont.*

Figure 18. The contour maps of land subsidence in 2050 under (**a**) TS1, (**b**) TS3, and (**c**) TS4 scenarios.

5. Conclusions

The northern part of Semarang City that is formed by alluvial deposits is prone to land subsidence, which has been exacerbated by over-drafting groundwater from deep wells in recent decades. To quantitatively analyze and model this problem, a systematic investigation of land subsidence caused by excessive groundwater abstraction was carried out in the current study using MODFLOW's flow and subsidence packages integrated with the recorded data obtained from groundwater level observation, leveling, GPS, and InSAR. The precise calibration and validation of our model were achieved excellent agreement with field measurements, and it was then applied to project future developments of subsidence and groundwater level in 2050.

To mitigate the negative consequences of land subsidence in the future, two groundwater regulatory strategies are proposed. Our results indicate that the subsidence rate can fall from 26% to 59% in different regions with the introduction of these two measures. It is also shown that although Genuk sub-district (LIK Kaligawe) has a land subsidence rate of 36.8 mm/year, these measures can reduce the subsidence rate by up to 59%. The maximum reduction in the affected area of land subsidence was demonstrated to occur in Region 1 (around Prawiro Jaya Baru), where the affected area can be decreased by up to 75%. Since the northern part of Semarang City is experiencing land subsidence and is located in the coastal area, we strongly suggest that evaluation and prediction of the risk of coastal flooding due to land subsidence and global sea-level rise should be at the forefront of future research.

Author Contributions: Conceptualization, W.L. and S.N.P.; methodology, W.L. and S.N.P.; software, S.N.P.; validation, W.L. and S.N.P.; formal analysis, S.N.P.; investigation, S.N.P. and B.G.D.; resources, W.L.; data curation, D.S. and S.; writing—original draft preparation, S.N.P.; writing—review and editing, W.L. and D.S.; visualization, S.N.P.; supervision, W.L.; funding acquisition, W.L. All authors have read and agreed to the published version of the manuscript.

Funding: This research received no external funding.

Data Availability Statement: DEM data can be found here: http://tides.big.go.id/DEMNAS/ (accessed on 19 November 2019). InSAR data is Sentinel-1 data from 2015–2020 (https://comet.nerc.ac.uk/comet-lics-portal/, accessed on 25 June 2020), with frame ID 076D_09725_121107 covering Central Java Province. Rainfall data can be obtained on request from the Water Resource Agency, Public Work Department of Indonesia. Soil property data can be obtained on request from the Mining and Energy Agency of Central Java Province. Well locations and discharge data can be obtained on request from the Mining and Energy Agency of Central Java Province. Observation well locations and groundwater level data can be obtained on request from the Center for Groundwater and Environmental Geology, Ministry of Energy, and Mineral Resources. Hydrogeological data can be obtained on request from BBWS Pemali–Juana, Ministry of Public Work, Mining and Energy Agency of Central Java Province, Department of Geology, Diponegoro University.

Acknowledgments: The author would like to thank the Research and Community Service Institution (LPPM) Jenderal Sudirman University for technical support. We also appreciate the Water Resource Agency of the Public Work Department of Indonesia, the Center for Groundwater and Environmental Geology of the Ministry of Energy and Mineral Resources of Indonesia, and the Potential Analysis of Groundwater of the Department of Energy and Mineral Resources of Central Java Province for providing their data in the study area for analysis.

Conflicts of Interest: The authors declare no conflict of interest.

References

1. Ezquerro, P.; Guardiola-Albert, C.; Herrera, G.; Fernández-Merodo, J.A.; Béjar-Pizarro, M.; Bonì, R. Groundwater and Subsidence Modeling Combining Geological and Multi-Satellite SAR Data over the Alto Guadalentín Aquifer (SE Spain). *Geofluids* **2017**, *2017*, 1359325. [CrossRef]
2. Keith, J.; Larson, K.J.; Marino, M. *Numerical Simulation of Land Subsidence in the Los Banos-Kettleman City Areal California Environmental Engineering*; The UC Center for Water Resources: Berkeley, CA, USA, 2001.

3. Sarah, D.; Satriyo, N.A.; Mulyono, A. Preliminary Study of Estimating Physical Losses Due to Land Subsidence in Semarang City (in Bahasa Indonesia). In Proceedings of the Presentation of Research Results of the Geological Research Center LIPI (Prosiding Pemaparan Hasil Penelitian Pusat Penelitian Geologi LIPI), Bandung, Indonesia, 2–3 December 2014; pp. 37–45.
4. Carminati, E.; Di Donato, G. Separating Natural and Anthropogenic Vertical Movements in Fast Subsiding Areas: The Po Plain (N. Italy) Case. *Geophys. Res. Lett.* **1999**, *26*, 2291–2294. [CrossRef]
5. Gambolati, G.; Teatini, P.; Tomasi, L.; Gonella, M. Coastline Regression of the Romagna Region, Italy, Due to Natural and Anthropogenic Land Subsidence and Sea Level Rise. *Water Resour. Res.* **1999**, *35*, 163–184. [CrossRef]
6. Hu, R.L.; Yue, Z.Q.; Wang, L.C.; Wang, S.J. Review on Current Status and Challenging Issues of Land Subsidence in China. *Eng. Geol.* **2004**, *76*, 65–77. [CrossRef]
7. Teatini, P.; Tosi, L.; Strozzi, T. Quantitative Evidence That Compaction of Holocene Sediments Drives the Present Land Subsidence of the Po Delta, Italy. *J. Geophys. Res. Solid Earth* **2011**, *116*, 1–10. [CrossRef]
8. Zoccarato, C.; Minderhoud, P.S.J.; Teatini, P. The Role of Sedimentation and Natural Compaction in a Prograding Delta: Insights from the Mega Mekong Delta, Vietnam. *Sci. Rep.* **2018**, *8*, 11437. [CrossRef] [PubMed]
9. Lo, W.-C.; Sposito, G.; Chu, H. Poroelastic Theory of Consolidation in Unsaturated Soils. *Vadose Zone J.* **2014**, *13*, vzj2013.07.0117. [CrossRef]
10. Gambolati, G.; Teatini, P. *Coastline Evolution of the Upper Adriatic Sea Due to Sea Level Rise and Natural and Anthropogenic Land Subsidence*; Singh, V.P., Ed.; Kluwer Academic Publishers: Dordrecht, The Netherlands, 1998.
11. Tran, D.-H.; Wang, S.-J. Land subsidence due to groundwater extraction and tectonic activity in Pingtung Plain, Taiwan. *Proc. Int. Assoc. Hydrol. Sci.* **2020**, *382*, 361–365. [CrossRef]
12. Lo, W.-C.; Lee, J.-W. Effect of water content and soil texture on consolidation in unsaturated soils. *Adv. Water Resour.* **2015**, *82*, 51–69. [CrossRef]
13. Lo, W.-C.; Sposito, G.; Lee, J.-W.; Chu, H. One-dimensional consolidation in unsaturated soils under cyclic loading. *Adv. Water Resour.* **2016**, *91*, 122–137. [CrossRef]
14. Lo, W.C.; Chao, N.C.; Chen, C.H.; Lee, J.W. Poroelastic Theory of Consolidation in Unsaturated Soils Incorporating Gravitational Body Forces. *Adv. Water Resour.* **2017**, *106*, 121–131. [CrossRef]
15. Lo, W.; Borja, R.I.; Deng, J.-H.; Lee, J.-W. Analytical solution of soil deformation and fluid pressure change for a two-layer system with an upper unsaturated soil and a lower saturated soil under external loading. *J. Hydrol.* **2020**, *588*, 124997. [CrossRef]
16. Lo, W.C.; Borja, R.I.; Deng, J.H.; Lee, J.W. Poroelastic Theory of Consolidation for a Two-Layer System with an Upper Unsaturated Soil and a Lower Saturated Soil under Fully Permeable Boundary Conditions. *J. Hydrol.* **2021**, *596*, 125700. [CrossRef]
17. Yang, Q.; Ke, Y. Relationship between Urban Construction and Land Subsidence in Beijing Region. In Proceedings of the 22nd International Congress on Modelling and Simulation, MODSIM 2017, Tasmania, Australia, 3–8 December 2017; pp. 1020–1026.
18. USGS. *Land Subsidence in the United States*; Galloway, D., David, R., Jones, D.R., Ingebritsen, S.E., Eds.; USGS Circular: Denver, CO, USA, 1999.
19. Hung, W.-C.; Hwang, C.; Liou, J.-C.; Lin, Y.-S.; Yang, H.-L. Modeling aquifer-system compaction and predicting land subsidence in central Taiwan. *Eng. Geol.* **2012**, *147–148*, 78–90. [CrossRef]
20. Al-Sittawy, M.; Gad, S.; Fouad, R.; Nofal, E. Assessment of soil subsidence due to long-term dewatering, Esna city, Egypt. *Water Sci.* **2019**, *33*, 40–53. [CrossRef]
21. Zeitoun, D.G.; Wakshal, E. *Land Subsidence Analysis in Urban Areas*; Springer: New York, NY, USA, 2013.
22. Othman, A.; Abotalib, A.Z. Land subsidence triggered by groundwater withdrawal under hyper-arid conditions: Case study from Central Saudi Arabia. *Environ. Earth Sci.* **2019**, *78*, 243. [CrossRef]
23. Mahmoudpour, M.; Khamehchiyan, M.; Nikudel, M.R.; Ghassemi, M.R. Numerical simulation and prediction of regional land subsidence caused by groundwater exploitation in the southwest plain of Tehran, Iran. *Eng. Geol.* **2016**, *201*, 6–28. [CrossRef]
24. Shen, S.-L.; Xu, Y.-S. Numerical evaluation of land subsidence induced by groundwater pumping in Shanghai. *Can. Geotech. J.* **2011**, *48*, 1378–1392. [CrossRef]
25. Zhang, Y.; Liu, Y.; Jin, M.; Jing, Y.; Liu, Y.; Liu, Y.; Sun, W.; Wei, J.; Chen, Y. Monitoring Land Subsidence in Wuhan City (China) Using the SBAS-INSAR Method with Radarsat-2 Imagery Data. *Sensors* **2019**, *19*, 743. [CrossRef]
26. Lanari, R.; Casu, F.; Manzo, M.; Lundgren, P. Application of the SBAS-DInSAR Technique to Fault Creep: A Case Study of the Hayward Fault, California. *Remote Sens. Environ.* **2007**, *109*, 20–28. [CrossRef]
27. Motagh, M.; Djamour, Y.; Walter, T.; Wetzel, H.-U.; Zschau, J.; Arabi, S. Land subsidence in Mashhad Valley, northeast Iran: Results from InSAR, levelling and GPS. *Geophys. J. Int.* **2007**, *168*, 518–526. [CrossRef]
28. Aditiya, A.; Takeuchi, W.; Aoki, Y. Land Subsidence Monitoring by InSAR Time Series Technique Derived From ALOS-2 PALSAR-2 over Surabaya City, Indonesia. *IOP Conf. Ser. Earth Environ. Sci.* **2017**, *98*, 12010. [CrossRef]
29. Mateos, R.M.; Ezquerro, P.; Luque-Espinar, J.A.; Pizarro, M.B.; Notti, D.; Azañón, J.M.; Monserrat, O.; Herrera, G.; Fernández-Chacón, F.; Peinado, T.; et al. Multiband PSInSAR and long-period monitoring of land subsidence in a strategic detrital aquifer (Vega de Granada, SE Spain): An approach to support management decisions. *J. Hydrol.* **2017**, *553*, 71–87. [CrossRef]
30. Yi, S.; Wang, Q.; Sun, W. Predictability of Hydraulic Head Changes and Characterization of Aquifer System and Fault Properties from InSAR Derived Ground Deformation. *J. Geophys. Res. Solid Earth* **2014**, *119*, 6572–6590.
31. Rezaei, A.; Mousavi, Z.; Khorrami, F.; Nankali, H. Inelastic and elastic storage properties and daily hydraulic head estimates from continuous global positioning system (GPS) measurements in northern Iran. *Hydrogeol. J.* **2020**, *28*, 657–672. [CrossRef]

32. Bonì, R.; Cigna, F.; Bricker, S.; Meisina, C.; McCormack, H. Characterisation of hydraulic head changes and aquifer properties in the London Basin using Persistent Scatterer Interferometry ground motion data. *J. Hydrol.* **2016**, *540*, 835–849. [CrossRef]
33. Loáiciga, H.A. Consolidation Settlement in Aquifers Caused by Pumping. *J. Geotech. Geoenviron. Eng.* **2013**, *139*, 1191–1204. [CrossRef]
34. Chen, J.; Knight, R.; Zebker, H.A.; Schreüder, W.A. Confined aquifer head measurements and storage properties in the San Luis Valley, Colorado, from spaceborne InSAR observations. *Water Resour. Res.* **2016**, *52*, 3623–3636. [CrossRef]
35. Cao, Y.; Wei, Y.-N.; Fan, W.; Peng, M.; Bao, L. Experimental study of land subsidence in response to groundwater withdrawal and recharge in Changping District of Beijing. *PLoS ONE* **2020**, *15*, e0232828.
36. Gong, X.; Geng, J.; Sun, Q.; Gu, C.; Zhang, W. Experimental study on pumping-induced land subsidence and earth fissures: A case study in the Su-Xi-Chang region, China. *Bull. Int. Assoc. Eng. Geol.* **2020**, *79*, 4515–4525. [CrossRef]
37. Shen, S.-L.; Xu, Y.-S.; Hong, Z.-S. Estimation of Land Subsidence Based on Groundwater Flow Model. *Mar. Georesour. Geotechnol.* **2006**, *24*, 149–167. [CrossRef]
38. Galloway, D.; Sneed, M. Analysis and simulation of regional subsidence accompanying groundwater abstraction and compaction of susceptible aquifer systems in the USA. *Boletín Sociedad Geológica Mexicana* **2013**, *65*, 123–136. [CrossRef]
39. Thu, T.M.; Fredlund, D.G. Modelling Subsidence in the Hanoi City Area, Vietnam. *Can. Geotech. J.* **2000**, *37*, 621–637. [CrossRef]
40. Gumilar, I.; Abidin, H.Z.; Sidiq, T.P.; Andreas, H.; Maiyudi, R.; Gamal, M. Mapping and Evaluating the Impact of Land Subsidence In Semarang (Indonesia). *Indones. J. Geospat.* **2013**, *2*, 26–41.
41. Abidin, H.Z.; Andreas, H.; Gumilar, I.; Sidiq, T.P.; Gamal, M.; Murdohardono, D.; SUpriyadi, S.; Fukuda, Y. Studying Land Subsidence in Semarang (Indonesia) Using Geodetic Methods. In Proceedings of the FIG Congress, Facing the Challenges—Building the Capacity, Sydney, Australia, 11 April 2010.
42. Lubis, A.M.; Sato, T.; Tomiyama, N.; Isezaki, N.; Yamanouchi, T. Ground subsidence in Semarang-Indonesia investigated by ALOS–PALSAR satellite SAR interferometry. *J. Southeast Asian Earth Sci.* **2011**, *40*, 1079–1088. [CrossRef]
43. Marfai, M.A.; King, L. Monitoring land subsidence in Semarang, Indonesia. *Environ. Earth Sci.* **2007**, *53*, 651–659. [CrossRef]
44. Kuehn, F.; Albiol, D.; Cooksley, G.; Duro, J.; Granda, J.; Haas, S.; Hoffmann-Rothe, A.; Murdohardono, D. Detection of land subsidence in Semarang, Indonesia, using stable points network (SPN) technique. *Environ. Earth Sci.* **2010**, *60*, 909–921. [CrossRef]
45. Sarah, D. Natural Compaction of Semarang Demak Alluvial Deposit. Ph.D. Thesis, Institut Teknologi Bandung, Bandung, Indonesia, 2019, *Unpublished*. (In Bahasa Indonesia)
46. Abidin, H.; Andreas, H.; Gumilar, I.; Sidiq, T.; Fukuda, Y. Land subsidence in coastal city of Semarang (Indonesia): Characteristics, impacts and causes. *Geomat. Nat. Hazards Risk* **2013**, *4*, 226–240. [CrossRef]
47. Sarah, D.; Hutasoit, L.M.; Delinom, R.M.; Sadisun, I.A. Natural Compaction of Semarang-Demak Alluvial Plain and Its Relationship to the Present Land Subsidence. *Indones. J. Geosci.* **2020**, *7*, 273–289. [CrossRef]
48. Andreas, H.; Abidin, H.Z.; Sarsito, D.A.; Meilano, I.; Susilo, S. Investigating the Tectonic Influence to the Anthropogenic Subsidence along Northern Coast of Java Island Indonesia Using GNSS Data Sets. *E3S Web Conf.* **2019**, *94*, 04005. [CrossRef]
49. Lo, W.; Purnomo, S.N.; Sarah, D.; Aghnia, S.; Hardini, P. Groundwater Modelling in Urban Development to Achieve Sustainability of Groundwater Resources: A Case Study of Semarang City, Indonesia. *Water* **2021**, *13*, 1395. [CrossRef]
50. Thaden, R.E.; Sumardirdja, H.; Richards, P.W. *Geological Map of Semarang-Magelang Quadrangle, Java, Scale 1:100.000*; Geological Research and Development Centre: Bandung, Indonesia, 1996.
51. Rukayah, R.S.; Abdullah, M. The Glory of Semarang Coastal City in the Past, Multi-Ethnic Merchants and Dutch Commerce. *J. Southwest Jiaotong Univ.* **2019**, *54*, 54. [CrossRef]
52. Tobing, M.H.L.; Syarief, E.A.; Murdohardono, D. *Engineering Geological Investigation of Subsidence in Semarang and Its Surroundings, Central Java Province*; Geological Environment Center, Geological Agency: Bandung, Indonesia, 2000. (In Bahasa Indonesia)
53. Marsudi. Prediction of Land Subsidence Rate in Semarang Alluvial Plain, Central Java. Ph.D. Thesis, Institut Teknologi Bandung, Bandung, Indonesia, 2000, *Unpublished*. (In Bahasa Indonesia)
54. VanBemmelen, R.W. The Geology of Indonesia. General Geology of Indonesia and Adjacent Archipelagoes. *Gov. Print. Off. Hague* **1949**, 545–547, 561–562.
55. Badan Standardisasi Nasional. *SNI 2812: One-Dimensional Soil Consolidation Test Method*; National Standardization Agency of Indonesia: Jakarta, Indonesia, 2011. (In Bahasa Indonesia)
56. Badan Standardisasi Nasional. *SNI 03-6870: How to Test Water Passing in the Laboratory for Fine-Grained Soils with High Pressure Drops*; National Standardization Agency of Indonesia: Jakarta, Indonesia, 2002. (In Bahasa Indonesia)
57. Badan Standardisasi Nasional. *SNI 3423: How to Test Soil Grain Size Analysis*; National Standardization Agency of Indonesia: Jakarta, Indonesia, 2008; pp. 1–27. (In Bahasa Indonesia)
58. Mining and Energy Agency of Central Java Province and Directorate of Geological Environment and Mining Areas. Overview of Groundwater System Configuration. In *Final Report: Study on the Configuration and Zoning of Underground Water in the Semarang–Demak, Subah, and Karanganyar–Boyolali, Central Java Province*; Mining and Energy Agency of Central Java Province: Semarang, Indonesia, 2003. (In Bahasa Indonesia)
59. Mining and Energy Agency Central Java Province (DESDM Provinsi Jawa Tengah). *Recapitulation of Water Calculation and Water Acquisition Value 2001–2010. Internal Report*; Mining and Energy Agency: Banjarnegara, Central Java, Indonesia, 2012, *Unpublished*. (In Bahasa Indonesia)

60. Tirtomihardjo, H. *Groundwater Resource Potential in Indonesia and Their Management. Presentation Report*; Mining and Energy Agency: Jakarta, Indonesia, 2011, *Unpublished*. (In Bahasa Indonesia)
61. Holzer, T.L.; Galloway, D.L.; Ehlen, J.; Haneberg, W.C.; Larson, R.A. Impacts of land subsidence caused by withdrawal of underground fluids in the United States. *Hum. Geol. Agents* **2005**, *XVI*, 87–99.
62. CTI Engineering International Co., L.& A. *Final Report on Land Subsidence Survey: Part II River Improvement Works, Water Resources Development & Land Subsidence Survey under JICA Loan IP-534*; Directorate General of Water Resources, Ministry of Public Works and Housing od Central Java: Semarang, Indonesia, 2016.
63. Luo, Q.; Perissin, D.; Lin, H.; Zhang, Y.; Wang, W. Subsidence Monitoring of Tianjin Suburbs by TerraSAR-X Persistent Scatterers Interferometry. *IEEE J. Sel. Top. Appl. Earth Obs. Remote Sens.* **2014**, *7*, 1642–1650. [CrossRef]
64. Yan, S.; Liu, G.; Deng, K.; Wang, Y.; Zhang, S.; Zhao, F. Large deformation monitoring over a coal mining region using pixel-tracking method with high-resolution Radarsat-2 imagery. *Remote Sens. Lett.* **2016**, *7*, 219–228. [CrossRef]
65. Samsonov, S.; D'Oreye, N.; Smets, B. Ground deformation associated with post-mining activity at the French–German border revealed by novel InSAR time series method. *Int. J. Appl. Earth Obs. Geoinf.* **2013**, *23*, 142–154. [CrossRef]
66. Jia, H.; Liu, L. A technical review on persistent scatterer interferometry. *J. Mod. Transp.* **2016**, *24*, 153–158. [CrossRef]
67. Catalão, J.; Nico, G.; Lollino, P.; Conde, V.; Lorusso, G.; Silva, C. Integration of InSAR Analysis and Numerical Modeling for the Assessment of Ground Subsidence in the City of Lisbon, Portugal. *IEEE J. Sel. Top. Appl. Earth Obs. Remote Sens.* **2015**, *9*, 1663–1673. [CrossRef]
68. Berardino, P.; Fornaro, G.; Lanari, R.; Sansosti, E. A new algorithm for surface deformation monitoring based on small baseline differential SAR interferograms. *IEEE Trans. Geosci. Remote Sens.* **2002**, *40*, 2375–2383. [CrossRef]
69. Hooper, A.; Zebker, H.; Segall, P.; Kampes, B. A new method for measuring deformation on volcanoes and other natural terrains using InSAR persistent scatterers. *Geophys. Res. Lett.* **2004**, *31*, L23611. [CrossRef]
70. Morishita, Y.; Lazecky, M.; Wright, T.J.; Weiss, J.R.; Elliott, J.R.; Hooper, A. LiCSBAS: An Open-Source InSAR Time Series Analysis Package Integrated with the LiCSAR Automated Sentinel-1 InSAR Processor. *Remote Sens.* **2020**, *12*, 424. [CrossRef]
71. Wang, Q.; Yu, W.; Xu, B.; Wei, G. Assessing the Use of Gacos Products for Sbas-Insar Deformation Monitoring: A Case in Southern California. *Sensors* **2019**, *19*, 3894. [CrossRef]
72. Schmidt, D.A.; Bürgmann, R. Time-Dependent Land Uplift and Subsidence in the Santa Clara Valley, California, from a Large Interferometric Synthetic Aperture Radar Data Set Time-Dependent Land Uplift and Subsidence in the Santa Clara Valley, California, from a Large Interferomet. *J. Geophys. Res.* **2003**, *108*, 1–13.
73. Agram, P.; Jolivet, R.; Simons, M. Generic InSAR Analysis Toolbox (GIAnT)–User Guide, ed. Available online: http://earthdef.caltech.edu (accessed on 31 May 2021).
74. Andaryani, S.; Nourani, V.; Trolle, D.; Dehghani, M.; Asl, A.M. Assessment of land use and climate change effects on land subsidence using a hydrological model and radar technique. *J. Hydrol.* **2019**, *578*, 124070. [CrossRef]
75. Hooper, A.; Bekaert, D.; Spaans, K.; Arıkan, M. Recent advances in SAR interferometry time series analysis for measuring crustal deformation. *Tectonophysics* **2012**, *514–517*, 1–13. [CrossRef]
76. Hu, J.; Li, Z.-W.; Ding, X.; Zhu, J.; Zhang, L.; Sun, Q. Resolving three-dimensional surface displacements from InSAR measurements: A review. *Earth-Sci. Rev.* **2014**, *133*, 1–17. [CrossRef]
77. PUSGEN. *Peta Sumber Dan Bahaya Gempa Indonesia Tahun 2017 (Map of Indonesia Earthquake Sources and Hazards in 2017)*; National Earthquake Study Center, Center for Research and Development of Housing and Settlements: Jakarta, Indonesia, 2017. (In Bahasa Indonesia)
78. Zhang, Y.; Xue, Y.; Wu, J.; Wang, H.; He, J. Mechanical modeling of aquifer sands under long-term groundwater withdrawal. *Eng. Geol.* **2012**, *125*, 74–80. [CrossRef]
79. Purnomo, S.N.; Lo, W.C. Estimation of Groundwater Recharge in Semarang City, Indonesia. *IOP Conf. Ser. Mater. Sci. Eng.* **2020**, *982*, 012035. [CrossRef]
80. Taufiq Nz, A.; Solihin, I.; Wahyudin. *Map of the Groundwater Conservation Zone of the Semarang Demak Basin in 2010*; Geological Environment Center, Geological Agency: Bandung, Indonesia, 2010. (In Bahasa Indonesia)

Article

Water/Cement/Bentonite Ratio Selection Method for Artificial Groundwater Barriers Made of Cutoff Walls

Cristian-Ștefan Barbu, Andrei-Dan Sabău, Daniel-Marcel Manoli * and Manole-Stelian Șerbulea

Department of Geotechnical and Foundation Engineering, Technical University of Civil Engineering Bucharest, 020396 Bucharest, Romania; cs.barbu@gmail.com (C.-Ș.B.); andreisabau228@gmail.com (A.-D.S.); manole.serbulea@utcb.ro (M.-S.Ș.)
* Correspondence: daniel.manoli@utcb.ro; Tel.: +40-723-166-182

Abstract: As urban development requires groundwater table isolation of various historically polluted sources, the necessity of building effective, strong, flexible, and low-permeability cutoff walls raises the question of choosing optimum construction materials. Various authors have proposed water–cement–bentonite mixtures, which are often chosen by experience or a trial-and-error approach, using classical methods for testing (Marsh funnel) and representation of results (water–cement ratio, water–bentonite ratio). The paper proposes a more precise approach for assessing the viscosity and global representation of the three components. Moreover, this approached is exemplified with a better documented recipe for the choice of materials based on laboratory results. The representation of the mixtures was undertaken on a limited domain of a ternary diagram, where the components are given in terms of mass percentage. The derived properties (viscosity, permeability, and compressive strength) are presented on a grid corresponding to the physically possible mixtures. Based on this representation, the most efficient recipes are chosen. Because the mixture contains only fine aggregates, the viscosity was determined using a laboratory viscosimeter.

Keywords: cutoff walls; plastic concrete; cement-bentonite-water ratio

Citation: Barbu, C.-Ș.; Sabău, A.-D.; Manoli, D.-M.; Șerbulea, M.-S. Water/Cement/Bentonite Ratio Selection Method for Artificial Groundwater Barriers Made of Cutoff Walls. *Water* 2022, 14, 376. https://doi.org/10.3390/w14030376

Academic Editor: C. Radu Gogu

Received: 10 December 2021
Accepted: 24 January 2022
Published: 26 January 2022

Publisher's Note: MDPI stays neutral with regard to jurisdictional claims in published maps and institutional affiliations.

Copyright: © 2022 by the authors. Licensee MDPI, Basel, Switzerland. This article is an open access article distributed under the terms and conditions of the Creative Commons Attribution (CC BY) license (https://creativecommons.org/licenses/by/4.0/).

1. Introduction

Since the expansion and development of urban areas, there has been an increasing preoccupation with the quality of the groundwater table. In order to preserve the cleanliness of groundwater, polluting sites must be isolated by containing them via the use of perimetral impervious barriers. In addition to the main, water-retaining function, this type of structure may also have a structural influence on the protected buildings or sites.

This technology is being broadly used in hydrotechnical structures for seepage control, and official guidelines already exist in certain countries, without being mandatory. The issue is still far from being solved or regulated, and the behavior of cutoff walls depends on the design requirements. Cutoff walls made of plastic concrete have a lesser impact on the hydro-geological urban environment because, unlike reinforced concrete trench walls, this type of structure may be broken in-place at the end of construction work, thus resetting the groundwater flow.

The plastic concrete differentiation is mainly described by the water–cement–bentonite content, because the aggregates' role is well established. The approach of this paper is similar to the characterization of binders in regular concrete, namely, the use for testing of a standardized sand according to EN 196-1 in a ratio of 1:1 with the cement–bentonite mixture.

Depending on the design requirements, there are a range of materials or mixtures that are used for building the barriers: soil–bentonite mixtures, soil–cement–bentonite mixtures, or plastic concrete. Various authors have proposed a wide range of variation for the amount of water, cement, and bentonite. In this study, we attempted to cover the proposed range as much as possible, in correlation with our own studies, which showed that a large amount of bentonite tends to reduce the compressive strength and increases the shrinkage-swelling capacity during soil moisture variation. Hence, in this study we limited the amount of bentonite to about 30%.

The ratios used in this paper are given in Table 1, with a number of case histories documented in previous papers.

Table 1. Reference values from previous contributions.

Paper Reference	Water to Cement Ratio	Water to Bentonite Ratio	Bentonite to Cement Ratio
This paper	1.25–3.65	2.00–10.42	0.175–1.25
David Alos S. et al., 2020 [1]	3.30–10.00	-	0.1–0.24
Fadaie M.A., et al., 2019 [2]	1.60–2.00	8.00–20.00	0.00–0.40
Pisheh Y.P., et al., 2018 [3]	1.80	6.30–10.20	0.14–0.29
U.S. Dept. of Interior, 2014 [4]	1–2.78	6.67–13.90	0.1–0.22
Hinchberger S.D., et al., 2010 [5]	1.70–2.35	12.50–18.20	0.20–0.30
Bagheri A., et al., 2008 [6]	1.80–2.60	13.00	0.14–0.23

Because the ratio between the main components—water, cement, and bentonite—needs to be represented on a scale with three dimensions, with one linear dependency (percentage of cement + percentage of bentonite + percentage of water = 1), to place them into context, the three elements were represented on a ternary diagram using their mass percentages.

To highlight the area of interest, the proposed ternary diagram was limited to a subdomain in which the percentual ratios characterize mixtures that are feasible for all applications. A uniform mesh of percentage combinations was chosen because it provides a smooth representation of each component effect (Figure 1).

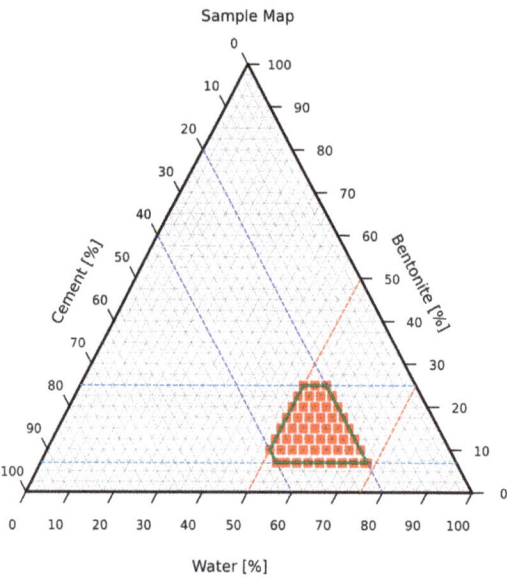

Figure 1. Mesh of cement–bentonite–water combinations.

The ternary subdomain is used as a planimetric representation of mixtures for which various parameters (such as viscosity and the coefficient of permeability) are represented. To provide a proper representation, the areas outside of the represented domain were cropped (Figure 2). The limit lines were set based on the values of the contributions listed in Table 1. Because the out-of-plane values may have exponential variations, where required for the readability of the plot, a logarithmic scale was employed.

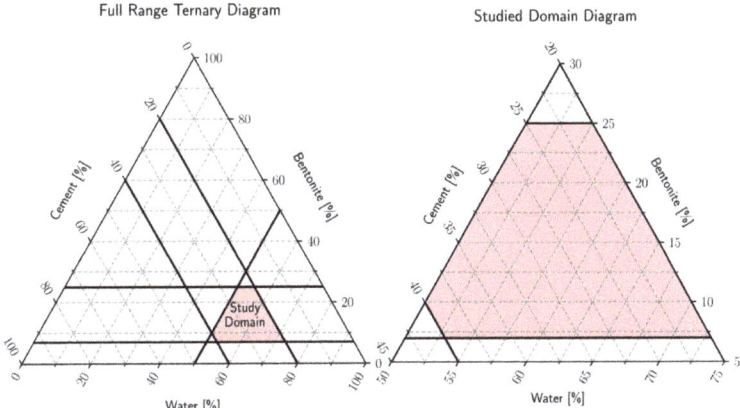

Figure 2. Ternary subdomain cropping.

2. Preparation of Samples

For the preparation of the samples, standard sand was used because it is recommended by norms that describe the method for determination of the compressive strength of cement mortar (EN 196-1:2016—Particle size distribution is given in Table 2). No other aggregate grading distribution was used to ensure the repeatability of tests and to keep the focus on the water active components. For each sample, the ratio of sand to hydraulic binder (cement and bentonite) was 1:1. The usage of sand was employed to prevent the cracking of the samples that may occur due to the lack of resistance. Even if the content of sand itself influences the behavior of the overall mixture, the relationship between the mass of sand and the sum of masses of bentonite and cement is bijective. Moreover, the amount of sand determines in a 4D-barycentric representation a plane due to the linearity of the variation (Figure 3). For tidiness of the representation, the axis of sand was disregarded.

Table 2. Particle size distribution of the CEN Standard sand.

Square Mesh Size (mm)	2.00	1.60	1.00	0.50	0.16	0.08
Cumulative Sieve Residue (%)	0	7 ± 5	33 ± 5	67 ± 5	87 ± 5	99 ± 1

The 1:1 ratio used herein is not generally challenged in the industry, and therefore was not among the objectives of this study.

The sand is not an active factor in the development of water–cement and water–bentonite gels. This contribution aimed at showcasing the reciprocal influence of hydration when the solids are simultaneously added in the mixture as normally happens in situ.

The cement chosen for sample preparation was Portland cement with high initial strength, having the minimum standard compressive strength of 42.5 MPa (CEM IIA 42.5R). Sodium bentonite was used as additive in the plastic concrete mix. The water introduced in the mixtures was regular tap water.

In previous studies, the mixing sequence implied the hydration of bentonite up to 24 h before adding the other components [5–7], although similar testing results were obtained by dry mixing all the parts of the mix [2]. Considering the actual site conditions, dry mixing is closer to the technological reality and more effective in terms of time and cost. Detailing the application of the method mentioned before, in this study the bentonite and the cement were added together with the aggregates, followed by the addition of water. The mixture was poured in 50 × 100 mm cylindrical plastic molds matching the triaxial test sample.

The laboratory testing program started 28 days after the mixtures were poured. The time interval was selected in order to ensure that the plastic concrete reached its strength.

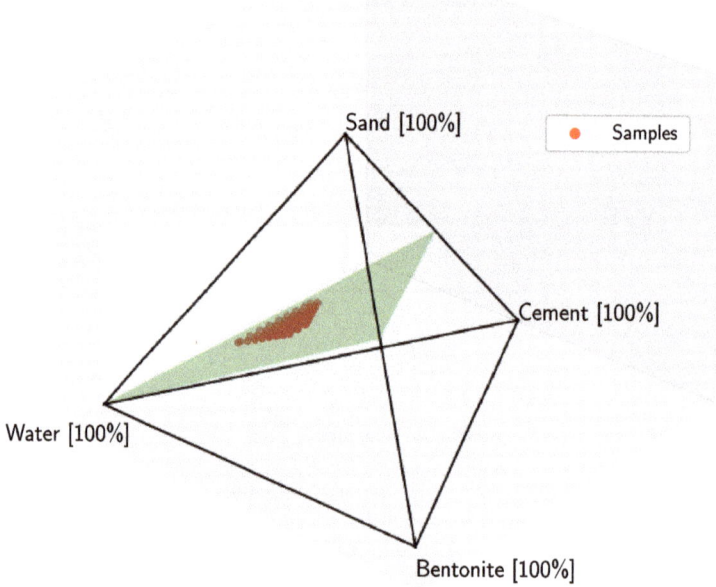

Figure 3. Ternary diagram in a 4D-barycentric representation in the plane defined by the equation $Mass_{Sand} = Mass_{Bentonite} + Mass_{Cement}$.

The determination of the permeability coefficient was undertaken using the constant head permeameter method, whereas the determination of the axial compressive strength was performed following the norm EN ISO 17892-7:2018.

The chosen range for different components proved to be sufficient because unfavorable effects could be noticed at the extremes. The high content of water, despite ensuring a good workability, led to contraction cracking of the samples (Figure 4) and sedimentation of sand; all the sedimented samples were tested and are marked as such in the Figure 5.

Figure 4. Contraction cracking caused by high water content.

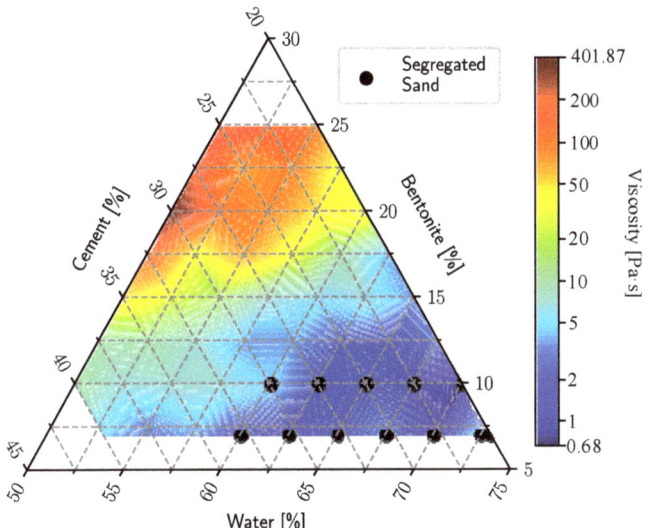

Figure 5. Viscosity variation of the fresh mixtures and marks of segregated sand samples.

An excessive amount of bentonite led to a crumbling behavior of the material, whereas large amounts of cement reduced the workability of fresh mixtures.

3. Test Results

The viscosity was measured using a rotational viscometer to measure the torque on rods of various geometries depending on the liquid. The slope of τ-γ' is the dynamic viscosity considering the fresh mix as being quasi-Newtonian. As expected, the variation in viscosity is quite important (between 0.68–401.87 Pa·s); thus, in order to provide a balanced representation of colors, a [0, 1] normalized logarithmic scale was employed as described in Equation (1) [8].

$$norm(\mu_i) = \frac{\frac{\ln(\eta_i)}{\ln(\prod_{i=1}^{n}\eta_i)} - min\left(\frac{\ln(\eta_i)}{\ln(\prod_{i=1}^{n}\eta_i)}\right)}{max\left(\frac{\ln(\eta_i)}{\ln(\prod_{i=1}^{n}\eta_i)}\right) - min\left(\frac{\ln(\eta_i)}{\ln(\prod_{i=1}^{n}\eta_i)}\right)} \qquad (1)$$

For graph plotting, Python 3.9.9 and numpy, matplotlib, mpltern, and ternary libraries were used.

As expected, the larger the ratio of water–solids, the smaller the viscosity. It should be noted that the Marsh funnel method refers to reference values for bentonite slurry of 32 to 60 s [8], corresponding to viscosities too low to be considered for workable plastic concrete, whereas the concrete viscosity is reported to be between 2–27 Pa·s [9,10], so that the measured values presented herein cover the latter reported range.

The variation is progressive and smooth, so Figure 5 may be used for assessing expected workability when preparing cement–bentonite–water mixtures with sand. It may be noted that limiting the lower viscosity to 5 Pa·s prevents the segregation. The advantage of the proposed plot consists also in setting the mixture without limiting the water content, and in choosing a proper ratio among components, and an increase in bentonite solves the sedimentation issue.

We consider the workability acceptable up to 100 Pa·s, but ideal under 50 Pa·s. For higher values, additives such as superplasticizers are required to attain proper flow for casting.

For choosing the best suited mixture, it is important to correlate workability with the targeted failure behaviors and permeabilities.

The quantity of bentonite negatively impacts the compressive strength when portions larger than 17% are used. In this case, the strength is decreased by factors of 2 and more with respect to lower bentonite amounts (Figure 6). This highlights a boundary of material behavior at values of bentonite content around 17%, from which the hardened mixtures shift from having predominant compressive strength to shearing strength; that is, switching from weak concrete to hard soil (Figure 7).

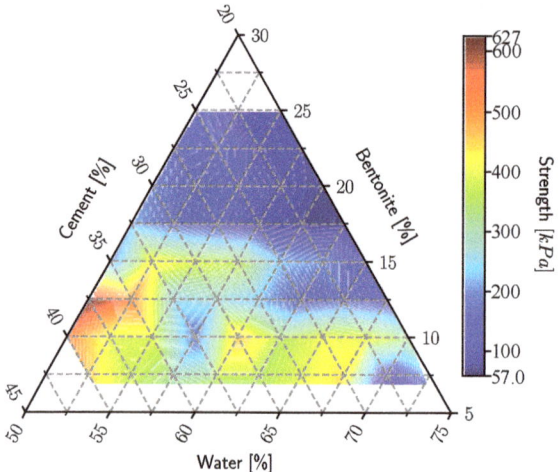

Figure 6. Axial compressive strength.

Figure 7. Failure patterns of plastic concrete samples. (**a**) "weak concrete"—compressive failure. (**b**) "strong soil"—shearing failure.

Conversely, the material permeability is also decreased by a factor of 5 or more (Figure 8) along the same boundary of about 17%, confirming the model of switching between the mechanical behavior of the hardened mixture.

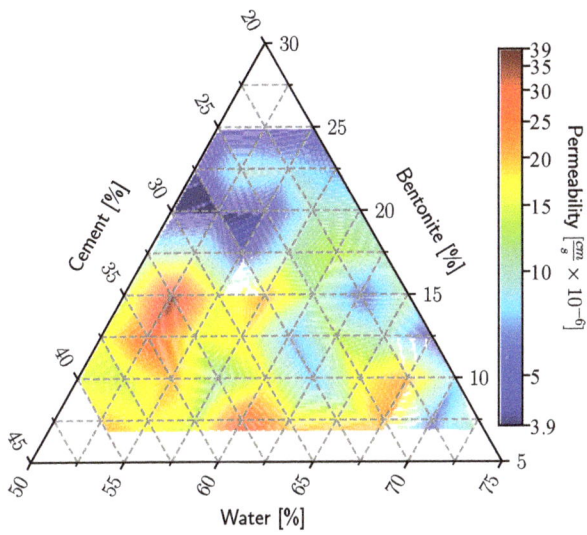

Figure 8. Permeability.

All data resulting from laboratory tests are given in Table 3. The mixtures are given both in terms of percentage of each component and as classical ratios.

Table 3. Test data table.

No.	Water [%]	Bent. [%]	Cem. [%]	w/c	w/b	c/b	K [cm/s]	c_u [kPa]	Segregated	Γ [kN/m³]	μ [Pa·s]
1	50.0	25.0	25.0	2.00	2.00	1.00	3.92×10^{-5}	162	No	13.04	237.73
2	52.5	25.0	22.5	2.33	2.10	0.90	3.90×10^{-5}	86	No	11.95	154.33
3	55.0	25.0	20.0	2.75	2.20	0.80	3.74×10^{-5}	82	No	10.98	103.48
4	50.0	22.5	27.5	1.82	2.22	1.22	3.69×10^{-5}	166	No	12.32	204.73
5	52.5	22.5	25.0	2.10	2.33	1.11	1.49×10^{-5}	156	No	12.02	74.04
6	55.0	22.5	22.5	2.44	2.44	1.00	1.70×10^{-5}	93	No	11.11	98.57
7	57.5	22.5	20.0	2.88	2.56	0.89	2.16×10^{-5}	79	No	10.49	36.13
8	50.0	20.0	30.0	1.67	2.50	1.50	3.82×10^{-5}	193	No	13.12	401.87
9	52.5	20.0	27.5	1.91	2.63	1.38	3.89×10^{-5}	145	No	12.58	132.92
10	55.0	20.0	25.0	2.20	2.75	1.25	3.62×10^{-5}	124	No	12.06	96.34
11	57.5	20.0	22.5	2.56	2.88	1.13	1.47×10^{-5}	93	No	11.31	49.51
12	60.0	20.0	20.0	3.00	3.00	1.00	1.53×10^{-5}	71	No	11.19	44.60
13	50.0	17.5	32.5	1.54	2.86	1.86	1.40×10^{-5}	219	No	13.45	156.56
14	52.5	17.5	30.0	1.75	3.00	1.71	1.32×10^{-5}	158	No	12.89	34.34
15	55.0	17.5	27.5	2.00	3.14	1.57	3.25×10^{-5}	124	No	12.09	80.29
16	57.5	17.5	25.0	2.30	3.29	1.43	1.40×10^{-5}	89	No	11.95	31.22
17	60.0	17.5	22.5	2.67	3.43	1.29	1.11×10^{-5}	72	No	11.38	10.17
18	62.5	17.5	20.0	3.13	3.57	1.14	1.44×10^{-5}	67	No	10.81	3.97
19	50.0	15.0	35.0	1.43	3.33	2.33	1.12×10^{-5}	194	No	13.86	48.62
20	52.5	15.0	32.5	1.62	3.50	2.17	3.89×10^{-6}	424	No	15.06	26.76
21	55.0	15.0	30.0	1.83	3.67	2.00	9.82×10^{-6}	388	No	14.62	15.08
22	57.5	15.0	27.5	2.09	3.83	1.83	6.91×10^{-6}	368	No	13.77	7.05
23	60.0	15.0	25.0	2.40	4.00	1.67	1.35×10^{-5}	228	No	12.76	5.71
24	62.5	15.0	22.5	2.78	4.17	1.50	2.71×10^{-5}	201	No	11.96	6.87
25	65.0	15.0	20.0	3.25	4.33	1.33	1.35×10^{-5}	120	No	12.16	3.64
26	50.0	12.5	37.5	1.33	4.00	3.00	9.45×10^{-6}	627	No	13.85	10.70
27	52.5	12.5	35.0	1.50	4.20	2.80	5.25×10^{-6}	524	No	14.69	10.35

Table 3. *Cont.*

No.	Water [%]	Bent. [%]	Cem. [%]	w/c	w/b	c/b	K [cm/s]	c_u [kPa]	Segregated	Γ [kN/m³]	μ [Pa·s]
28	55.0	12.5	32.5	1.69	4.40	2.60	9.75×10^{-6}	251	No	13.71	7.31
29	57.5	12.5	30.0	1.92	4.60	2.40	8.46×10^{-6}	264	No	13.4	3.81
30	60.0	12.5	27.5	2.18	4.80	2.20	2.29×10^{-5}	223	No	12.48	3.79
31	62.5	12.5	25.0	2.50	5.00	2.00	1.13×10^{-5}	136	No	11.01	2.03
32	65.0	12.5	22.5	2.89	5.20	1.80	1.54×10^{-5}	161	No	10.93	1.65
33	67.5	12.5	20.0	3.38	5.40	1.60	3.35×10^{-5}	120	No	10.16	2.94
34	50.0	10.0	40.0	1.25	5.00	4.00	9.98×10^{-6}	529	No	13.6	10.79
35	52.5	10.0	37.5	1.40	5.25	3.75	1.08×10^{-5}	389	No	13.84	7.14
36	55.0	10.0	35.0	1.57	5.50	3.50	7.32×10^{-6}	390	No	12.69	7.58
37	57.5	10.0	32.5	1.77	5.75	3.25	1.44×10^{-5}	173	No	12.18	7.40
38	60.0	10.0	30.0	2.00	6.00	3.00	9.44×10^{-6}	471	Yes	13.11	5.44
39	62.5	10.0	27.5	2.27	6.25	2.75	2.19×10^{-5}	360	Yes	12.19	1.90
40	65.0	10.0	25.0	2.60	6.50	2.50	8.71×10^{-6}	415	Yes	12.76	2.10
41	67.5	10.0	22.5	3.00	6.75	2.25	6.74×10^{-6}	402	Yes	13.37	1.43
42	70.0	10.0	20.0	3.50	7.00	2.00	1.90×10^{-5}	334	Yes	13.1	0.68
43	53.0	7.0	40.0	1.33	7.57	5.71	7.29×10^{-6}	443	No	14.24	4.68
44	55.0	7.0	38.0	1.45	7.86	5.43	8.71×10^{-6}	368	No	13.5	2.83
45	57.5	7.0	35.5	1.62	8.21	5.07	1.08×10^{-5}	319	No	12.95	3.93
46	60.0	7.0	33.0	1.82	8.57	4.71	4.70×10^{-6}	349	Yes	14.69	1.65
47	62.5	7.0	30.5	2.05	8.93	4.36	7.15×10^{-6}	353	Yes	13.75	1.34
48	65.0	7.0	28.0	2.32	9.29	4.00	7.03×10^{-6}	393	Yes	13.82	2.10
49	67.5	7.0	25.5	2.65	9.64	3.64	6.96×10^{-6}	449	Yes	14.01	3.48
50	70.0	7.0	23.0	3.04	10.00	3.29	2.84×10^{-5}	57	No	10.37	1.10
51	72.5	7.0	20.5	3.54	10.36	2.93	1.40×10^{-5}	161	No	10.82	5.93
52	73.0	7.0	20.0	3.65	10.43	2.86	1.57×10^{-5}	94	Yes	10.96	2.71

4. Discussion

This research focused mainly on covering all the possible mixture contents as described by various contributions. This led to a wide distribution of results in terms of the diverse mechanical behavior. In order to aid the visual identification of the mixtures, a colormap based on red–yellow–blue coding was derived, as shown in Figure 9.

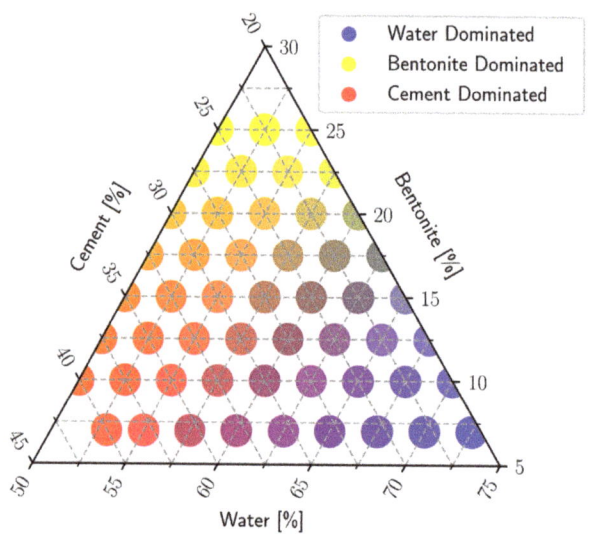

Figure 9. Component distribution colormap.

The three properties studied for each mixture were viscosity, strength, and permeability. A distinctive clustering of results induced by the feature tradeoff governed by the components may be noted in Figure 10.

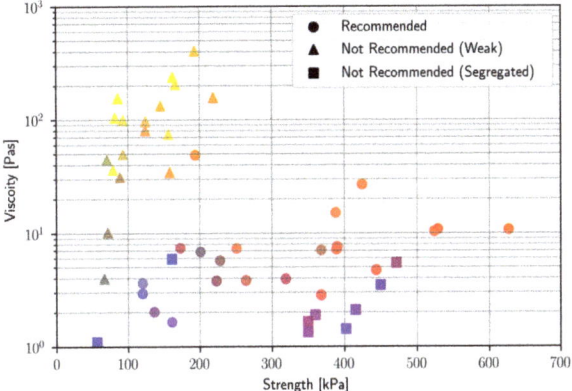

Figure 10. Viscosity vs. strength.

The increase in bentonite amount (yellower dots in the graphs) leads to the foreseeable loss of strength of about 1–5 times with respect to the cement-governed mixtures; however the viscosity increases by about 1–1.5 orders of magnitude. The large variation in strength for plastic concrete is documented in the literature, and the range found herein was confirmed by contributions such as [5,11,12]. Other authors, such as [6], obtained slightly overlapping results but with most of the strengths being an order of magnitude higher.

Studying the variation in the strength with respect to the commonly used ratios employed in engineering practice, namely water/cement and water/bentonite, the influence of the third component in the mixture may be noted.

The effect of strength reduction due to the bentonite may be readily noted in Figure 11, in the mixtures containing over 17.5% active clay (yellow shaded), or in Figure 12 with a water/bentonite ratio lower than 4. In Figure 11, the plastic concrete with high content of sodium bentonite is less prone to develop hydrated clinker bridges, so the strength drops under 200 kPa for this composition of samples. These results are also noticeable in the grouping of yellow-shaded markers in Figure 12.

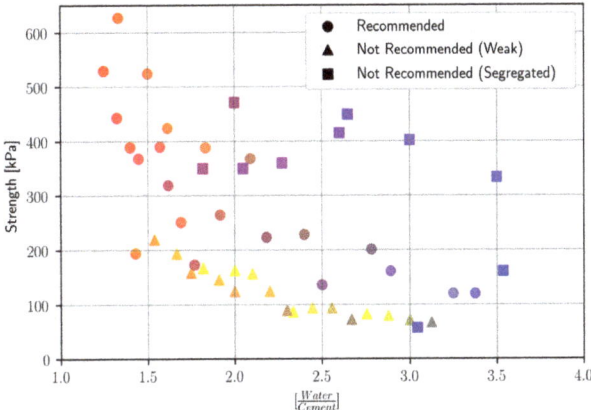

Figure 11. Strength vs. water/cement ratio.

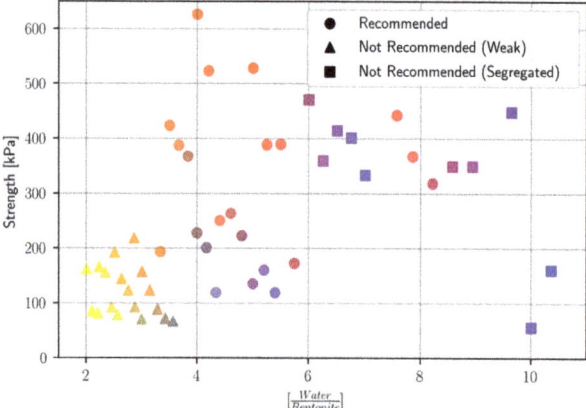

Figure 12. Strength vs. water/bentonite ratio.

We consider viscosity to be a very important feature of the fresh plastic concrete, because this property prevents the contamination of the cast material with water or soil from the trench walls when replacing the slurry used for excavation. The values obtained in our tests did not exceed the limit of proper workability.

In addition to the role of thickener for the fresh mixture, bentonite prevents the proper development of cement gel and is twice as expensive as the binder; however, the decrease in permeability of the wall for low amounts of bentonite is canceled or even worsened if this component is used in excess. This effect is also observed in the case of strength.

In Figure 13, the low scattering of test data in the graph indicates the fact that the viscosity of the mix is almost entirely governed by the bentonite and water content, with little to no influence from the cement (Figure 14).

Due to the large number of samples and long testing time required (employing triaxial test equipment), and the hardening time limitation on batching, only one sample was tested for each combination, resulting in some lack of accuracy, especially for permeability tests. However, a trend in the data emerged that helps to narrow the mixture variation range to a domain closer to the targeted purposes. A future stage in this research will be to employ three to five samples of each combination in order to be able to process the data on a statistical basis, yet for a narrower subdomain, as proposed in Figure 15.

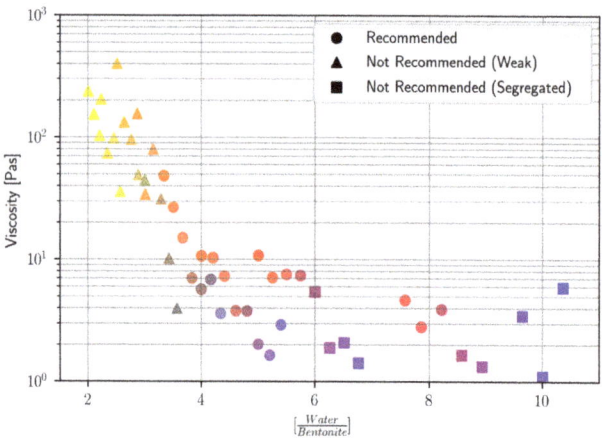

Figure 13. Viscosity vs. water/bentonite ratio.

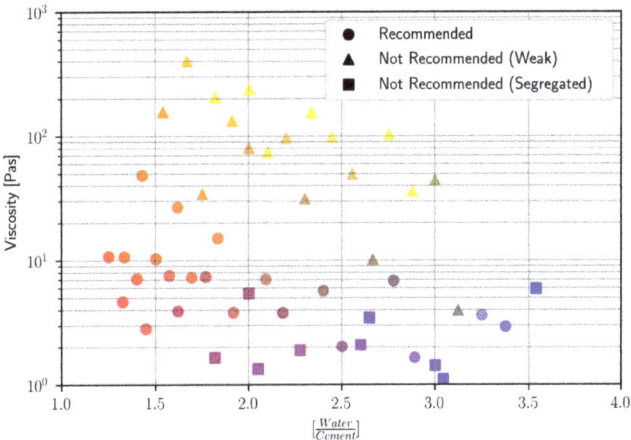

Figure 14. Viscosity vs. water/cement ratio.

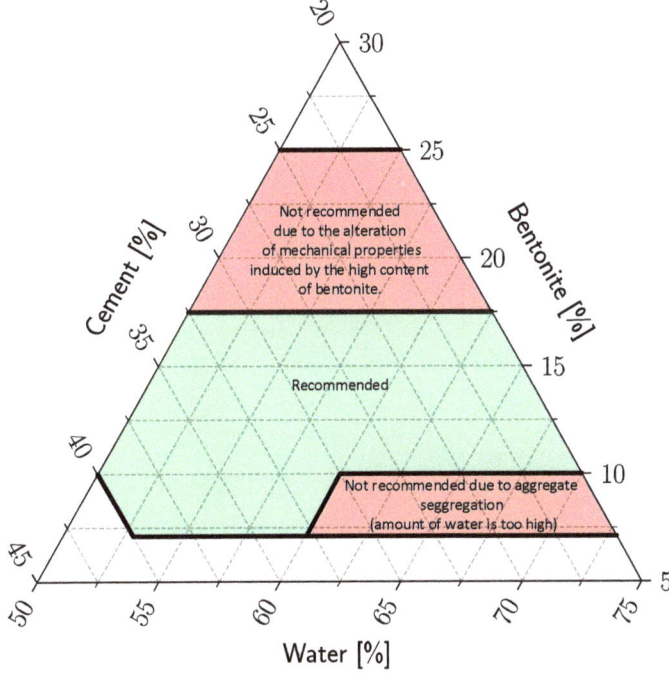

Figure 15. Boundaries of recommended water–bentonite–cement mixes based on the test results.

5. Conclusions

The plastic concrete used in urban hydrogeological barriers requires a set of properties that sometimes yields to contradictory requirements. The best example is the tradeoff between permeability and strength that involves using bentonite as a control component. As shown in this study, a large amount of bentonite decreases permeability and thus efficiency of the cutoff wall but, at the same time, also reduces compressive strength with consequences for the cracking control of the structure. The best approach should be to choose bentonite content within limits that provide satisfactory behavior in terms of both parameters. The recommended dosage is less than 17%, as shown herein.

Even if various authors proposed wide ranges of ratios of the mixture components, extreme values definitely lead to unfavorable effects, such as sand segregation, low workability, or long setting time. Because sodium bentonite and cements are well standardized, we consider it possible to predefine mixture recipes that are optimal for certain purposes.

This paper does not aim to set pre-determined mixture ratios but rather to guide the choice of plastic concrete recipe towards the targeted design purpose. As is readily noticeable, slight variations in components may induce an important improvement or decay in certain mechanical characteristics.

Author Contributions: Investigation, C.-Ș.B.; Data curation, A.-D.S., Methodology, D.-M.M.; Software, A.-D.S.; Supervision, M.-S.Ș. All authors have read and agreed to the published version of the manuscript.

Funding: This research received no external funding.

Institutional Review Board Statement: Not applicable.

Informed Consent Statement: Not applicable.

Data Availability Statement: Data is contained within the article in Table 3.

Conflicts of Interest: The authors declare no conflict of interest.

References

1. Shepherd, D.A.; Kotan, E.; Dehn, F. Plastic concrete for cut-off walls: A review. *Constr. Build. Mater.* **2020**, *255*, 119248. [CrossRef]
2. Fadaie, M.A.; Nekooei, M.; Javadi, P. Effect of Dry and Saturated Bentonite on Plastic Concrete. *KSCE J. Civ. Eng.* **2019**, *23*, 3431–3442. [CrossRef]
3. Pashang Pisheh, Y.; Mir Mohammad Hosseini, M. Experimental Investigation of Mechanical Behavior of Plastic Concrete in Cutoff Walls. *J. Mater. Civ. Eng.* **2018**, *31*, 04018355. [CrossRef]
4. U.S. Bureau of Reclamation. *Design Standards No.13: Embankment Dam, Chapter 16: Cutoff Walls*; U.S. Department of the Interior: Washington, DC, USA, 2014; Volume 7.
5. Hinchberger, S.; Weck, J.; Newson, T. Mechanical and hydraulic characterization of plastic concrete for seepage cut-off walls. *Can. Geotech. J.* **2010**, *47*, 461–471. [CrossRef]
6. Bagheri, A.; Alibabaie, M.; Babaie, M. Reduction in the permeability of plastic concrete for cut-off walls through utilization of silica fume. *Constr. Build. Mater.* **2008**, *22*, 1247–1252. [CrossRef]
7. EZavadkas, K.; Turskis, Z. A new logarithmic normalization method in games theory. *Informatica* **2008**, *19*, 303–314.
8. CEN, EN 1538+A1:2015; Execution of Special Geotechnical Work—Diaphragm Walls. European Committee for Standardization: Brussels, Belgium, 2015.
9. Banfill, P.F. Rheology of Fresh Cement and Concrete. *Mater. Sci.* **1991**, 61–130. [CrossRef]
10. Mahboubi, A.; Ali, A. Experimental study of the mechanical behavior of plastic concrete in triaxial compression. *Cem. Concr. Res.* **2005**, *35*, 412–419. [CrossRef]
11. Zhang, P.; Guan, Q.; Li, Q. Mechanical Properties of Plastic Concrete Containing Bentonite. *Res. J. Appl. Sci. Eng. Technol.* **2013**, *5*, 1317–1322. [CrossRef]
12. Kazemian, S.; Ghareh, S.; Torkanloo, L. To Investigation of Plastic Concrete Bentonite Changes on it's Physical Properties. *Procedia Eng.* **2016**, *145*, 1080–1087. [CrossRef]

Article

Spatial and Time Variable Long Term Infiltration Rates of Green Infrastructure under Extreme Climate Conditions, Drought and Highly Intensive Rainfall

Floris Cornelis Boogaard [1,2]

1 Research Centre for Built Environment NoorderRuimte, Hanze University of Applied Sciences, 9747 AS Groningen, The Netherlands; floris@noorderruimte.nl; Tel.: +31-641852172
2 Deltares Daltonlaan 600, 3584 BK Utrecht Postbus, 85467 3508 AL Utrecht, The Netherlands

Abstract: Swales are widely used Sustainable Urban Drainage Systems (SuDS) that can reduce peak flow, collect and retain water and improve groundwater recharge. Most previous research has focused on the unsaturated infiltration rates of swales without considering the variation in infiltration rates under extreme climate events, such as multiple stormwater events after a long drought period. Therefore, fieldwork was carried out to collect hydraulic data of three swales under drought conditions followed by high precipitation. For this simulation, a new full-scale infiltration method was used to simulate five rainfall events filling up the total storage volume of the swales under drought conditions. The results were then compared to earlier research under regular circumstances. The results of this study show that three swales situated in the same street show a variation in initial infiltration capacity of 1.6 to 11.9 m/d and show higher infiltration rates under drought conditions. The saturated infiltration rate is up to a factor 4 lower than the initial unsaturated rate with a minimal rate of 0.5 m/d, close to the minimum required infiltration rate. Significant spatial and time variable infiltration rates are also found at similar research locations with multiple green infrastructures in close range. If the unsaturated infiltration capacity is used as the design input for computer models, the infiltration capacity may be significantly overestimated. The innovative method and the results of this study should help stormwater managers to test, model, plan and schedule maintenance requirements with more confidence, so that they will continue to perform satisfactorily over their intended design lifespan.

Keywords: infiltration of stormwater; green infrastructure; nature-based solutions; bioretention; hydrologic performance; full-scale testing; drought

Citation: Boogaard, F.C. Spatial and Time Variable Long Term Infiltration Rates of Green Infrastructure under Extreme Climate Conditions, Drought and Highly Intensive Rainfall. *Water* 2022, 14, 840. https://doi.org/10.3390/w14060840

Academic Editors: C. Radu Gogu, Andrzej Witkowski and Ian Prosser

Received: 2 January 2022
Accepted: 7 March 2022
Published: 8 March 2022

Publisher's Note: MDPI stays neutral with regard to jurisdictional claims in published maps and institutional affiliations.

Copyright: © 2022 by the author. Licensee MDPI, Basel, Switzerland. This article is an open access article distributed under the terms and conditions of the Creative Commons Attribution (CC BY) license (https://creativecommons.org/licenses/by/4.0/).

1. Introduction

Urbanisation and climate change effect the water balance in our cities, resulting in challenges such as flooding, droughts and heat stress. The development and urbanization of watersheds increases impervious land cover and leads to an increase in stormwater runoff volume [1,2]. Stormwater management has shifted to include techniques that reduce runoff volumes and improve runoff water quality in addition to reducing the peak flow rate. Sustainable Urban Drainage System (SuDS), green infrastructure (GI), nature-based solutions (NBS) and bio-retention practices are typically designed to reduce runoff through infiltration and have been used for decades globally to provide infrastructure conveyance and water quality treatment [2–4]. Swales are typical landscape surface-drainage system vegetated (generally grass-lined) channels that receive stormwater runoff through gentle side-slopes and convey this stormwater downstream by way of longitudinal slopes [4–7]. Water quality treatment in a swale occurs through the process of sedimentation, filtration, infiltration and biological and chemical interactions with the soil. Swales have been shown to be very efficient in removing sediment particles from urban runoff [8–10]. The embankment slopes of swales can provide a bonus for volume retention, which in turn is positively impacted by the infiltration capacity of the filter media, soil moisture deficit, and side-slope length [11–14]. The type of vegetation, such as deep-rooted grass species and proliferation of root and earthworm channels, also improve infiltration [15–17]. Regarding soil composition in general, bioswales are composed of loamy sands, loams, or sandy loams resulting in variation in the infiltration rates of bio-swales and bio filters [14,16]. Several studies show that the performance of swales and sustainable urban drainage systems in general can be influenced by (human) failures in the design, implementation and maintenance of swales [4,18,19]. Besides these factors, there is a general consensus on the increase in extreme events in Europe in the 21st century [20–22], fostered by a temperature rise in the context of global warming that will affect the performance of green infrastructure. Drought impacts recorded in the recent past in Europe [23–25] will become more substantial in the future, making the identification of areas where droughts are projected to become more frequent and severe an important subject. In many European countries, the year 2018 broke drought records, with growing concerns in The Netherlands [26,27] after a 2-month period of no rain. The dry visual state of green infrastructure (Figure 1) in 2018 compared to 2017 raised many questions from Dutch municipalities and water authorities on the efficiency of swales being the most commonly used Sustainable Urban Drainage Systems in The Netherlands.

Figure 1. (a) Swale at study site in the municipality Dalfsen showing normal condition in 2017 and (b) the same swale after a 7-weeks dry period in 2018. The swales in Dalfsen is the case study presented in this paper.

A large variation in the hydraulic performance of swales can be expected by the several discussed factors as different filter media, soil moisture content, side-slope length, type of vegetation, soil composition and (human) errors in the design, implementation and maintenance phase. There are several international studies that determined the variation in mean volume reduction in swales from 11 to 75% [10,28–32] and peak flow rate reductions from 10 to 74% with detention provided by infiltration or check dams improving this mitigation [33,34]. In The Netherlands, check dams are hardly applied; the hydraulic efficiency of swales rely on infiltration and retention capacity. The infiltration capacity of swales is usually estimated by measuring the rate at which water soaks away from small test pits or boreholes [4,35–37] or ring infiltrometer tests [38–40]. A number of studies have demonstrated a high degree of spatial variability between different infiltration measurements since the results were based on the infiltration rate through a very small area that is used to represent the total infiltration area [19,41–43]. Studies show large spatial variation in infiltration rates with individually measured infiltration values varying by a factor of 100, concluding that about 20 measurements at each swale is needed to reduce the uncertainty [44].

No studies have been found on the effect of long dry periods on the infiltration capacity of (Dutch) swales. Since swales are the most implemented nature-based climate adaptation method in the Netherlands (Figure 2a), research on this topic is advised. It is estimated that over 5000 swales are implemented in The Netherlands and more will be constructed to replace grey infrastructure in the near future. Urban planners and stakeholders need to have an understanding of the spatial and temporal variability of swales under normal and extreme circumstances, such as the severe drought in 2018. Therefore, this study set out to answer the following questions:

- Which variation of the (un)saturated infiltration capacity can be expected under extreme weather conditions of drought followed by severe rainfall?
- Do the swales empty their storage volume within 2 days under all circumstances according to the Dutch guidelines [45]?

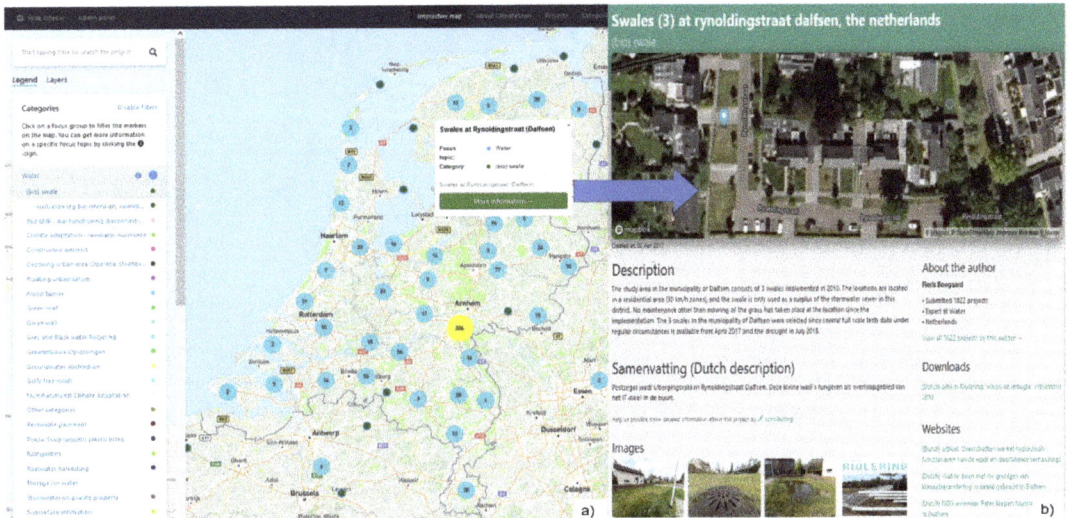

Figure 2. (**a**) Aggregated map of The Netherlands with over 1000 locations installed Dutch swales (Source: ClimateScan. Available online: https://www.climatescan.nl/map#filter-1-1 (accessed on 1 January 2022)). (**b**) Exact location of 3 swales in the Reynoldstraat in Dalfsen (more information: ClimateScan. Available online: https://www.climatescan.nl/projects/1114/detail (accessed on 1 January 2022)).

While previous research on swales used small scale (ring infiltrometer) tests to gain insight into hydraulic performance, in this study, a full-scale test is applied where the total volume of 3 swales is repeatedly filled up with a tanktruck.

Study Area and Data

To answer the research questions, Dalfsen was chosen as the study area (Figure 1), since full-scale research on variation in infiltration capacity was conducted here in 2017 and was repeated at the ending of the drought in 2018. It is a unique location as three swales are constructed next to each other in the same street at the same time, built by the same design characteristics under similar geo-hydraulic circumstances in 2010. At this location, the variation in hydraulic performance can be studied and compared to earlier research results, with the permission of the municipality and residents.

The open-source online platform "Climatescan" presented in Figure 2a shows over 1000 locations where swales have been installed in The Netherlands from 1999 to the present [46]. Figure 2b shows the exact location of the study area in the municipality of Dalfsen, where three swales were implemented.

Dutch swales are shallow (often <0.3 m deep), dry, vegetated, and generally grass-lined, receiving stormwater runoff through drainage pipes and gullies from the connected surface area of roads and houses [45], see Figure 3. The figure shows the basic green infrastructure with the road surface and the grassed swale with engineered soil and inlet and outlet.

Figure 3. The swale construction in the municipality Dalfsen, The Netherlands.

A number of swale infiltration systems have been in use in The Netherlands for years (Figure 2b), or even decades, but despite their long-term operation there is limited documentation on their hydraulic performance over their expected lifetime.

Dutch guidelines dictate a maximum emptying time of 48 h for swales with a depth of 30 to 50 cm [45]. Guidelines in The Netherlands are based on several factors such as the limited availability of space in urban areas, the low permeability of the soil, high groundwater tables and limited public health concerns (drowning of kids and mosquito nuisance) as a safety factor, given that the infiltration capacity of swales may reduce over time by clogging [19,45]. Similar guidelines on the hydraulic performance of swales can be found in Germany, where advice has been given on a design that enables the swale to infiltrate stormwater with a minimum infiltration rate of 0.864 m/d [47].

The study area in the municipality of Dalfsen is located in a residential area (30 km/h zones) and the three individual swales are used as a surplus of the stormwater sewer in this district. Figures 3 and 4 show the impact of drought on vegetation in 2018 compared to 2017. The data of the full-scale infiltration test used in this study was executed on 20 April 2017, which is compared to the data of infiltration tests on 27 July 2018 after 7 weeks of drought.

Figure 4. Inlet of the swale in Dalfsen—(**a**) dry period in 2018, (**b**) during full-scale test in 2017.

The dimensions of the swales are presented in Figure 5. All swales have a maximum depth of 50 cm to street level. Swales 1 and 2 are similar (both an average storage volume 7.6 m^3 up to the overflow level of appr. 10 cm height); swale 3 is smaller (storage volume 1.7 m^3 up to the overflow level).

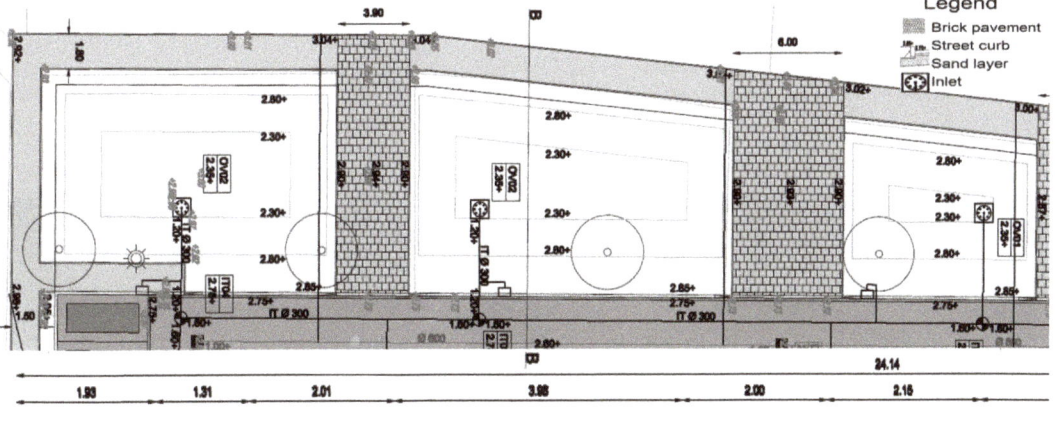

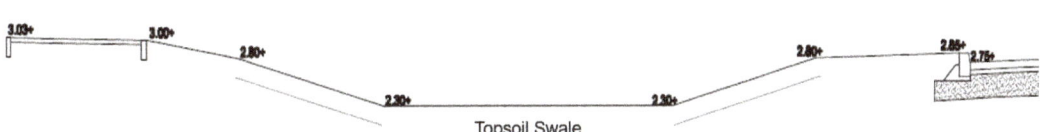

Figure 5. Swale 1 (**left**) to 3 (**right**) and cross-section B-B (swale 2).

2. Materials and Methods

For the case study Dalfsen, a full-scale infiltration test was used [19] where the total volume of the swales was submerged (simulating high intensive rainfall) and the emptying time was measured by several measurements during regular and extreme drought conditions during 2018 (7 weeks of drought). A number of issues were considered using the full-scale test method in the municipality Dalfsen:

1. Selection swale for testing;
2. Water supply alternatives;
3. Accurate determination of surface infiltration rate.

(1) The swales in the municipality of Dalfsen are a unique research location as the three swales are situated next to each other and were all built by the same contractor and with the same materials in 2010, under the same geo-hydraulic conditions. All swales have a confined space which can be filled by a tank truck up to the water level of outflow, without any additional constructions to prevent water leaving the swale during the full-scale infiltration test.

(2) For this full-scale infiltration test in Dalfsen, a tanktruck was used with a larger storage volume of 10 m^3 than the volume of the swales (swale 1 and 2 have a storage volume of 8 m^3 up till the overflow level of 10 cm), so every individual swale was filled with water continuously with the tanktruck (Figures 6 and 7). The time interval between refilling the swales was within 10 min.

(3) Wireless, self-logging, pressure transducer loggers (Minidiver. Available online: https://www.eijkelkamp.com/producten/sensors-monitoring_nl/ (accessed on 1 January 2022)) were used in the study as the primary method of measuring and recording the reduction in water levels over time. Two loggers were installed at the lowest points of the swale. The transducers continuously monitored the static water pressures at those locations, logging the data in internal memory. Three different measurement methods were used in conjunction with the pressure transducers in order to verify the transducer readings, as shown in Figure 7. The three methods were: hand measurements, underwater camera (Figure 7) and time-lapse photography (movies available at ClimateScan. Available online: https://www.climatescan.nl/projects/1114/detail (accessed on 1 January 2022)).

Figure 6. The full-scale infiltration test with a tank truck executed repetitively for monitoring the hydraulic behaviour of swales in Dalfsen—(**a**) under regular conditions in 2017, (**b**) Same monitoring after 7 weeks of drought in 2018.

Figure 7. The methods for monitoring the water height in the swales during the infiltration tests were (**a**) hand measurements on a measuring tape and underwater video camera, continuous time-lapse photography (**b**) and pressure transducer loggers.

3. Results and Discussion

Three swales were tested repeatedly with the full-scale infiltration tests on two occasions under different (simulated) weather conditions. The results show the linear emptying curves of the three swales in Dalfsen (Figures 8 and 9). The number of tests at individual swales depended on the duration of each test and the availability of the tanktruck as the water source. Swale 2 could be tested with a maximum of five full-scale tests (Figure 8). For comparative reasons, all graphs display the measuring interval 8 to 4.5 cm of the water column; 8 cm is the height just below the outlet, thereby ruling out any error due to leakage in the data (Figure 8). Below 4.5 cm the infiltration surface of the swale was not regular, with the water being just above the "bumpy" bottom of the swale, allowing water to flow horizontally from a small puddle to puddle. By truncating the data, any subsurface flow due to depressions effecting the emptying curve was ruled out (Figures 8 and 9).

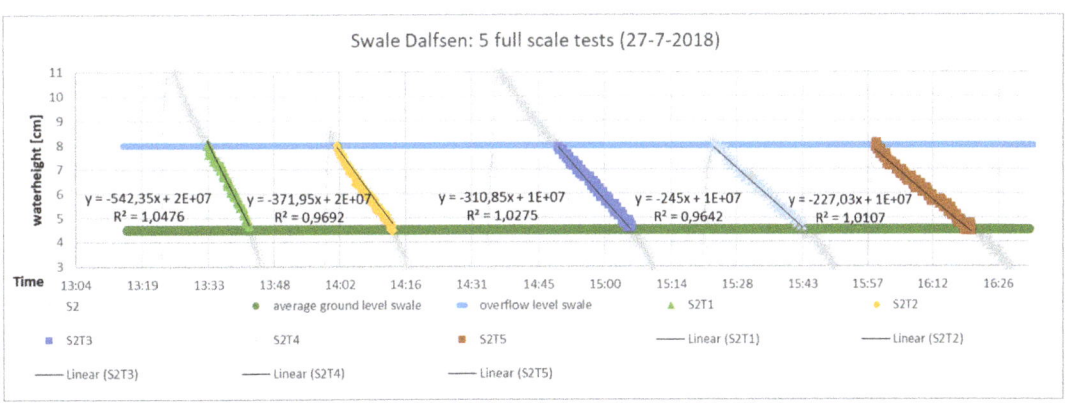

Figure 8. Empty curves swale 2 (S2) and 5 tests (T1–T5) in drought situation.

The maximum standard deviation of the pressure transducer loggers is in the range of +/−0.5 cm, which is reflected in the thickness of the lines in Figure 9. All the emptying curves for the three swales under normal conditions show similar patterns; rapid emptying time with steep linear curves for the first infiltration test and gradually increasing in time with lower infiltration rates (Figure 9).

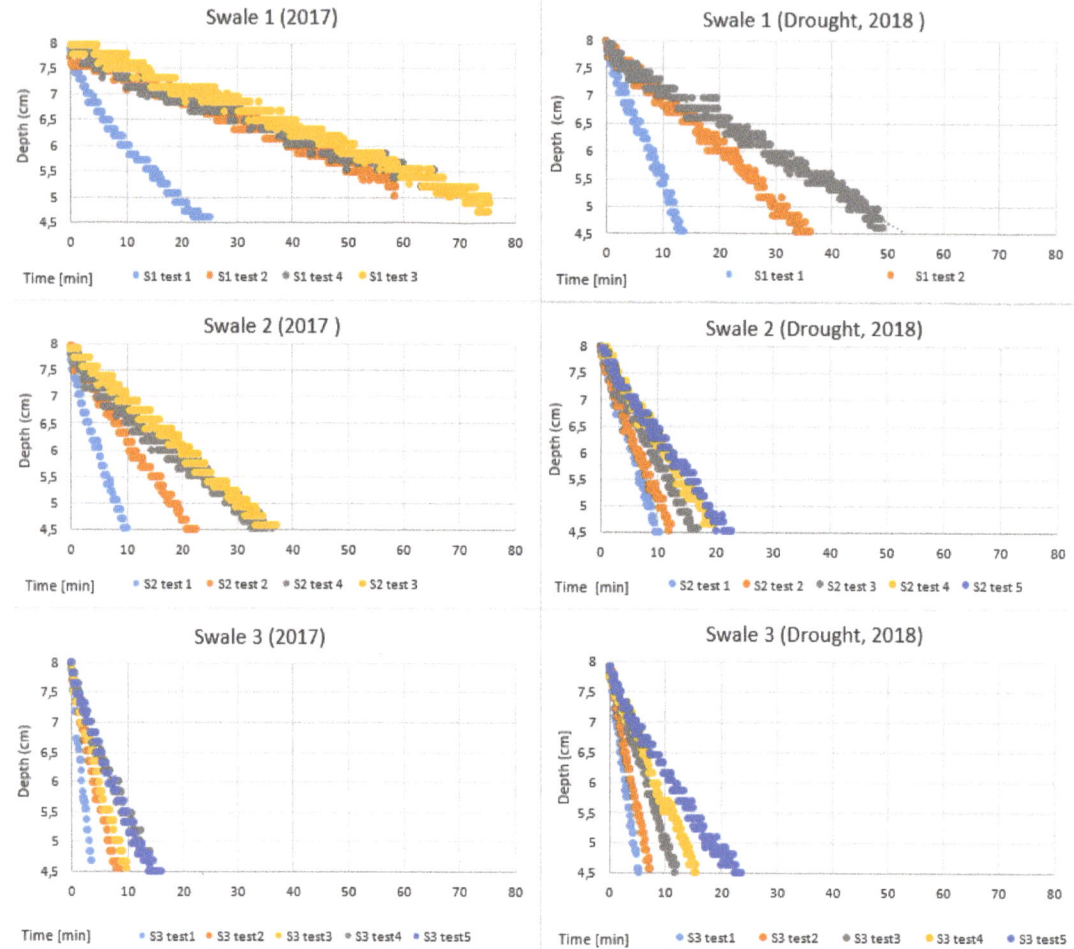

Figure 9. Emptying time graphs of swale 1, 2 and 3 under regular (2017, **left**) and drought conditions (2018, **right**).

Figure 9 shows a high variation in the infiltration rates of the swales. Simple linear regression analysis was used to generate lines of best fit for the transducer readings from each swale. A factor 7 difference in initial infiltration rate between swale 1 and 3 can be observed in 2017 (factor 3 in 2018), while the design, construction and maintenance of the swales are the same. Repeating the infiltration test can lead to reduction in infiltration rates in the order of 50% (swale 2), Figure 10.

The equations of the linear regression lines were then used to calculate the average infiltration rate in m/d. Figure 11 shows the calculated infiltration rates of all tests. The infiltration rate after five tests of swale 3 in 2017 and 2018 was 25% of its initial infiltration capacity. After five tests, a saturated steady state infiltration rate was not yet reached.

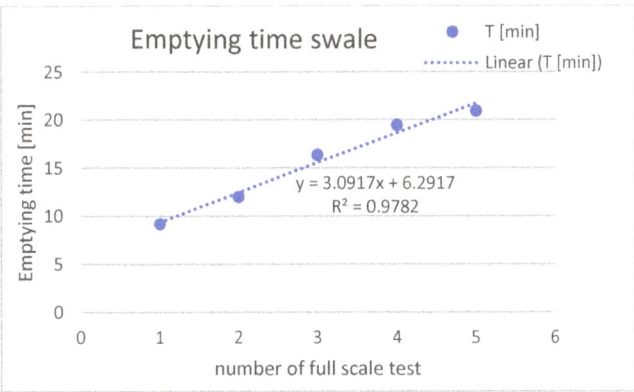

Figure 10. Emptying time of swale 2 show a factor of about 50% decline after 5 full-scale tests.

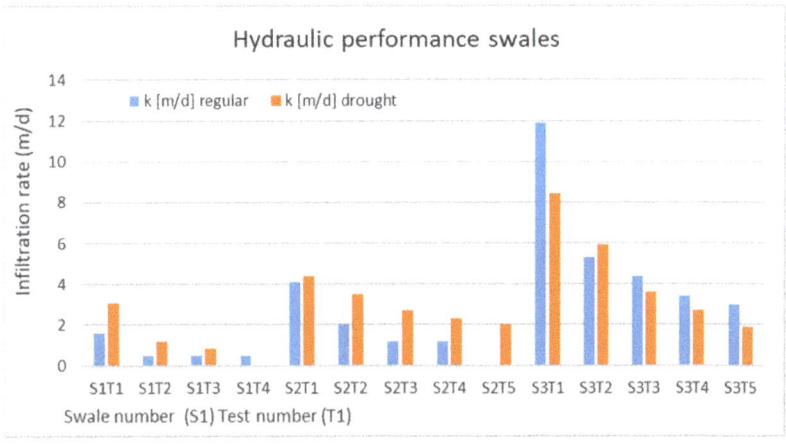

Figure 11. Infiltration rate in m/d (results of linear regression analysis).

In drought conditions, all swales did not reach the saturated steady state infiltration rate, even after five full-scale tests in contrast to swale 1 and 2 under regular circumstances in 2017. Swales 1 and 2 showed increased infiltration capacity during the drought period. The relatively small swale 3 (Figure 5) has a different pattern and the emptying time is lower in the drought situation.

4. Discussion

The largest swales 1 and 2 have a higher infiltration rate during the drought period in contrast to the smallest swale 3. Note that for swale 3 this difference in emptying time is only in the order of some minutes due to the much higher infiltration rate compared to swale 1 and 2 (the difference in emptying time is 47 min for swale 1). The municipality expected a lower infiltration rate during drought (seeing the dry baren state of the swales in Figure 1b) and feared intensive rainfall after the drought period, resulting in flooding. A possible explanation for the increased infiltration rate during drought is preferential flow [15,48] due to cracks in the soil and micro-pores as the result of dying roots, enabling the vertical flow of air and water (Figure 12).

Figure 12. Significant bubbles during the filling process indicating (preferential) flow during the drought period: air-filled pores are filled with water forcing out the trapped air.

The variation in initial infiltration capacity of 1.6 to 11.9 m/d of the three swales situated in the same street was received by the municipality with more surprise. However, same observations were recorded in other municipalities, such as Tilburg and Arnhem, where the same full-scale method was used on several green infrastructures on one street (Figure 13) showing a high variation in infiltration rate (Figure 14).

Figure 13. Measurements at two streets with multiple green infrastructure in The Netherlands: Azuurweg Tilburg (**left**, ClimateScan. Available online: https://www.climatescan.nl/projects/6052/detail (accessed on 1 January 2022)) and Reestraat Arnhem (**right**, ClimateScan. Available online: https://www.climatescan.nl/projects/1109/detail (accessed on 1 January 2022)).

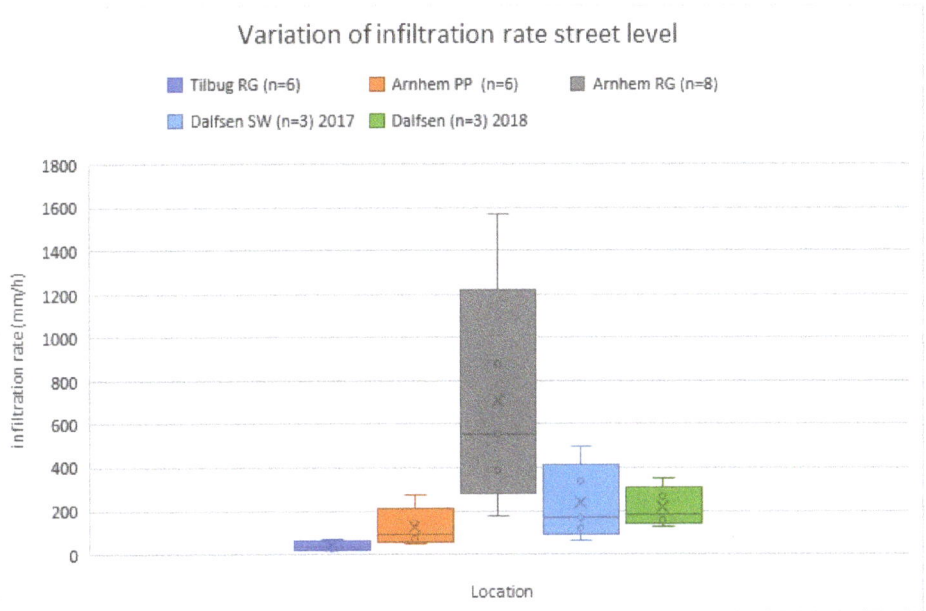

Figure 14. Range of infiltration rates in streets with multiple green infrastructure [19,49,50]. SW = Swale, RG = Raingarden, PP = Permeable Pavement, n = number of green infrastructures.

Surprising to the municipality is also the difference between the saturated and unsaturated infiltration capacity: the measured saturated infiltration rates are up to a factor of 4 lower than the initial unsaturated capacity, with a minimal rate of 0.5 m/d. Other studies, where small-scale tests were used, such as (Modified) Philip–Dunne (MPD) or infiltrometer tests [4,35–40,51], show a variation in (un)saturated infiltration rates. Research locations with green infrastructure, where the full-scale test method was applied under similar Dutch circumstances, also show a significant reduction in infiltration rates between the unsaturated and saturated tests, ranging from 27% to 48% [52]. The most important outcome of this research, particularly for public stakeholders, was that all swales empty their storage volume within two days under all circumstances, according to the Dutch guidelines [45]. These research results should raise awareness about the variation in infiltration rates among stormwater managers engaged with the planning, testing, and modelling of green infrastructure about the spatial and time variable long-term infiltration capacity rates of green infrastructure.

5. Conclusions

The full-scale infiltration testing method was applied in this study to determine the variation in the hydraulic performance of three swales under different (simulated) weather conditions. The results from this study show that the tested swales empty their storage volume within 48 h under regular and extreme drought (un)saturated conditions, even after five full-scale tests (lowest value 0.5 m/d) after 8 years in operation, and without maintenance other than mowing the grass. Stormwater managers are satisfied that their green infrastructure demonstrates the minimum required infiltration capacity, so any tangible improvement to their design and characterization, such as additional drainage, is not needed.

The individual swales in Dalfsen show a variation in the initial infiltration capacity of 1.6 to 11.9 m/d. Most previous research focused on unsaturated infiltration rates. However, the results of this study show that the difference in infiltration rates between saturated and unsaturated can differ by a factor 4 (after 5 full-scale tests). These large differences by a factor of 4 in (un)saturated infiltration rate in one street is also determined in other research with the full-scale test in The Netherlands. Therefore, if the unsaturated infiltration capacity is used as the design input for computer models, the infiltration capacity may be significantly overestimated. It is therefore recommended that an appropriate degree of pre-saturation is accounted for in the design or a linear increase in the emptying time, as shown in this paper. This could be applied by a simple reduction in the design unsaturated infiltration rate. The results of this study should help stormwater managers with the modelling, planning, testing and scheduling of maintenance requirements for swales with more confidence, so that they will continue to perform satisfactorily over their intended design lifespan.

Funding: This research was funded by SIA, grant number SVB/ RAAK.PUB07.015 project Groenblauwe oplossingen, kansen en risico's.

Data Availability Statement: Not applicable.

Acknowledgments: This research is supported through Raak GroenBlauw, kansen en risicos co-funded by SIA. Thanks to the municipality Dalfsen, Tilburg and Arnhem for support for this work. Photos by Thomas Klomp.

Conflicts of Interest: The authors declare no conflict of interest.

References

1. Marsalek, J.; Jiménez-Cisneros, B.E.; Malmquist, P.-A.; Karamouz, M.; Goldenfum, J.; Chocat, B. *Urban Water Cycle Process and Interactions*; UNESCO Publishing and Taylor and Francis Group: Paris, France, 2006.
2. Fletcher, T.; Andrieu, H.; Hamel, P. Understanding, management and modelling of urban hydrology and its consequences for receiving waters: A state of the art. *Adv. Water Resour.* **2013**, *51*, 261–279. [CrossRef]
3. Fletcher, T.D.; Shuster, W.; Hunt, W.F.; Ashley, R.; Butler, D.; Arthur, S.; Trowsdale, S.; Barraud, S.; Semadeni-Davies, A.; Bertrand-Krajewski, J.-L.; et al. SUDS, LID, BMPs, WSUD and more—The evolution and application of terminology sur-rounding urban drainage. *Urban Water J.* **2015**, *12*, 525–542. [CrossRef]
4. Ballard, B.W.; Wilson, S.; Udale-Clarke, H.; Illman, S.; Scott, T.; Ashley, R.; Kellagher, R. *The SuDS Manual*; CIRIA C753-The SuDS Manual, CIRIA Research Project (RP)992CIRIA; Department for Environment Food & Rural Affairs, CIRIA: London, UK, 2015; ISBN 978-0-86017-760-9.
5. Davis, A.P.; Jamil, E. Field Evaluation of Hydrologic and Water Quality Benefits of Grass Swales with Check Dams for Managing Highway Runoff. In Proceedings of the International Low Impact Development Conference 2008, Seattle, WA, USA, 16–19 November 2008. [CrossRef]
6. Davis, A.P.; Stagge, J.H.; Jamil, E.; Kim, H. Hydraulic performance of grass swales for managing highway runoff. *Water Res.* **2012**, *46*, 6775–6786. [CrossRef] [PubMed]
7. Stagge, J.H.; Davis, A.P.; Jamil, E.; Kim, H. Performance of grass swales for improving water quality from highway runoff. *Water Res.* **2012**, *46*, 6731–6742. [CrossRef]
8. Winston, R.J.; Hunt, W.F.; Kennedy, S.G.; Wright, J.D.; Lauffer, M.S. Field Evaluation of Storm-Water Control Measures for Highway Runoff Treatment. *Environ. Eng.* **2012**, *138*, 101–111. [CrossRef]
9. Barrett, M.E.; Walsh, P.M.; Malina, J.F.; Charbeneau, R.J. Performance of Vegetative Controls for Treating Highway Runoff. *Environ. Eng.* **1998**, *124*, 1121–1128. [CrossRef]
10. Deletic, A. Sediment transport in urban runoff over grassed areas. *J. Hydrol.* **2005**, *301*, 108–122. [CrossRef]
11. García-Serrana, M.; Gulliver, J.S.; Nieber, J.L. Infiltration capacity of roadside filter strips with non-uniform over-land flow. *J. Hydrol.* **2017**, *545*, 451–462. [CrossRef]
12. Blanco-Canqui, H.; Gantzer, C.J.; Anderson, S.H.; Alberts, E.E.; Thompson, A.L. Grass Barrier and Vegetative Filter Strip Effectiveness in Reducing Runoff, Sediment, Nitrogen, and Phosphorus Loss. *Soil Sci. Soc. Am. J.* **2004**, *68*, 1670–1678. [CrossRef]
13. Lucke, T.; Mohamed, M.A.K.; Tindale, N. Pollutant Removal and Hydraulic Reduction Performance of Field Grassed Swales during Runoff Simulation Experiments. *Water* **2014**, *6*, 1887–1904. [CrossRef]
14. Le Coustumer, S.; Fletcher, T.; Deletic, A.; Barraud, S.; Lewis, J.F. Hydraulic performance of biofilter systems for stormwater management: Influences of design and operation. *J. Hydrol.* **2009**, *376*, 16–23. [CrossRef]
15. Abu-Zreig, M.; Rudra, R.P.; LaLonde, M.N.; Whiteley, H.R.; Kaushik, N.K. Experimental investigation of runoff reduction and sediment removal by vegetated filter strips. *Hydrol. Process.* **2004**, *18*, 2029–2037. [CrossRef]
16. Le Coustumer, S.; Fletcher, T.D.; Deletic, A.; Barraud, S.; Poelsma, P. The influence of design parameters on clogging of stormwater biofilters: A large-scale column study. *Water Res.* **2012**, *46*, 6743–6752. [CrossRef]

17. Fort, F.; Jouany, C.; Cruz, P. Root and leaf functional trait relations in Poaceae species: Implications of differing re-source-acquisition strategies. *J. Plant Ecol.* **2012**, *6*, 211–219. [CrossRef]
18. De Bilt, V.; Nieuwenhuis, E.; van de Ven, F.; Langeveld, J. Root causes of failures in sustainable urban drainage systems (SUDS): An exploratory study in 11 municipalities in The Netherlands. *Blue-Green Syst.* **2021**, *3*, 31. [CrossRef]
19. Boogaard, F. Stormwater Characteristics and New Testing Methods for Certain Sustainable Urban Drainage Systems in The Netherlands. Ph.D. Thesis, Technische Universiteit Delft, Delft, The Netherlands, July 2015; p. 149.
20. IPCC. 2021: Climate Change 2021: The Physical Science Basis. In *Contribution of Working Group I to the Sixth Assessment Report of the Intergovernmental Panel on Climate Change*; Masson-Delmotte, V., Zhai, P., Pirani, A., Connors, S.L., Péan, C., Berger, S., Caud, N., Chen, Y., Goldfarb, L., Gomis, M.I., et al., Eds.; Cambridge University Press: Cambridge, UK, 2021.
21. Forzieri, G.; Feyen, L.; Russo, S.; Vousdoukas, M.; Alfieri, L.; Outten, S.; Migliavacca, M.; Bianchi, A.; Rojas, R.; Cid, A. Multi-hazard assessment in Europe under climate change. *Clim. Change* **2016**, *137*, 105–119. [CrossRef]
22. Spinoni, J.; Vogt, J.V.; Naumann, G.; Barbosa, P.; Dosio, A. Will drought events become more frequent and severe in Europe? *Int. J. Climatol.* **2018**, *38*, 1718–1736. [CrossRef]
23. Naumann, G.; Spinoni, J.; Vogt, J.; Barbosa, P. Assessment of drought damages and their uncertainties in Europe. *Environ. Res. Lett.* **2015**, *10*, 124013. [CrossRef]
24. Blauhut, V.; Stahl, K.; Stagge, J.H.; Tallaksen, L.M.; De Stefano, L.; Vogt, J. Estimating drought risk across Europe from re-ported drought impacts, hazard indicators and vulnerability factors. *Hydrol. Earth Syst. Sci. Discuss.* **2015**, *12*, 12515–12566.
25. Stahl, K.; Kohn, I.; Blauhut, V.; Urquijo, J.; De Stefano, L.; Acácio, V.; Dias, S.; Stagge, J.H.; Tallaksen, L.; Kampragou, E.; et al. Impacts of European drought events: Insights from an international database of text-based reports. *Nat. Hazards Earth Syst. Sci.* **2016**, *16*, 801–819. [CrossRef]
26. Eertwegh, G.A.P.H.; van den Bartholomeus, R.P.; Louw, P.; de Witte, J.P.M.; Dam, J.; van Deijl, D.; van Hoefsloot, P.; Clevers, S.H.P.; Hendriks, D.; Al, E. Droogte in Zandgebieden van Zuid-, Midden-en Oost-Nederland. Rapportage Fase 1: Ontwikkeling van Uniforme Werkwijze Voor Analyse van Droogte en Tussentijdse Bevindingen. KnowH$_2$O. 2019. Available online: https://research.wur.nl/en/publications/droogte-in-zandgebieden-van-zuid-midden-en-oost-nederland-rapport (accessed on 1 January 2022).
27. Knight, E.M.P.; Hunt, W.F.; Winston, R.J. Side-by-side evaluation of four level spreader-vegetated filter strips and a swale in eastern North Carolina. *J. Soil Water Conserv.* **2013**, *68*, 60–72. [CrossRef]
28. Deletic, A. Modelling of water and sediment transport over grassed areas. *J. Hydrol.* **2001**, *248*, 168–182. [CrossRef]
29. COASTAR. Available online: https://www.deltares.nl/nl/publication/coastar-nationaal-regionale-en-nationale-coastar-toepassingen-in-beeld/ (accessed on 1 January 2022).
30. Rushton, B.T. Low-impact parking lot design reduces runoff and pollutants loads. *J. Water Resour. Plan. Manag.* **2001**, *127*, 172–179. [CrossRef]
31. Barrett, M.E. Comparison of BMP Performance Using the International BMP Database. *J. Irrig. Drain. Eng.* **2008**, *134*, 556–561. [CrossRef]
32. Ackerman, D.; Stein, E. Evaluating the effectiveness of best management practices using dynamic modeling. *J. Environ. Eng.* **2008**, *134*, 628–639. [CrossRef]
33. Davis, A.P.; Traver, R.G.; Hunt, W.F.; Lee, R.; Brown, R.A.; Olszewski, J.M. Hydrologic Performance of Bioretention Storm-Water Control Measures. *J. Hydrol. Eng.* **2012**, *17*, 604–614. [CrossRef]
34. Fassman, E.A.; Liao, M. Monitoring of a series of swales within a stormwater treatment train. In Proceedings of the 32nd Hydrology and Water Resources Symposium, Newcastle, Australia, 30 November–3 December 2009; Engineers Australia: Barton, Australia, 2009; pp. 368–378.
35. Bettess, R. *Infiltration Drainage-Manual of Good Practice*; CIRIA R156; CIRIA: London, UK, 1996; ISBN 978-0-86017-457-8.
36. Palhegyi, G.E. Designing Storm-Water Controls to Promote Sustainable Ecosystems: Science and Application. *J. Hydrol. Eng.* **2010**, *15*, 504–511. [CrossRef]
37. BRE Digest 365. In *Soakway Design*; Buildings Research Establishment: Bracknell, UK, 1991; ISBN 0-85125-502-7.
38. *ASTM D3385-09*; Standard Test Method for Infiltration Rate of Soils in Field Using Double-Ring Infiltrometer. American Society for Testing and Materials (ASTM): West Conshohocken, PA, USA, 2009.
39. Nestingen, R.S. The Comparison of Infiltration Devices and Modification of the Philip-Dunne Permeameter for the Assessment of Rain Gardens. Master's Thesis, Department of Civil Engineering, University of Minnesota, Minneapolis, MN, USA, 2007.
40. *DIN 19682-7*; Soil Quality—Field Tests—Part 7: Determination of Infiltration Rate by Double Ring Infiltrometer. German Institute for Standardization: Berlin, Germany, 2015.
41. Lucke, T.; Beecham, S. Field investigation of clogging in a permeable pavement system. *Build. Res. Inf.* **2011**, *39*, 603–615. [CrossRef]
42. Pezzaniti, D.; Beecham, S.; Kandasamy, J. Influence of clogging on the effective life of permeable pavements. *Water Manag.* **2009**, *162*, 211–220. [CrossRef]
43. Fassman, E.A.; Blackbourn, S. Urban Runoff Mitigation by a Permeable Pavement System over Impermeable Soils. *J. Hydrol. Eng.* **2010**, *15*, 475–485. [CrossRef]
44. Ahmed, F.; Gulliver, J.; Nieber, J. Field infiltration measurements in grassed roadside drainage ditches: Spatial and temporal variability. *J. Hydrol.* **2015**, *530*, 604–611. [CrossRef]
45. RIONED. *Swales: Recommendations for Design, Implementation and Maintenance*; Stichting RIONED: Ede, The Netherlands, 2006.

46. Restemeyer, B.; Boogaard, F.C. Potentials and Pitfalls of Mapping Nature-Based Solutions with the Online Citizen Science Platform ClimateScan. *Land* **2021**, *10*, 5. [CrossRef]
47. Arbeitsblatt DWA-A 138-Planung, Bau und Betrieb von Anlagen zur Versickerung von Niederschlagswasser-Verlag: DWA, Ausgabe: 04/2005, ISBN Print: 978-3-937758-66-4, ISBN E-Book: 978-3-96862-062-6, 32 April 2005; Stand: Korrigierte Fassung März 2006. Available online: https://webshop.dwa.de/en/dwa-a-138-versickerungsanlagen-4-2005.html (accessed on 1 January 2022).
48. Allaire, S.E.; Roulier, S.; Cessna, A.J. Quantifying preferential flow in soils: A review of different techniques. *J. Hydrol.* **2009**, *378*, 179–204. [CrossRef]
49. Boogaard, F.; Roelofs, R.; Laurentzen, E. Time and Spatial Depending Infiltration Rates (In Dutch: Infiltratiecapaciteit Verharding Verschilt in Ruimte en Tijd). Land en Water nr. 5-mei 2021. Available online: https://research.hanze.nl/en/publications/infiltratiecapaciteit-verharding-verschilt-in-ruimte-en-tijd (accessed on 1 January 2022).
50. Boogaard, F.; Schilder, S.; Lekkerkerk, J. Variation in Infiltration Rates Swales (In Dutch: Infiltratiecapaciteit Wadi's Varieert in Ruimte en Tijd); Land en Water, nr 1/2- Februari 2021. Available online: https://research.hanze.nl/nl/publications/infiltratiecapaciteit-wadis-varieert-in-ruimte-en-tijd (accessed on 1 January 2022).
51. Ebrahimian, A.; Sample-Lord, K.; Wadzuk, B.; Traver, R. Temporal and spatial variation of infiltration in urban green infrastructure. *Hydrol. Process.* **2019**, *34*, 1016–1034. [CrossRef]
52. Boogaard, F.; Lucke, T. Long-Term Infiltration Performance Evaluation of Dutch Permeable Pavements Using the Full-Scale Infiltration Method. *Water* **2019**, *11*, 320. [CrossRef]

Article

Groundwater and Urban Planning Perspective

Alina Radutu [1], Oana Luca [2,*] and Constantin Radu Gogu [2]

[1] Romanian Space Agency, 010362 Bucharest, Romania; alina.radutu@rosa.ro
[2] Groundwater Engineering Research Center, Technical University of Civil Engineering, 020396 Bucharest, Romania; radu.gogu@utcb.ro
* Correspondence: oana.luca@utcb.ro; Tel.: +40-746961119

Abstract: An analysis of 17 Romanian cities' Urban General Plans showed that urban planning documents do not satisfactorily rely on groundwater information. The associated hydrogeological supporting studies include only general recommendations. However, they should include specifications to improve water-balance and detail the need to implement monitoring systems to monitor groundwater levels. The studies do not recommend special construction measures to be implemented for future infrastructure elements and do not include maps delimiting the particular geotechnical and hydrogeological characteristics. A study conducted on an urban river corridor using satellite remote sensing and a methodology characterizing the chosen zone clearly shows a major concordance between the groundwater level and vertical displacements. In addition, the presence of urban anthropogenic strata associated with the groundwater level fluctuations showed amplified vertical displacements of the ground when compared to the areas where the natural deposits exist. The methodology combines subsidence occurrence, land-cover changes, hydrogeological, geological, and hydrological characteristics, climatic aspects, the location, the extension of old quarries, and the last 100 years of topographical changes. These observations emphasize the need for accurate studies to properly discriminate between phenomena and processes generating subsidence, which must be used systematically to support the general urban plans of cities as the documentation of future developments

Keywords: urban groundwater; urban planning; hydrogeology; sustainable development

Citation: Radutu, A.; Luca, O.; Gogu, C.R. Groundwater and Urban Planning Perspective. *Water* 2022, 14, 1627. https://doi.org/10.3390/w14101627

Academic Editor: Yuanzheng Zhai

Received: 24 March 2022
Accepted: 16 May 2022
Published: 18 May 2022

Publisher's Note: MDPI stays neutral with regard to jurisdictional claims in published maps and institutional affiliations.

Copyright: © 2022 by the authors. Licensee MDPI, Basel, Switzerland. This article is an open access article distributed under the terms and conditions of the Creative Commons Attribution (CC BY) license (https://creativecommons.org/licenses/by/4.0/).

1. Introduction

Urban planners, innovators, and researchers are increasingly working on sustainable and smart initiatives for urban transformation [1]. Various perspectives and disciplines related to radical change towards sustainable and smart urban systems are being integrated into urban plans and strategies. Despite the growing body of evidence highlighting the importance of groundwater to the support of urban living, the impact of urbanism on natural groundwater systems is rarely considered in city systems planning [2], even in such cases where groundwater dynamics could modify the buildings structural behavior. Discrepancies in communication between the scientific community and city administrations add to the difficulties in addressing urban resilience and hydrogeology issues. Reliable subsurface management must be based on relevant knowledge and an understanding of the phenomena, processes, and data, which must be accessible, as well as knowledge of urban planning procedures to minimize the impact of urban developments on the water cycle. Experts should use robust datasets on urban fabric, infrastructure networks, groundwater, and geothermal energy systems at the urban scale [3].

The influence of urban planning on hydrology has been explored by Carneiro et al. [4]. The authors [4] identify the need for building accurate hydrogeological models that can be used as tools for urban planning decisions. These models must be based on accurate geological modeling which takes into account both vertical and lateral facies variations. Consequently, accurate local studies are needed to characterize hydraulically anthropogenic heterogeneous deposits and embankments. The models should englobe the influence of the underground structures and consider the drains, dewatering systems, and the losses from the water supply transportation pipes. In Canada and the United States, the required level of groundwater information for land use planning purposes has been described by Roxane et al. [5], and demonstrates that the centralization of groundwater data leads to an increased accessibility for land planners. The paper also confirmed the existing gap between groundwater information production and regional planning. The urgent need to guide urban planners on how to manage urban development with minimal damage to groundwater resources has also been demonstrated in several works [6,7]. Carmon et al. [6] provided solutions on how to protect groundwater for the development of water-sensitive urban planning. The paper highlighted the effects of urban development on the quantity and quality of rainwater that infiltrates into the soil on its way to recharge the aquifer. The authors analyzed the effect of urban development on runoff and infiltration processes, focusing on the effect of certain patterns of urban development and identifying options to mitigate the negative aspects through relatively simple and inexpensive means. Components of urban planning which influence runoff and infiltration were presented. As an aid to understanding the urbanization process and to the designing of policies and regulations, Patra et al. [7] studied the impacts of urbanization on land cover changes and its probable implications for the local climate and groundwater level. The authors pointed out that the increase in the urban built-up area over the last two decades led to fluctuations of the groundwater level in the Howrah Municipal Corporation (HMC) area of the Indian state of West Bengal. In the city of Erbil, in the Kurdistan Region of Iraq, Hameed et al. [8] identified a significant correlation between urbanization areas and groundwater levels. As the urban expansion increased by 278% between 2004 and 2014, a decline in groundwater levels was detected in the study area [8].

A study focusing [9] on the situation in the coastal peatlands of Netherlands detailed where careful land use planning was adopted with special attention paid to groundwater. This was due to the long history of subsidence followed by adaptation strategies over a period of more than nine centuries.

Focusing the use of satellite remote sensing techniques, another study [10] proposed the use of the satellite radar interferometry system InSAR to define potential ground fractures linked to aquifer system compaction, in Toluca Valley, Mexico. The ground fractures map was validated through a field campaign. The authors highlighted the great benefit of this map to providing operational support to urban planning. It was suggested that the ground fracture mapping shows a higher reliability when derived from InSAR than solely based on field observations.

The long-term urban development plan established in 2011 in Hanoi, Vietnam (Hanoi Master Plan to year 2030 and vision to 2050), with the purpose of a more sustainable development of the city, considers various environmental aspects, such as soil pollution, air quality, and water quality [11]. As groundwater is one of its main water supply resources, the continuous rapid urbanization of Hanoi city since the 1990s has led to the acceleration of groundwater withdrawal and the emergence of land subsidence [12]. Several other studies were carried out using satellite radar interferometry for land subsidence monitoring since 2007 [12–14], where the authors mention numerous foundation failures of different constructions in the 2000s [12]. The authors highlight the crucial role that a knowledge of surface deformation plays for urban planning to minimize the severe influences of surface subsidence [14].

The continuous monitoring of land subsidence is recognized as a critical instrument for the detection of potential hazards and for the designing of compensation strategies and urban planning activities, as in the case of Beijing (China) [15,16].

Furthermore, the lack of accurate data on urban groundwater represents a major problem in many cities and metropolitan areas, and more aspects should be addressed as mentioned below. The first aspect is the reduced number of monitoring wells for a certain urban area. This was mentioned in the scientific literature for Mexico City [10], Hanoi city [14], Lisbon, Portugal [17], Merton area, London [18], and Glasgow, UK [3]. The second aspect arises from the unknown number of users, wells, and corresponding pumping rates, as mentioned in [19] in a study for Gioia Tauro, Italy, in [20] for several Indonesian urban areas including Jakarta, and in [9] for Ho-Chi-Minh City (Vietnam). Management problems [21] and the issue of proprietary data production might additionally lead to a lack of data accuracy, as stated in [22] for several urban areas in the USA.

The connection between urban planning/city development and subsurface hydrology is therefore a work in progress, and our study undertakes an initial step to understanding this connection from a multi-perspective approach. Our study brings into discussion the necessity of hydrogeological data to the documentation of city development (buildings and infrastructure) as a complete and coherent picture of the underground environment's behavior with respect to supporting the healthy evolution of cities.

Objectives of the Paper

The first interdisciplinary exploratory study on the connection between urban planning, urban development, and urban subsurface hydrology in Romania is provided (Figure 1). Three research objectives are proposed: (1) analyze the urban plans in 17 large cities in Romania to examine the urban strategic and planning frameworks and thus understand the connection between the cities' development and urban groundwater; (2) analyze the land instability map of Bucharest to discern the role groundwater plays in main subsidence triggering factors; (3) discuss the results obtained from the previous objectives and formulate recommendations for the local and national authorities, urban planners, and water companies.

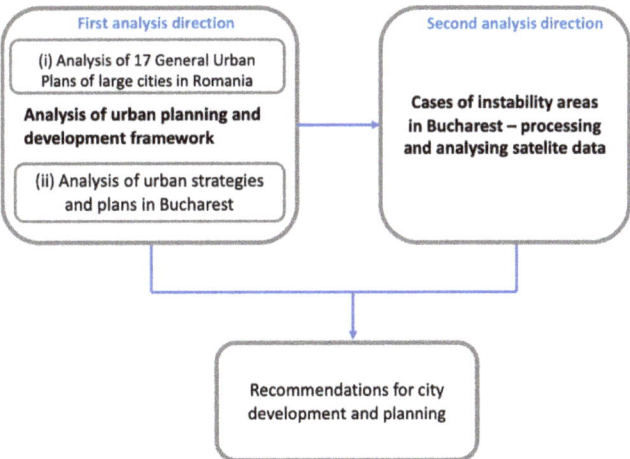

Figure 1. Research methodology for exploring the connection between urban development and urban subsurface hydrology. Source: elaborated by the authors.

2. Exploring the Connection between Urban Development and Urban Subsurface Hydrology

The direction of the first analysis includes two steps outlined in Figure 1: (i) an analysis of the urban planning documents and support studies for 17 large cities in Romania, exploring the reliance of the General Urban Plan (GUP) on groundwater documentation; (ii) the analysis of strategic documentation [23] for the city of Bucharest, namely the Integrated Urban Development Strategy for Bucharest (strategy for absorbing EU funds, elaborated in 2021) and of the strategic concept for Bucharest 2030 (elaborated in 2011); in addition, the analysis of urban planning documentation, namely the current general urban plan (drafted in 1998, with extended validity until the new GUP is completed).

Based on the land instability map of Bucharest, the second direction presented in Figure 1 focuses on discerning the role of groundwater within the main subsidence triggering factors and analyzing its relation to the urban planning framework. By monitoring vertical ground displacement it is possible to indirectly estimate the generated subsidence phenomena, with one of the most important of these phenomena being related to the dynamics of urban groundwater. The recent method for monitoring urban subsidence is the satellite remote sensing technique that uses Persistent Scatterer Interferometry (PSI), which allows for vertical ground displacements to be determined to the millimeter order [24] by measurements made along the Line of Sight (LoS) of the satellite [25]. PSI is a radar-based technique from the Interferometric Synthetic Aperture Radar (InSAR) family [26]. The InSAR operates by generating an interferometric pair using two radar scenes, acquired approximately from the same look angle, and over the same area, at different times [27]. If more interferograms are generated, their phase difference is used for estimating the displacements in the monitored area along the Line of Sight (LoS) of the satellite [28]. Considering the available radar data and the particularities of the studied area, different methods are used for estimating the vertical ground displacements from LoS displacements. For the PSI technique, the ground displacements are determined in temporally stable highly reflective ground features known as persistent scatterers (PS) [25] using multiple SAR interferograms generated from a time series of satellite radar data. The main advantages of the SAR techniques consist of its large area monitoring capability and high temporal resolution [27]. As an example, Sentinel-1, the current European SAR missions, provides a large amount of satellite data and has a revisit time of 6 days when both satellites of the constellation are considered. The Sentinel-1 Interferometric Wide Swath (IW) products, which are used for PSI, have a swath of 250 km, [29] facilitating an analysis at city level, or even at regional level. One disadvantage of the SAR technique is represented by the temporal and spatial decorrelation, which leads to a lack of ground displacement information for land surfaces covered with vegetation or for areas where land changes occur during the studied time interval [30].

2.1. Current Urban Planning and Development Framework

Romanian legislation requires localities to have a General Urban Plan (GUP) that has a directive character and includes operational regulations, thus being the legal support for the implementation of development programs and actions [23]. A GUP includes several written documents such as a summary report, a general report (the main results of the studies justifying the chosen solutions), local planning regulations, and graphic representations: the city or commune plan, an analysis illustrating the identified dysfunctions, planning regulations and zoning, the land ownership, and the legal circulation of land. Geotechnical or sometimes hydrogeological studies are part of the support studies.

A GUP follows the structure of a guide approved by a Ministerial Order. This indicates that the content and topics covered by the support studies must be decided by the contracted service provider along with the municipality. However, by default, some sections are considered mandatory, optional, or non-required, depending on the locality category.

An amount of 17 large cities (Figure 2) with more than 170,000 inhabitants have been selected (Table 1). It has been examined whether or not groundwater characteristics are considered in their General Urban Plans.

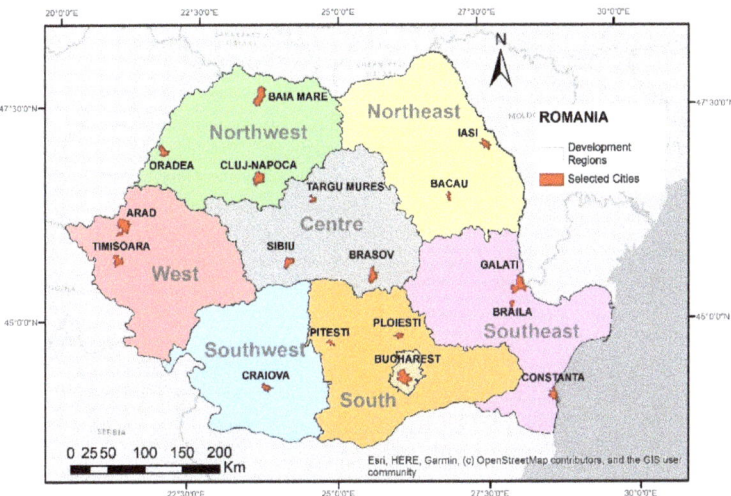

Figure 2. The location of the chosen 17 cities on the Romanian map. Map generated in Esri® ArcMap™ 10.3. Base map source: ESRI. Source: elaborated by the authors.

Table 1. Summary of the selection criteria.

Region	Name	Built Area Density (Inhabitants/sq km)	Population (2019)	Function	GUP Elaborated in
Northeast	Iași	6347.08	382,767	County residence municipality (CRM)	1999
Northeast	Bacău	5819.05	197,097	CRM	2009
Southeast	Constanța	5372.78	312,250	CRM	2010
Southeast	Galați	5172.35	305,386	CRM	2017
Southeast	Brăila	7017.42	201,414	CRM	2021
South	Ploiești	4293.23	226,133	CRM	2017
South	Pitești	7043.76	173,537	CRM	2021
Southwest	Craiova	5286.90	300,375	CRM	2000
West	Timișoara	5333.53	326,636	CRM	2021
West	Arad	3165.74	176,455	CRM	1998
Northwest	Cluj Napoca	4810.47	326,145	CRM	2018
Northwest	Oradea	2987.65	221,301	CRM	2021
Northwest	Baia Mare	3844.24	145,220	CRM	1999
Centre	Brașov	3223.75	289,190	CRM	2010
Centre	Sibiu	5211.81	168,477	CRM	2010
Centre	Târgu Mureș	5442.24	147,305	CRM	2010
Bucharest-Ilfov	Bucharest	8903.9	2,139,493	Country capital	1998

The selection criteria included: the region (minimum one town in each existing development region), the number of inhabitants (2019 as the year of reference) according to the National Institute of Statistics, the population density (more than 3000 inhabitants/sq km), and the year the GUP had been finalized (before 2000, between 2010–2015, and after 2015).

2.2. Land Instability in Bucharest City

Located in southeastern Romania, with an area of approximately 240 km^2 and a population of over 2.1 million inhabitants [31], Bucharest is crossed by two rivers: the Dâmbovița and Colentina. The first is extensively canalized and the second was reshaped at the beginning of the 20th century in a series of chained lakes [32], which have a direct hydraulic connection to the shallow aquifer.

Bucharest is a dynamic city with an expanding population and surface coverage, and thus its infrastructure shows considerable development and subsurface changes. Aspects of urban development in connection to groundwater dynamics have been highlighted. Two parallel directions of study have followed: the examination of the current urban development and planning framework and a land instability analysis (Figure 1).

Considering the particularities of the subsidence phenomena, two time periods have been considered, for which data sets from two European satellite radar missions were available. Thus, for the 2004–2010 period, data from the ENVISAT ASAR mission have been used and processed using the Persistent Scatterer Interferometry (PSI) technique. For the 2014–2018 period, data from the current European radar mission (Sentinel-1) has been used. Maps of the vertical displacements of the terrain for the entire urban area were generated using these data sets for the two time intervals mentioned above. The consistency of the results for the two time intervals has been observed by analyzing two corresponding land displacement maps. Numerous areas of instability that were identified on the land displacements map for the 2004–2010 interval show the same trends as for the 2014–2018 interval. Alongside the PSI technique, a land surface change detection technique was used for a better interpretation of the changes and displacements at the land's surface level. A new cartographic product has been obtained through combining the map of the vertical land displacements with the surface terrain change detection map [28]. Therefore, the potential areas that locate the source of vertical displacements have been established accurately.

The chosen area is located along the Colentina River corridor (Figure 3). For the 2014–2018 period, a methodology to characterize the area has been proposed. It considers specific elements of the studied area, such as: subsidence occurrence; land-cover changes; hydrogeological, geological, and hydrological characteristics; the history of the study area (considering the location and the extension of old quarries for building materials); the last 100 years of topographical changes to the Colentina River chained lakes; and climatic aspects.

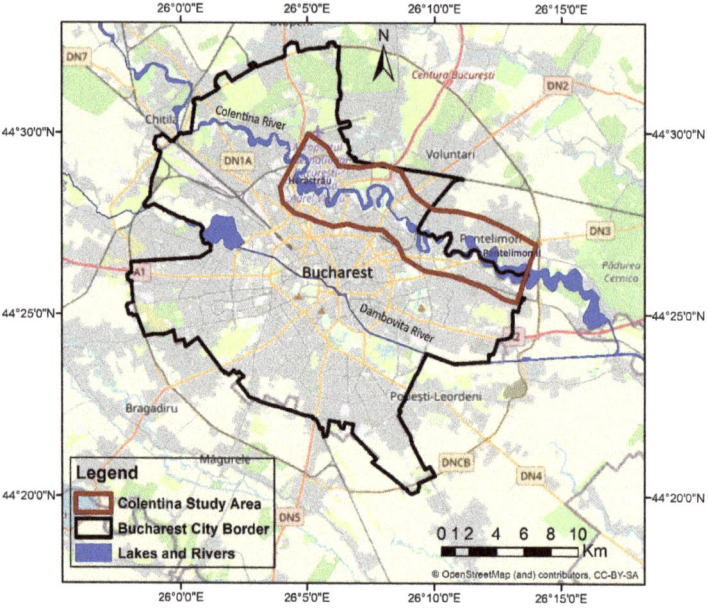

Figure 3. Colentina River corridor study area. Map generated in Esri® ArcMap™ 10.3. Base map source: OpenStreetMap. Source: elaborated by the authors.

Short Overview of the Processing and Analysis of Satellite Data for Identification and an Outline of the Main Subsidence Triggering Factors

To generate the PSI maps for the 2004–2010 and 2014–2018 time intervals, two different data sets and methodologies were used. Based on the data coming from the radar European mission ENVISAT ASAR, a land displacement map was generated for the June 2004–July 2010 time interval using the ENVI SARScape software (ENVI SARScape version 5.5, L3Harris, purchased from ESRI Romania (Bucharest)-authorised distributor of Harris Geospatial products (ENVI and SARScape) for Romania) for data processing [33] and by applying the PSI technique. ENVISAT ASAR data are available in the archive of the European Space Agency (ESA) along with data from other historical missions of ESA [34]. For this study, some of the SAR ENVISAT ASAR data used for processing was purchased through the ESA C1P 6050 proposal, using the Romanian Space Agency as coordinator.

The main PSI processing steps for the ENVISAT ASAR data in ENVI SARScape include [33]:

- ENVISAT ASAR images were imported to ENVI SARScape;
- Application of the Orbit File: precise orbit files were downloaded from ESA archive while the SAR images were downloaded;
- Connection Graph Generation: the SAR pair combination (Master and Slaves) are defined together with the connection network, used for the generation of the multiple differential interferograms. The Master image that was used was from 24 February 2007;
- Area of Interest Definition: Bucharest city;
- Interferometric process, including the coregistration, interferogram generation, and the flattening and Amplitude dispersion index. Here, the digital elevation model (DEM) is specified. The STRM-3 v4 DEM was used for this processing;
- Inversion: First Step—estimation of the first model inversion to derive the residual height and the displacement velocity;
- Inversion: Second Step—estimation of the atmospheric phase components, considering the first linear model product coming from the previous step;
- Geocoding—the PS products are geocoded, and the displacements can be displayed.

During satellite data processing, the baseline was considered between the master and the slave scenes to have values lower than 300 m. The coherence threshold established for the ENVISAT ASAR data series is 0.8.

For 2014–2018 surface displacements in Bucharest, data from the SAR Sentinel-1 sensor have been used. The data that were used, available on the "Copernicus Open Access Hub" website [35], cover the period between October 2014–April 2018 and consider acquisitions from both a descending (109D) and an ascending orbit (131A). The PSI maps for both ascending and descending orbits were produced by the Norwegian Geological Survey within the Norwegian Ground Motion Service [36], by applying the classic PSI technique. The PSI processing of the Sentinel-1 data was carried out as an activity within the INXCES-INnovations for eXtreme Climatic EventS project [37].

Based on the ascending and descending PSI maps, where the land displacements are measured along the Line of Sight (LoS), a PSI map of vertical displacements was computed considering the methodology described in [38].

Several areas showing subsidence on the PSI map have been outlined: even the general trend of the city of Bucharest for the period 2014–2018 showed stability. Most of these areas are located along the corridors of each the two rivers crossing the city. Figure 4 illustrates the PSI map of vertical land displacements obtained using Sentinel-1 images. They are classified according to several criteria, depending on the main triggering factor, as it is shown in Table 2.

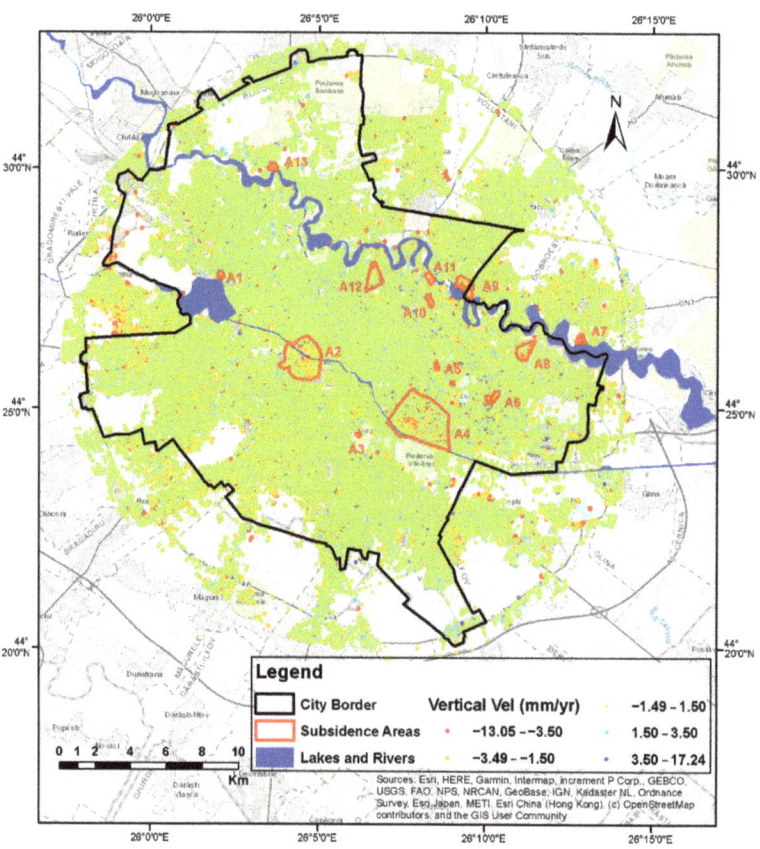

Figure 4. PSI map of vertical land displacements in Bucharest city (obtained from Sentinel-1 images) and the identified subsidence areas (modified after [38]). Map generated in Esri® ArcMap™ 10.3. Source: elaborated by the authors.

Table 2. Areas with identified displacements (modified after [28]).

Areas Where Displacements Have Been Identified, Classified by Land Use Category	Areas on the Map
Areas with small above-ground constructions (residential, office, or commercial buildings) and a small number of floors, where previous demolition took place	A4, A12, A10
Urban areas where land has shifted from green space to construction during the last 10–15 years	A6, A9, A13
Peri-urban areas in the pasture or forest category whose use has passed into the construction category during the last 10–15 years	A7
Green areas where works have been carried out for the practice of leisure sports (bicycle courts, tennis courts, or skateboarding)	A3, A5
Areas where the land use category has not changed, but there are specific factors producing displacements—e.g., the presence of an anthropogenic superficial layer of construction debris	A1, A8

To understand main the phenomena and processes that take place along the Colentina River corridor, several types of data sources were used. These include data describing natural environment of the area, geographical and urban conditions, and hydrogeological, hydrological, and geotechnics information. Table 3 mentions data types and information along with their sources used in the analysis.

Table 3. Data types and sources (modified after [28]).

Data	Sources
Hydrological data	Water level data of the Colentina chained lakes for the 2014–2017 period were provided by several institutions: the National Administration "Romanian Waters", the Argeș Water Basin Administration, the Vedea-Ilfov-Bucharest Water Management System, and the Administration of Lakes, Parks, and Leisure of Bucharest city. We used the series of data for the lakes Herastrau, Floreasca, Tei, Plumbuita, and Pantelimon I.
Geographical characteristics	The maps comprising the geographical characteristics of the area (landscape units, morphological maps, hypsometric maps), geological data representing the thicknesses of the loess formation, and geotechnical data on the location of former construction quarries were taken from the Geo-Atlas of Bucharest City [39]. The Digital Surface Model (DSM) that was used, which represents the land surface including buildings, infrastructure, and vegetation, is the digital model developed under the European Copernicus Earth Observation Program [40].
Geological data	The geological maps of Bucharest and Romania, scale 1: 200,000, sheet 44, were used [41]
Hydrogeological data	The hydrogeological data (Hydrogeological map of Bucharest city, groundwater levels during the analyzed period, location of the wells, results of existing urban hydrogeological models, and others) have been extracted from the databases of the Groundwater Engineering Research Centre of the Technical University of Civil Engineering, Bucharest [41].
Geotechnical data	Geotechnical data on anthropogenic soil areas and information on the geotechnical monitoring boreholes extracted from the Groundwater Engineering Research Centre of the Technical University of Civil Engineering, Bucharest [41]. Other geotechnical information was extracted from the Geotechnical Zoning Map of Bucharest, made by the Institute of Studies and Design in Land Improvements (ISPIF) in the 1970s.
Meteorological data	Extracted from Romania's statistical yearbooks, published by the National Institute of Statistics [42].

In addition to the SAR products used for radar interferometry, which use phase information, SAR images also contain amplitude information that can be used for other types of analysis. Thus, through the Interferometric Wide Swath (IW) mode of operation, in addition to the SLC (Single Look Complex) products used for interferometry, Sentinel-1 sensors provide GRD (Ground Range Detected) products. By using this type of product, it is possible to analyze the land-cover surface changes between the times of acquisition of the first image and second images. For consistency with previous land displacement analyzes, in which Sentinel-1 SLC data were acquired between October 2014 and April 2018, two Sentinel-1 GRD images were used to detect changes: the first from 25 October 2014, and the second from 25 April 2018 [38]. The data were downloaded free of charge from the Copernicus Open Access Hub [35].

Figure 5 shows the land-cover changes detection map generated for Bucharest. The areas where changes occurred in 2018 in respect to 2014 are marked in red. Areas where no changes occurred between the two moments of time are symbolized in gray tones. Water bodies have a high reflectivity, and the backscatter towards the radar is low; therefore, the lakes and rivers appear dark on the map. Ground changes are encountered mainly in areas where new residences (e.g., the western border of Bucharest city) or commercial centers are built. More details on the methodology and change detection trends in Bucharest city, considering the 2014–2018 time interval, are presented in [43].

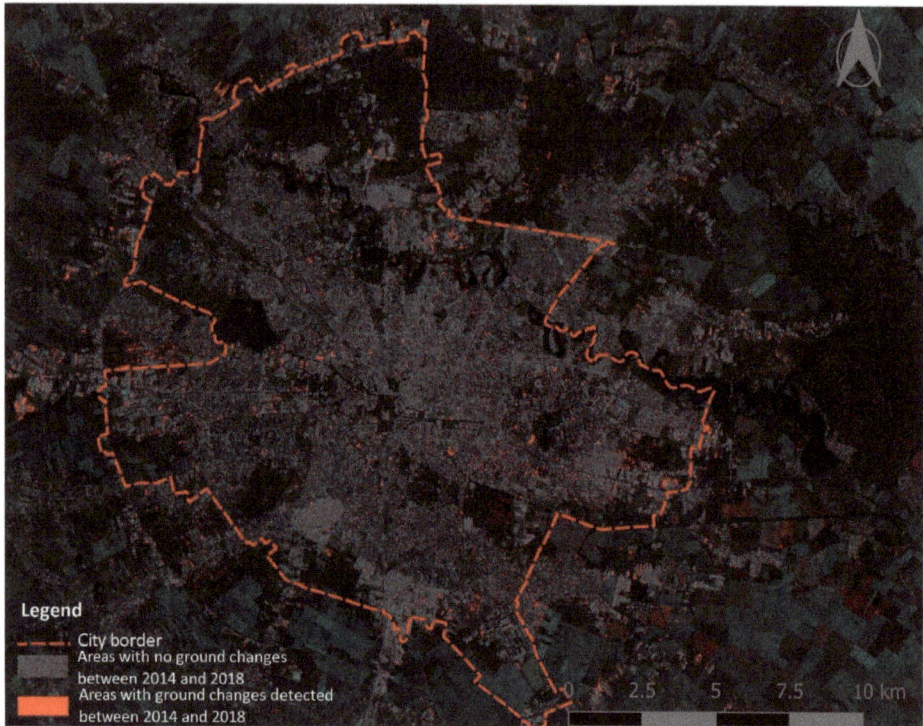

Figure 5. Detection map of land surface changes for Bucharest (2014–2018), obtained from Sentinel-1 GRD images (modified after [28]). Gray tones symbolize areas where no changes were detected between 2014 and 2018. Areas where ground changes occurred between 2014 and 2018 are symbolized in red. Water bodies are shown on the map in dark tones. Map generated using ESA SNAP (Sentinel Application Platform) and QGIS open-source software. Source: elaborated by the authors.

3. Discussion of Urban Planning and Subsidence

3.1. Analysis of the Current Urban Planning and Development Plans in Romania

The analysis performed by the authors on the urban planning and development documentation emphasizes that groundwater information is not regularly mentioned in the urban planning documentation of the 17 cities [44–60]. According to the Romanian legislation, the plans are based on support studies, but only the main planner decides which studies are required. The information from the support studies is incorporated into the general report. Only in some cases such as Iasi, Braila, Bacau, Ploiesti, and Pitesti [44,45,48,50] do the plans include geotechnical or hydrogeological analysis. In some cases, even if these studies include the hydrological component, they only refer to surface water (for example, the city of Bacau [45]), thus outlining the areas that are unsuitable for construction due to the risk of floods or erosion.

In other cases, for example the city of Pitesti [50], the local planning regulation accompanying the plan, which takes the information from the support study, provides measures to encourage building on difficult foundation sites. For example, underground parking areas are allowed to be built if the beneficiary can prove that the terrain has difficult foundation conditions, including high groundwater levels. The proof is presented by the significant comparative costs estimated for both underground and surface options, supported by geotechnical or hydrogeological studies.

In some cases, for example the Manta Roșie district of Iasi city (where new high buildings have been constructed) [44], although the problem of a high groundwater level flooding the cellars of old houses is acknowledged in the general report of the urban planning documentation, the associated regulation does not prescribe specific construction restrictions nor recommend specific measures to improve the urban hydrogeological conditions. Although recently elaborated, other general urban plans for the cities of Braila, Galati, Timisoara, and Sibiu [47,48,52,58] incorporate only generic statements of the current legislation (General Urban Planning Regulation adopted by Government Decision 525/1996, updated in 2002), namely that, in the "constructions are allowed in the buffer protection zone of the rail transport infrastructure if all measures triggering landslides, subsidence, or altering the groundwater table, are provided".

For Bucharest, the analysis of the studied strategies and General Urban Plan [60–62] concluded that information related to groundwater is not considered.

3.2. Analysis of the Subsidence Trigerring Factors in Colentina River Corridor

The study area has been defined by using a buffer zone of 1 km on both banks of the Colentina River. It starts from the northwest area of Bucharest, from the entrance to Herăstrău Lake, and ends at the exit of Pantelimon II Lake, in the southeastern part of the municipality (Figure 3).

Based on the vertical displacement PSI maps of the land surface for the 2004–2010 and 2014–2018 intervals, it was possible to analyze the evolution of vertical displacements in the Colentina study area. Figure 6 shows the evolution of the terrain instability in the studied area. The areas that presented subsidence during the period from 2004–2010 are highlighted in red. Areas with negative vertical displacements during the period from 2014–2018 are outlined in purple [28]. In most of the cases, in the immediate vicinity, the areas overlap during the two time intervals. This is the case for the areas situated on the right shore of the Tei, Colentina, and Pantelimon I Lakes, as well as some of those situated on the left shore of Floreasca and Colentina Lakes. For the subsidence areas situated on the right shore of Tei Lake, a detailed study is presented in [38], the main factors involved in the ground displacement process being the presence of a thick anthropic soil, the dynamics of the water supply losses in the area from 2006 to 2014, and the decrease in the area groundwater hydraulic head. For the 2014–2018 period, there are several areas showing subsidence without having a correspondent to the previous period from 2004–2010. These new subsidence areas occurred due to construction work extensions in areas where other categories of use were previously involved (e.g., the pasture area replaced by residential area in the area situated on the left shore of Pantelimon II Lake; the industrial area replaced by a commercial center in the area situated on the right shore of Fundeni Lake, near the border of the study area). The current extension of the lakes has been delineated using Open Street Map data [63].

From the land-cover change detection map generated from Sentinel-1 GRD data for the Colentina River buffer zone, the areas where land use changes occurred during 2014–2018 have been identified. Figure 7 shows the map for determining the land-cover changes, along with these areas. The identified areas are marked in blue.

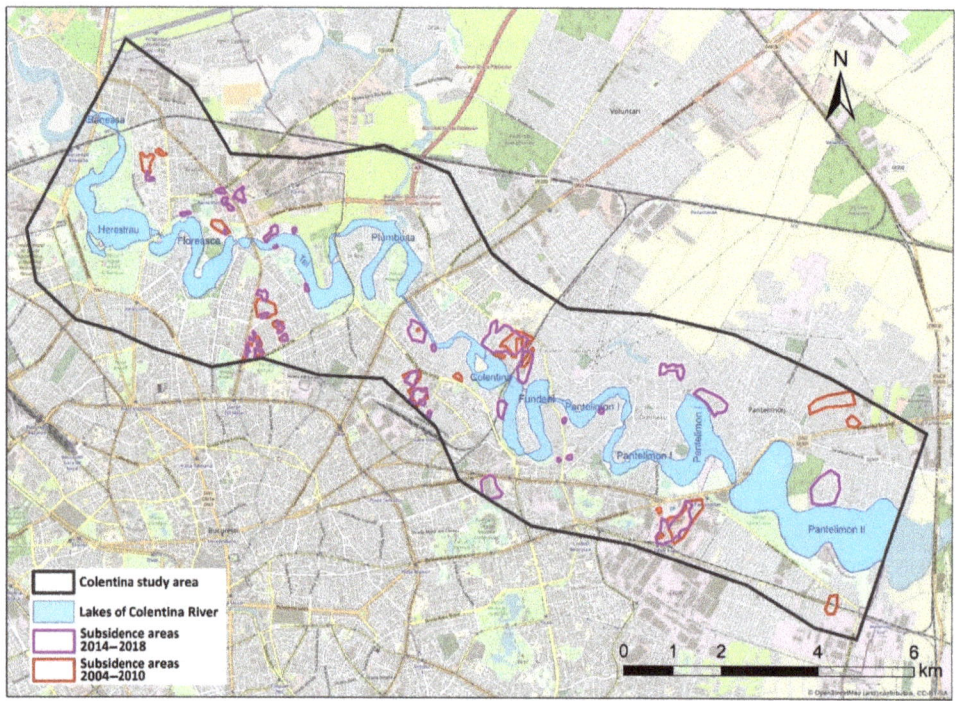

Figure 6. Subsidence areas in the Colentina River buffer zone, for the time intervals 2004–2010 and 2014–2018 [28]. Map generated in Esri® ArcMap™ 10.3. Base map source: OpenStreetMap. Source: elaborated by the authors.

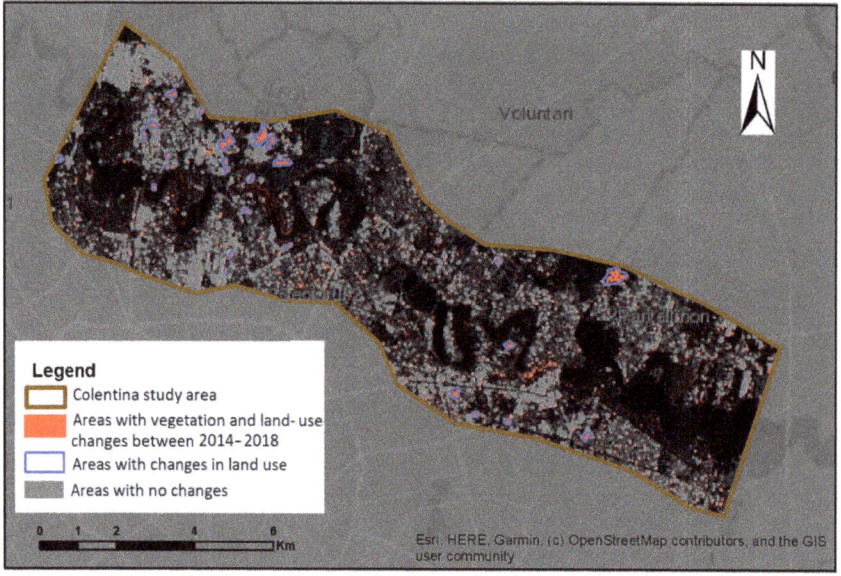

Figure 7. Land-cover changes for the Colentina River buffer zone modified after [28]. Map generated in Esri® ArcMap™ 10.3. Base map source: ESRI. Source: elaborated by the authors.

These areas include:

- Areas covered with vegetation in 2014 on which high buildings (residential complexes) were built until 2018;
- Areas covered with vegetation in 2014, which were transformed into residential districts with houses until 2018;
- Areas where infrastructure elements have been developed (e.g., parking extensions, construction of new parking lots for already built buildings);
- Areas with buildings that have been demolished and where new buildings have been built;
- Areas covered with construction sites where the works were completed by 2018 [28].

Figure 8 shows the areas with land-cover changes during 2014–2018 (marked in brown) as well as the areas showing vertical displacement during the two studied time intervals, 2004–2010 (red color) and 2014–2018 (purple). Some areas with land-cover changes are independent of areas with subsidence or uplift. However, it is evident that there are also areas where the two types of phenomena occur in the immediate vicinity. In many cases, this happens due to the extension of construction work into distinct residential areas. The combined analysis of the two types of phenomena can help when looking for the factors generating subsidence and their localization. As an example, for the Intrarea Chefalului area, situated on the left shore of Tei Lake, near the area indicating subsidence, on the northeast, there is an area where changes were detected. The values of negative subsidence are higher on the border of the two types of areas. From the other available information, it was evident that in the area with land changes detected, construction work took place and was finalized in mid-2019. This work can contribute to the ground displacement values in the neighboring area.

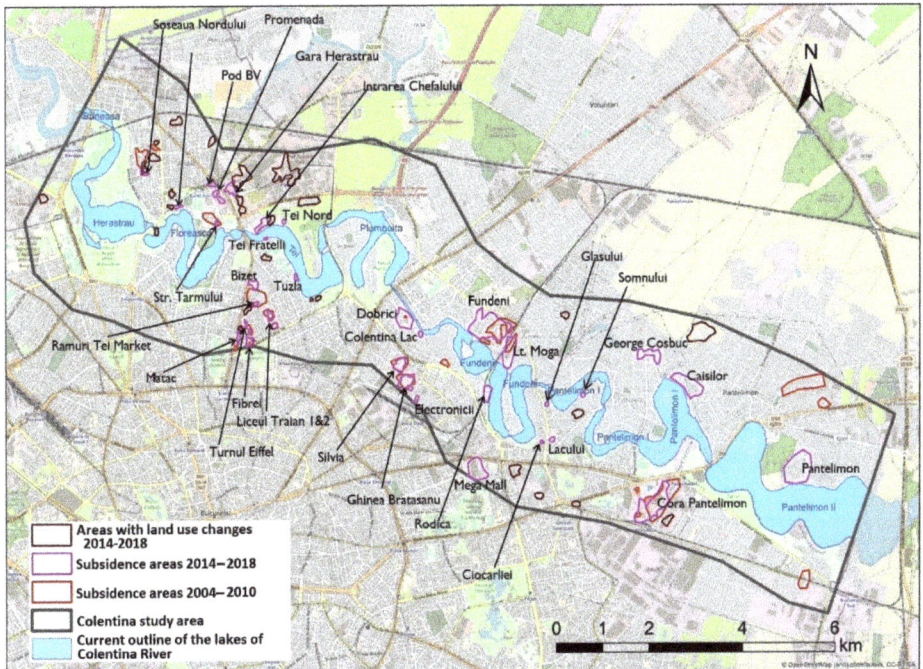

Figure 8. Areas showing land-cover changes and vertical displacements for the periods 2004–2010 and 2014–2018. Map generated in Esri® ArcMap™ 10.3. Base map source: OpenStreetMap. Source: elaborated by the authors.

The map presented by Figure 9 brings together both areas with potential instability and areas showing land-cover changes, and areas with negative vertical displacements determined from PSI maps generated from Sentinel-1 data for 2014–2018 and ENVISAT ASAR data for 2004–2010. Most of the subsidence areas, both for the 2004–2010 and 2014–2018 periods, are in regions that have been determined to have the potential for instability, generated by construction activity and outlined by the land-cover changes or historical activities in terms of quarries for building materials or similar. In the vicinity of the lakes of Colentina, an important land-cover change yet to be mentioned is the drainage of some areas which in the past were part of the lakes of Colentina and are now urban areas. These are highlighted in shades of red in Figure 9. In some areas (e.g., the embankment zone of Șoseaua Nordului), partial stabilization of the area could occurred between the analyzed time intervals due to the interruption of the construction works or by completion of the buildings settlement processes.

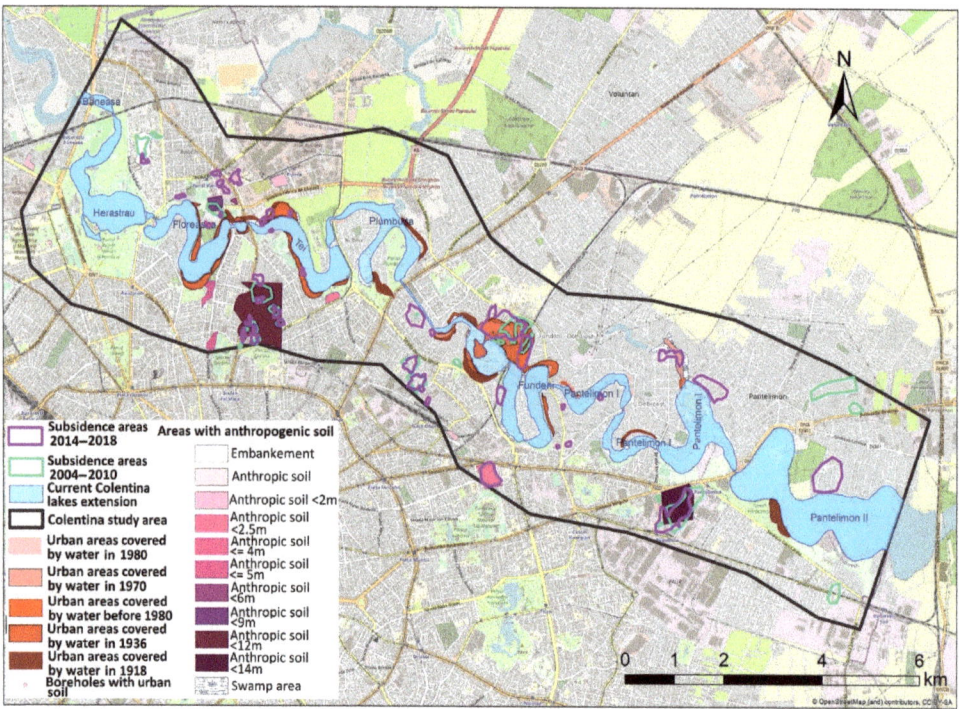

Figure 9. Areas showing subsidence and instability potential. Map generated in Esri® ArcMap™ 10.3. Base map source: OpenStreetMap. Source: elaborated by the authors.

Four of the 37 instability areas are further discussed, namely Promenada, Pod BV, Gara Herastrau, and Soseaua Nordului. Table 4 provides the processed information used in the analysis: the name of the area; Sentinel-1 orbit used to generate the PSI map; the behavior of the area during the period 2004–2010; the average annual velocity along the line of sight (LOS); the distance between the centroid of the area and the lakes; the presence of anthropogenic soil and its thickness; the thickness of the loess layer; the previous area situation (water or swamp cover identified in the historical maps); the presence of a career of construction materials used in the past; the geology of the area; the monitoring of borehole locations nearby each area; the depth of the shallow aquifer strata (Colentina gravels); the depth of the water table of the shallow aquifer (unconfined). Changes of the shallow aquifer groundwater hydraulic head matches the chained lake's water level variation as their bed shows a direct hydraulic connection to the shallow aquifer.

Table 4. Processed information for Promenada, Pod BV, Gara Herastrau, and Soseaua Nordului areas. Used data sources [28,38,41,64].

Name of the Area	PSI SI	Subsidence during 2004–2010	Average Displacement (LOS) [mm/year]	Distance Area-Lake [m]	Anthropogenic deposits [m]	Loess Presence [m]	Dried Area	Previous Career of Construction Materials	Geology	Depth to the Shallow Aquifer Strata [m]	Depth to the Water Table of the Shallow, Unconfined Aquifer [m]
Promenada	109D	No	−1.3	460	No	No	No	Yes	qp2/3-Dd	5 to 20	0.3
Gara Herastrau	109D	No	−2.1	611	<2	15–17	No	No	qp2/3-Dd	10 to 20	15.8
Pod BV	109D	No	−1.1	547	No	2.5–6 partially	No	Partially	qp2/3-Dd	10 to 20	8.3
Soseaua Nordului	131A	Partially	−1.4	513		2.5–6	No	No	qp2/3-Dd	0 to 5	4.27

Located on the left bank of Lake Floreasca, at a distance of over 500 m, the areas Promenada, Pod BV, and Gara Herastrau (Figure 8) are characterized by the presence of anthropogenic deposits or as being used as the location for previous construction materials quarries. The construction work that carried on during the period preceding the 2014–2018 interval in these areas as well as in neighboring ones represent the identified land-cover changes. Most of these land-cover changes are related to the construction of high office buildings. Table 4 shows the comparison between the evolution of the water level in the Floreasca Lake and the evolution of the values of the vertical terrain displacement. Even for both, there is a general downward trend during the period when the water level of the Floreasca Lake decreased at a sharp pace; a steeper downward trend for vertical terrain displacements has been registered (Figure 10). Therefore, when the lake's water level decreased 2 m from November 2014 to April 2016, a negative displacement of more than 20 mm occurred at the ground surface. An offset of a few months can be observed between the water level and ground displacement trends. This is due to the different behavior of the surface water and of the groundwater. Considering the linear trendline for Floreasca Lake and for the ground displacements, for medium decrease of 1.5 m in the water lake, 15 mm of subsidence is registered.

The Soseaua Nordului area has similar characteristics to the aforementioned ones, more precisely, it has neighboring areas with anthropogenic deposits and areas with land-cover changes in the vicinity due to construction activity. The main difference is that the construction work in the Soseaua Nordului area includes office buildings along with apartment buildings. A peculiarity of the area is the fact that it was partially affected by land displacements during 2004–2010. Figure 11 shows the comparison between the evolution of Herastrau Lake water level and the land displacement trend of this area. Herastrau Lake is the only lake of Colentina River where the water level increases periodically (controlled). It is evident that there is a slight reverse correspondence between the two data series. Although the general trend of the vertical displacements shows subsidence, during the periods when the lake water level registers minimum values, the values of the terrain surface displacements have a slight ascending trend followed by a descending trend. During periods when the lake water level has maximum values, the terrain surface shows a descending trend followed by a slight rising trend.

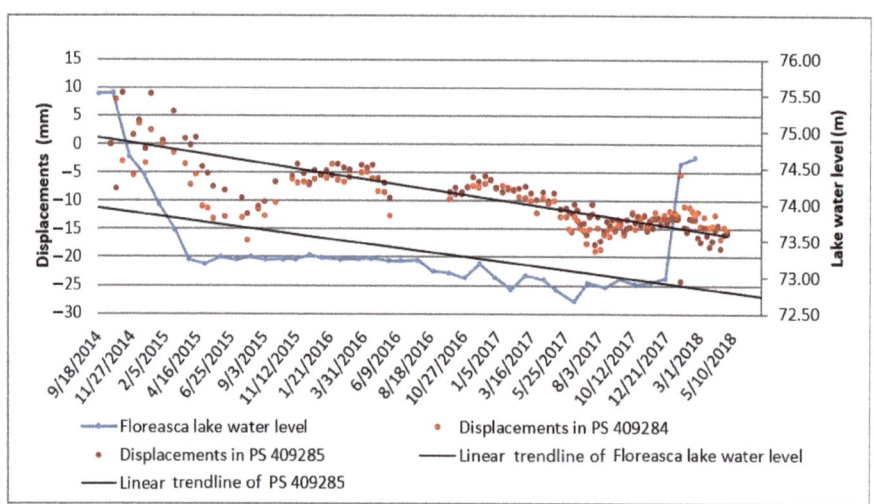

Figure 10. Comparison between the Floreasca Lake water level and the evolution of the land vertical displacements in the Gara Herăstrău area. Source: elaborated by the authors.

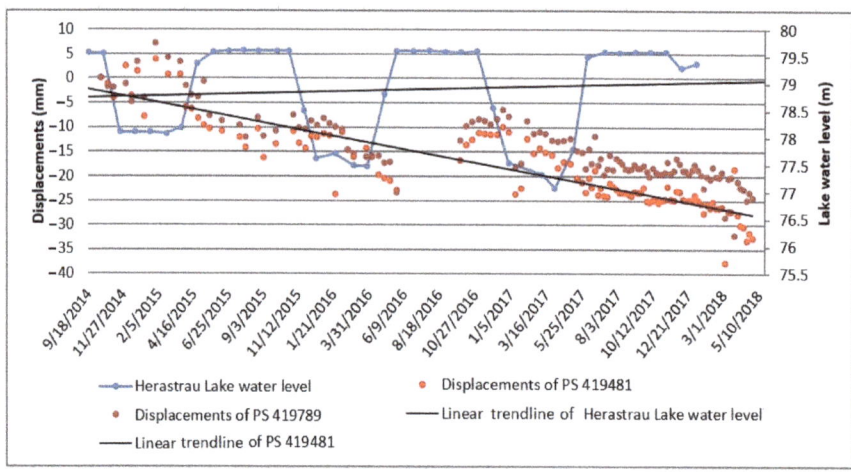

Figure 11. Comparison between the Herăstrău Lake water level and the evolution of the land vertical displacements in Soseaua Nordului area. Source: elaborated by the authors.

The obtained results show a major concordance between the water level of the lakes and the vertical displacements of the surrounding areas, both with decreasing tendencies, in the context of a strong hydraulic connection between the Colentina riverbed and the shallow aquifer strata. The presence of anthropogenic deposits associated with groundwater fluctuations showed amplified ground vertical displacements, compared to the areas where natural deposits exist. Another association that can produce land movements is provided by the presence of marshy soils as a base layer for heterogeneous anthropogenic deposits. In all the areas mentioned above, these observations lead to the necessity for further local studies [65] in order to properly discriminate between the phenomena and processes generating subsidence. Civil engineers and urban planners should use these studies to further investigate local geology and possible anthropogenic strata, the ground geotechnical properties, and the local hydrogeological settings. These studies, providing an accurate overview of the ground conditions, should be included as support for the general urban plans of the cities, documenting future developments.

4. Conclusions and Recommendations

The analysis of the support studies of the Urban General Plans for 17 Romanian cities showed that urban planning documents do not sufficiently rely on groundwater information. Corresponding hydrogeological studies include only general recommendations, and these do not include specifications targeting reliable water-balance preservation or detailing the necessity of implementing monitoring systems to survey groundwater levels. These do not recommend either special constructive measures to be implemented for future infrastructure elements and do not include maps delineating particular geotechnical and hydrogeological characteristics.

The analysis of Bucharest clearly shows that anthropogenic influences severely alter the water cycle in urban areas. The change in the water cycle induced by urban areas directly affects the groundwater level. Increasing groundwater levels can lead to flooding while lowering groundwater levels can lead to subsidence.

The construction of high-rise offices or residential buildings, made in several locations in the Colentina study area, create basements on several levels, which makes the extension of the buildings exceed 10 m in depth. This leads to the use of dewatering systems required to locally reduce the hydraulic head of the shallow aquifer to ease the construction work and later allow for the optimal use of the basements. In the case of the intensive development of the subsurface constructions, the associated procedures of lowering or raising the groundwater level would have a strong impact on the neighboring infrastructure elements or networks. Depending on the pumped volume of the groundwater, this operation can have effects on the aquifer dynamics and can lead to displacement of the land surface, water level decrease in neighboring lakes, and the instability of infrastructure elements or changes of their hydraulic interaction.

The vertical displacement of the earth's surface generally has a combined cause, involving problems of geotechnics, hydrology, hydrogeology, and geological aspects. Since elements in urban areas both underground and aboveground are exposed to more factors than in the extra-urban area, it is very difficult to accurately identify the causes and effects of certain phenomena that occur. The historical component of this problem is that in the past various works were carried out without keeping coherent and concrete records of the work process. In the rare cases where this information exists, it is not sufficiently detailed.

Increasing awareness of the strong linkage between urban development and groundwater is necessary for urban planners and decision makers because of the urgent need for applying coherent and correct city guidelines that consider both aboveground and underground development, including groundwater management.

Accurate analysis of urban groundwater should be performed with two considerations: (1) the resilience of city development, and (2) groundwater resource protection. The first consideration focuses on the unfailing development of city infrastructure in terms of its ecological resilience as well as land stability, preventing the delay of construction projects and cost overrun, and of subsequent damage to urban infrastructure [3]. The second consideration emphasizes the qualitative and quantitative protection of the entire aquifer system the city is laying on.

Underground city management could be performed by implementing a service within the municipalities in charge of data and decision-making concerning geology, groundwater, and the subsurface infrastructure elements and networks. This management should directly cooperate with the main actors represented by the city institutions and professional groups. This inter-institutional framework should regroup the (a) public administration including the municipality's environmental department; the institutions in charge of underground transportation networks, utility operators, the management of Lakes and Parks, and others; (b) urban planners; (c) water operators and infrastructure developers—those who make underground constructions: subways, railways, warehouses, buildings with deep foundations; and (d) academic research centers.

The directions that should be followed are: (1) The development of a guide for urban hydrology data to be collected and used in urban planning; (2) defining a pallet of constructive solutions to allow a steady urban water cycle, for example, measures to allow aquifers to recharge or reduce the barrier effect; (3) integrating urban hydrogeology technical guides in the cities' urban planning documents, thus focusing the low impact development in terms of minimizing the hydrological urban balance disruption.

As future research continues, it must be mentioned that even if the time interval in which the subsidence analyses are performed is relevant (a period of four years), the follow-up over a longer period may allow for a broader analysis. This will facilitate a more accurate identification of the subsidence triggers which are mainly related to geotechnical, hydrogeological, or geological data. Although the time frame is relevant, the consistency of the analyses may be affected by the heterogeneity of the data acquisition intervals, which differ for each data set. In the case of climate and satellite data there are daily or a few days' step measurements, however, in the case of the hydraulic head measurements performed for this case study, measurements were at longer intervals. By improving the hydraulic head data acquisition system, a coherent overlap with other types of data can be obtained.

Author Contributions: Conceptualization, O.L. and C.R.G.; methodology, A.R., O.L. and C.R.G.; software, A.R.; validation, O.L. and C.R.G.; formal analysis, A.R., O.L. and C.R.G.; investigation, A.R., O.L. and C.R.G.; resources, A.R., O.L. and C.R.G.; data curation, A.R. and O.L.; writing—original draft preparation, A.R. and O.L.; writing—review and editing, A.R., O.L. and C.R.G.; visualization, A.R., O.L. and C.R.G.; supervision, C.R.G.; project administration, C.R.G. All authors have read and agreed to the published version of the manuscript.

Funding: The PSI processing of Sentinel-1 data is part of the INXCES (INnovations for eXtreme Climatic EventS, https://inxces.eu/) project 2016–2019 funded by WaterJPI WaterWorks 2014. It was partly funded by the Romanian National Authority for Scientific Research and Innovation, CCCDI-UEFISCDI grant number 48/2013 Cofound-202-INXCES, with PNCDI III. A part of the SAR ENVISAT ASAR data used for processing was purchased through the ESA C1P 6050 proposal, having Romanian Space Agency as coordinator.

Institutional Review Board Statement: Not applicable.

Informed Consent Statement: Not applicable.

Data Availability Statement: Not applicable.

Conflicts of Interest: The authors declare no conflict of interest.

References

1. Hölscher, K.; Frantzeskaki, N. Perspectives on urban transformation research: Transformations *in, of,* and *by* cities. *Urban Transform.* **2021**, *3*, 2. [CrossRef]
2. Bricker, S.H.; Banks, V.J.; Galik, G.; Tapete, D.; Jones, R. Accounting for groundwater in future city visions. *Land Use Policy* **2017**, *69*, 618–630. [CrossRef]
3. Foster, S.; Gogu, R.; Gathu, J. Urban groundwater-mobilizing stakeholders to improve monitoring. *Source 2019*, *2019*, 58–62.
4. Carneiro, J.; Carvalho, J.M. Groundwater modelling as an urban planning tool: Issues raised by a small-scale model. *Geol. Soc. Spec. Publ.* **2010**, *43*, 157–170. [CrossRef]
5. Roxane, L.; Lebel, A.; Joerin, F.; Rodriguez, M.J. Integration of groundwater information into decision making for regional planning: A portrait for North America. *J. Environ. Manag.* **2013**, *114*, 496–504. [CrossRef]
6. Carmon, N.; Shamir, U.; Meiron-Pistiner, S. Water-sensitive Urban Planning: Protecting Groundwater. *J. Environ. Plan. Manag.* **1997**, *40*, 413–434. [CrossRef]
7. Patra, S.; Sahoo, S.; Mishra, P.; Mahapatra, S.C. Impacts of urbanization on land use/cover changes and its probable implications on local climate and groundwater level. *J. Urban Manag.* **2018**, *7*, 70–84. [CrossRef]
8. Hameed, H.M.; Faqe, G.R.; Qurtas, S.S.; Hashemi, H. Impact of Urban Growth on Groundwater Levels using Remote Sensing—Case study: Erbil City, Kurdistan Region of Iraq. *J. Nat. Sci. Res.* **2015**, *5*, 72–84.
9. Erkens, G.; Bucx, T.; Dam, R.; de Lange, G.; Lambert, J. Sinking coastal cities. *Proc. IAHS* **2015**, *372*, 189–198. [CrossRef]
10. Castellazzi, P.; Garfias, J.; Martel, R.; Brouard, C.; Rivera, A. InSAR to support sustainable urbanization over compacting aquifers: The case of Toluca Valley, Mexico. *Int. J. Appl. Earth Obs. Geoinf.* **2017**, *63*, 33–44. [CrossRef]

11. Kubota, T.; Lee, H.S.; Trihamdani, A.R.; Phuong, T.T.T.; Tanaka, T.; Matsuo, K. Impacts of land use changes from the Hanoi Master Plan 2030 on urban heat islands: Part 1. Cooling effects of proposed green strategies. *Sustain. Cities Soc.* **2017**, *32*, 295–317. [CrossRef]
12. Dang, V.K.; Doubre, C.; Weber, C.; Gourmelen, N.; Masson, F. Recent land subsidence caused by the rapid urban development in the Hanoi region (Vietnam) using ALOS InSAR data. *Nat. Hazards Earth Syst. Sci.* **2014**, *14*, 657–674. [CrossRef]
13. Nguyen, M.; Chang, C.P.; Tseng, K.H. Using Sentinel-1 Tops SAR and SBAS for land subsidence monitoring in Hanoi, Vietnam. In Proceedings of the International Symposium on Remote Sensing, Pyeongchang, Korea, 9–11 May 2018.
14. Luyen, K.B.; Le, P.V.V.; Dao, P.D.; Nguyen, Q.L.; Hai, V.P.; Hong, H.T.; Xie, L. Recent land deformation detected by Sentinel-1A InSAR data (2016–2020) over Hanoi, Vietnam, and the relationship with groundwater level change. *GIsci Remote Sens.* **2021**, *58*, 161–179. [CrossRef]
15. Chen, M.; Tomás, R.; Li, Z.; Motagh, M.; Li, T.; Hu, L.; Gong, H.; Li, X.; Yu, J.; Gong, X. Imaging Land Subsidence Induced by Groundwater Extraction in Beijing (China) Using Satellite Radar Interferometry. *Remote Sens.* **2016**, *8*, 468. [CrossRef]
16. Deng, Z.; Ke, Y.; Gong, H.; Li, X.; Li, Z. Land subsidence prediction in Beijing based on PS-InSAR technique and improved Grey-Markov model. *GIsci. Remote Sens.* **2017**, *54*, 797–818. [CrossRef]
17. Heleno, S.I.N.; Oliveira, L.G.S.; Henriques, M.J.; Palcao, A.P.; Lima, J.N.P.; Cooksley, G.; Ferretti, A.; Lobo-Ferreira, J.P.; Fonseca, J.F.B.D. Persistent Scatterers Interferometry detects and measures ground subsidence. *Remote Sens. Environ.* **2011**, *115*, 2152–2167. [CrossRef]
18. Aldiss, D.; Burke, H.; Chacksfield, B.; Bingley, R.; Teferle, N.; Williams, S.; Blackman, D.; Burren, R.; Press, N. Geological interpretation of current subsidence and uplift in the London area, UK, as shown by high precision satellite-based surveying. *Proc. Geol. Assoc.* **2014**, *125*, 1–13. [CrossRef]
19. Righini, G.; Raspini, F.; Moretti, S.; Cigna, F. Unsustainable Use of Groundwater Resources in Agricultural and Urban Areas: A Persistent Scatterer Study of Land Subsidence at the Basin Scale. *WIT Trans. Ecol. Environ.* **2011**, *144*, 81–92. [CrossRef]
20. Chaussard, E.; Amelung, F.; Abidin, H.; Hong, S.H. Sinking cities in Indonesia: ALOS PALSAR detects rapid subsidence due to groundwater and gas extraction. *Remote Sens. Environ.* **2013**, *128*, 150–161. [CrossRef]
21. Ruiz-Constan, A.; Ruiz-Armenteros, A.M.; Lamas-Fernandez, F.; Martos-Rosillo, S.; Delgado, J.M.; Bekaert, D.P.S.; Sousa, J.J.; Gil, A.; Cuenca, M.C.; Hanssen, R.; et al. Multi-temporal InSAR evidence of ground subsidence induced by groundwater withdrawal: The Montellano aquifer (SW Spain). *Environ. Earth Sci.* **2016**, *75*, 242. [CrossRef]
22. Semple, A.; Pritchard, M.; Lohman, R. An Incomplete Inventory of Suspected Human-Induced Surface Deformation in North America Detected by Satellite Interferometric Synthetic-Aperture Radar. *Remote Sens.* **2017**, *9*, 1296. [CrossRef]
23. Luca, O.; Gaman, F.; Răuță, E. Towards a National Harmonized Framework for Urban Plans and Strategies in Romania. *Sustainability* **2021**, *13*, 1930. [CrossRef]
24. Ferretti, A.; Prati, C.; Rocca, F. Permanent Scatterers in SAR Interferometry. *IEEE Trans. Geosci. Remote. Sens.* **2001**, *39*, 8–20. [CrossRef]
25. Galloway, D.L.; Hoffman, J. The application of the satellite differential SAR interferometry-derived ground displacements in hydrogeology. *Hydrogeol. J.* **2006**, *15*, 133–154. [CrossRef]
26. Crosetto, M.; Monserrat, O.; Cuevas-Gonzales, M.; Devanthery, N.; Crippa, B. Persistent Scatter Interferometry: A Review. *ISPRS J. Photogramm. Remote Sens.* **2016**, *115*, 78–89. [CrossRef]
27. Radutu, A.; Nedelcu, I.; Gogu, C. An overview of ground surface displacements generated by groundwater dynamics, revealed by InSAR techniques. *Procedia Eng.* **2017**, *209*, 119–126. [CrossRef]
28. Radutu, A. Use of Remote Sensing Techniques for Monitoring the Dynamics of Urban Groundwater. Ph.D. Thesis, Technical University of Civil Engineering Bucharest, Bucharest, Romania, 2021.
29. European Space Agency, Sentinel Online, Interferometric Wide Swath. Available online: https://sentinelesa.int/web/sentinel/user-guides/sentinel-1-sar/acquisition-modes/interferometric-wide-swath (accessed on 28 April 2020).
30. Ferretti, A.; Fumagalli, A.; Novali, F.; Prati, C.; Rocca, F.; Rucci, A. A New Algorithm for Processing Interferometric Data-Stacks: SqueeSAR. *IEEE Trans. Geosci. Remote Sens.* **2011**, *49*, 3460–3470. [CrossRef]
31. INS-Direcția Regională de Statistică a Municipiului București. Available online: http://www.bucuresti.insse.ro/despre-bucuresti/ (accessed on 20 July 2019).
32. Gaitanaru, S.D.; Gogu, R.; Boukhemacha, M.A.; Litescu, L.; Zaharia, V.; Moldovan, A.; Mihailovici, M. Bucharest city urban groundwater monitoring system. *Procedia Eng.* **2017**, *209*, 143–147. [CrossRef]
33. L3Harris Geospatial. ENVI SARscape. Available online: https://www.l3harrisgeospatial.com/Software-Technology/ENVI-SARscape (accessed on 14 March 2022).
34. ESA. Earth Online. Available online: https://earth.esa.int (accessed on 17 July 2020).
35. Copernicus Open Access Hub. Available online: https://scihub.copernicus.eu (accessed on 14 March 2022).
36. Dehls, J.F.; Larsen, Y.; Marinkovic, P.; Lauknes, T.R.; Stodle, D.; Moldestad, D.A. INSAR.NO: A National InSAR Deformation Mapping/Monitoring Service in Norway-From Concept to Operations. In Proceedings of the IGARSS 2019-IEEE International Geoscience and Remote Sensing Symposium, Yokohama, Japan, 28 July–2 August 2019; IEEE: Piscataway, NJ, USA; pp. 5461–5464. [CrossRef]
37. INnovations for eXtreme Climatic Events Project. Available online: https://inxces.eu/ (accessed on 14 March 2022).

38. Radutu, A.; Venvik, G.; Ghibus, T.; Gogu, C.R. Sentinel-1 Data for Underground Processes Recognition in Bucharest City, Romania. *Remote Sens.* **2020**, *12*, 4054. [CrossRef]
39. Lacatusu, R.; Anastasiu, N.; Popescu, M.; Enciu, P. *Geo-Atlasul Municipiului București*; Estfalia Publishing House: Bucharest, Romania, 2008.
40. Copernicus Space Component Data Access. Available online: https://spacedata.copernicus.eu/web/cscda/dataset-details?articleId=394198 (accessed on 21 April 2021).
41. Boukhemacha, M.A.; Gogu, R.C.; Serpescu, I.; Gaitanaru, D.; Bica, I. A hydrogeological conceptual approach to study urban groundwater flow in Bucharest city, Romania. *Hydrogeol. J.* **2015**, *23*, 437–450. [CrossRef]
42. National Institute of Statistics. Romanian Statistical Yearbook. Available online: https://insse.ro/cms/ro/tags/anuarul-statistic-al-romaniei (accessed on 1 March 2021).
43. Radutu, A.; Vlad Sandru, M.I.; Nedelcu, I.; Poenaru, V. Change detection trends in urban areas with remote sensing and socio-economic diagnosis in Bucharest city. In Proceedings of the 21st International Multidisciplinary Scientific GeoConference SGEM 2021, Albena, Bulgaria, 14–22 August 2021; STEF92 Technology: Sofia, Bulgaria; pp. 255–266. [CrossRef]
44. Iași General Urban Plan. Available online: http://www.primaria-iasi.ro/portal-primaria-municipiului-iasi/pug/10452/dezvoltare-urbana (accessed on 14 February 2022).
45. Bacau General Urban Plan. Available online: http://arhiva.municipiulbacau.ro/subpagini_fisiere/RLU_BACAU_2012.pdf (accessed on 14 February 2022).
46. Constanța General Urban Plan. Available online: http://www.primaria-constanta.ro/primarie/urbanism/regulamentul-local-de-urbanism (accessed on 14 February 2022).
47. Galati General Urban Plan. Available online: https://ro.scribd.com/document/510417220/RLU-PUG-GALATI (accessed on 14 February 2022).
48. Braila General Urban Plan. Available online: http://www.primariabr.ro/directii-proprii-2/directia-arhitect-sef/pug (accessed on 15 February 2022).
49. Ploiesti General Urban Plan. Available online: http://www.ploiesti.ro/18.02.2015.php (accessed on 15 February 2022).
50. Pitesti General Urban Plan. Available online: https://www.primariapitesti.ro/pug-p762 (accessed on 31 January 2022).
51. Craiova General Urban Plan. Available online: https://www.primariacraiova.ro/pozearticole/userfiles/files/01/17929.pdf (accessed on 31 January 2022).
52. Timisoara General Urban Plan. Available online: https://www.primariatm.ro/urbanism/planuri-urbanistice/planuri-urbanistice-aprobate/pug/ (accessed on 31 January 2022).
53. Arad General Urban Plan. Available online: https://portal1.primariaarad.ro/info.php?page=/temp/PUG-2015/index_PUG_Arad2015.html&newlang=ron&theme=th1-ron (accessed on 31 January 2022).
54. Cluj Napoca General Urban Plan. Available online: https://gis.primariaclujnapoca.ro/Public/ (accessed on 26 January 2022).
55. Oradea General Urban Plan. Available online: https://www.oradea.ro/subpagina/plan-urbanistic-general (accessed on 26 January 2022).
56. Baia Mare General Urban Plan. Available online: https://www.baiamare.ro/ro/Servicii-publice/Serviciul-Dezvoltare-Urbana-si-Amenajarea-Teritoriului/Regulamentul-Local-de-Urbanism-(R.L.U)-si-Planul-Urbanistic-General-(P.U.G.)---Aprobat-cu-HCL-3491999/ (accessed on 26 January 2022).
57. Brasov General Urban Plan. Available online: https://visumbrasov.org/planul-urbanistic-general-al-municipiului-brasov-varianta-iii/ (accessed on 26 January 2022).
58. Sibiu General Urban Plan. Available online: https://sibiu.ro/primaria/urbanism/pug-municipiul-sibiu-aprobat-cu-hcl-165-28.04.2011 (accessed on 26 January 2022).
59. Targu Mures General Urban Plan. Available online: https://www.tirgumures.ro/index.php?option=com_content&view=article&id=197&Itemid=196&lang=ro (accessed on 2 March 2022).
60. Bucharest General Urban Plan. Available online: https://www2.pmb.ro/servicii/urbanism/pug/pug.php (accessed on 17 January 2022).
61. Bucharest Urban Concept 2035. Available online: https://issuu.com/almihai/docs/concept_strategic_bucuresti_2035 (accessed on 17 January 2022).
62. Integrated Urban Development Strategy. Available online: https://estibucuresti.org/sidu (accessed on 17 January 2022).
63. Open Street Map. Available online: https://www.openstreetmap.org (accessed on 21 April 2021).
64. Gogu, C.R.; Gaitanaru, D.; Boukhemacha, M.A.; Serpescu, I.; Litescu, L.; Zaharia, V.; Moldovan, A.; Mihailovici, M.J. Urban Hydrogeology studies in Bucharest City, Romania. *Procedia Eng.* **2017**, *209*, 135–142. [CrossRef]
65. Lyu, H.M.; Shen, S.L.; Wu, Y.X.; Zhou, A.N. Calculation of groundwater head distribution with a close barrier during excavation dewatering in confined aquifer. *Geosci. Front.* **2021**, *12*, 791–803. [CrossRef]

MDPI
St. Alban-Anlage 66
4052 Basel
Switzerland
Tel. +41 61 683 77 34
Fax +41 61 302 89 18
www.mdpi.com

Water Editorial Office
E-mail: water@mdpi.com
www.mdpi.com/journal/water

www.ingramcontent.com/pod-product-compliance
Lightning Source LLC
LaVergne TN
LVHW070405100526
838202LV00014B/1393